AUTODESK

AUTOCAD MECHANICAL

400 PRACTICE DRAWINGS

SACHIDANAND JHA

cadin360°
Learning Tutorials

Dear Reader,

Thank you for choosing **AUTOCAD MECHANICAL** book. This book is part of a family of premium-quality CADin360 books, all of which are written by Outstanding author who combine practical experience with a gift for teaching.

CADin360 was founded in 2016. More than 3 years later, we're still committed to producing consistently exceptional books. With each of our titles, we're working hard to set a new standard for the industry. From the paper we print on, to the authors we work with, our goal is to bring you the best books available.

I hope you see all that reflected in these pages. I'd be very interested to hear your comments and get your feedback on how we're doing. Feel free to let me know what you think about this or any other CADin360 book by sending me an email at contactus@cadin360.com.

If you think you've found a technical error in this book, please visit https://cadin360.com/contact-us/. Customer feedback is critical to our efforts at CADin360.

Best regards,

Sachidanand Jha
Founder & CEO, CADin360

AUTOCAD MECHANICAL

Published by
CADin360
cadin360.com
Copyright © 2019 by CADin360, ALL Rights Reserved

Limit of Liability/Disclaimer of Warranty:

Examination Copies

Electronic Files

Disclaimer:

Preface

AUTOCAD MECHANICAL

- ❖ This book contain 200 2D CAD practice drawings & 200 3D exercises for practice.
- ❖ This book does not provide step by step tutorial to design 3D models.
- ❖ S.I Units is used.
- ❖ Predominantly used Third Angle Projection.
- ❖ This book is for **AUTOCAD MECHANICAL** and Other Feature-Based Modeling Software such as Inventor, Catia, SolidWorks, NX, Solid Edge, AutoCAD, PTC Creo etc.
- ❖ It is intended to provide Drafters, Designers and Engineers with enough 2D & 3D CAD exercises for practice on **AUTOCAD MECHANICAL.**
- ❖ It includes almost all types of exercises that are necessary to provide, clear, concise and systematic information required on industrial machine part drawings.
- ❖ Third Angle Projection is intentionally used to familiarize Drafters, Designers and Engineers in Third Angle Projection to meet the expectation of world wide Engineering drawing print.
- ❖ Clear and well drafted drawing help easy understanding of the design.
- ❖ This book is for Beginner, Intermediate and Advance CAD users.
- ❖ These exercises are from Basics to Advance level.
- ❖ Each exercises can be assigned and designed separately.
- ❖ No Exercise is a prerequisite for another. All dimensions are in mm.
- ❖ Note: Assume any missing dimensions.

SKETCHING

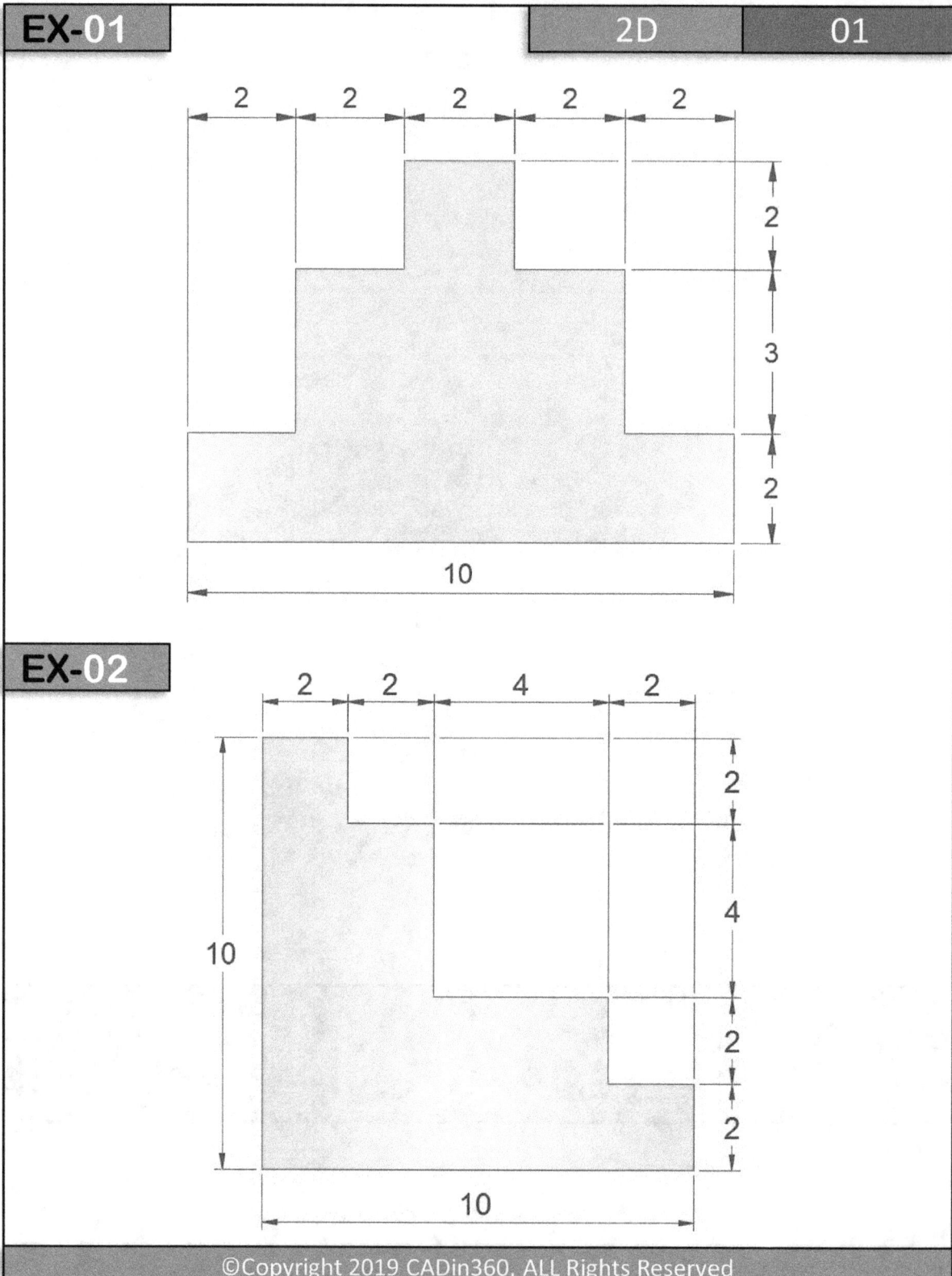

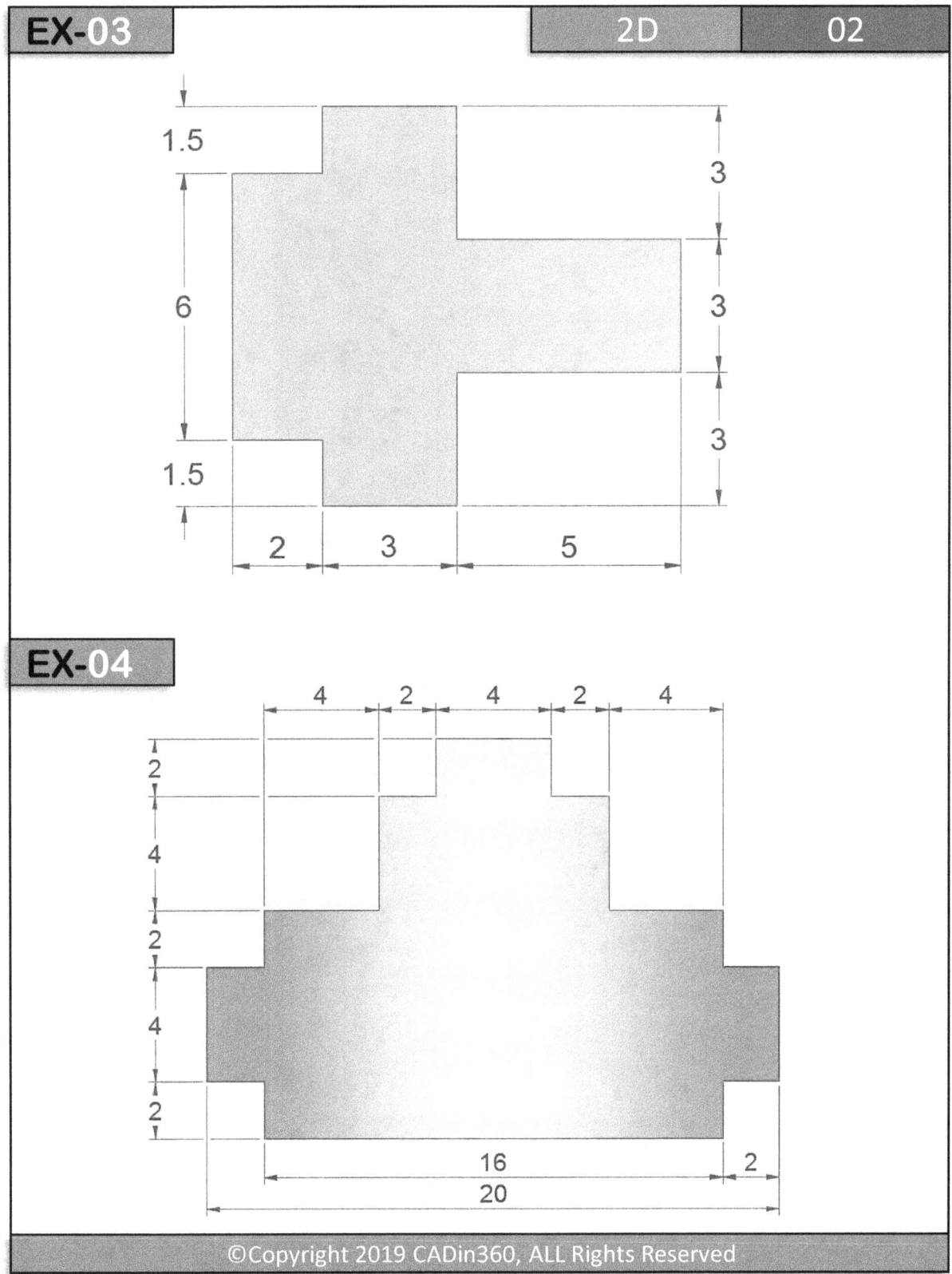

EX-04

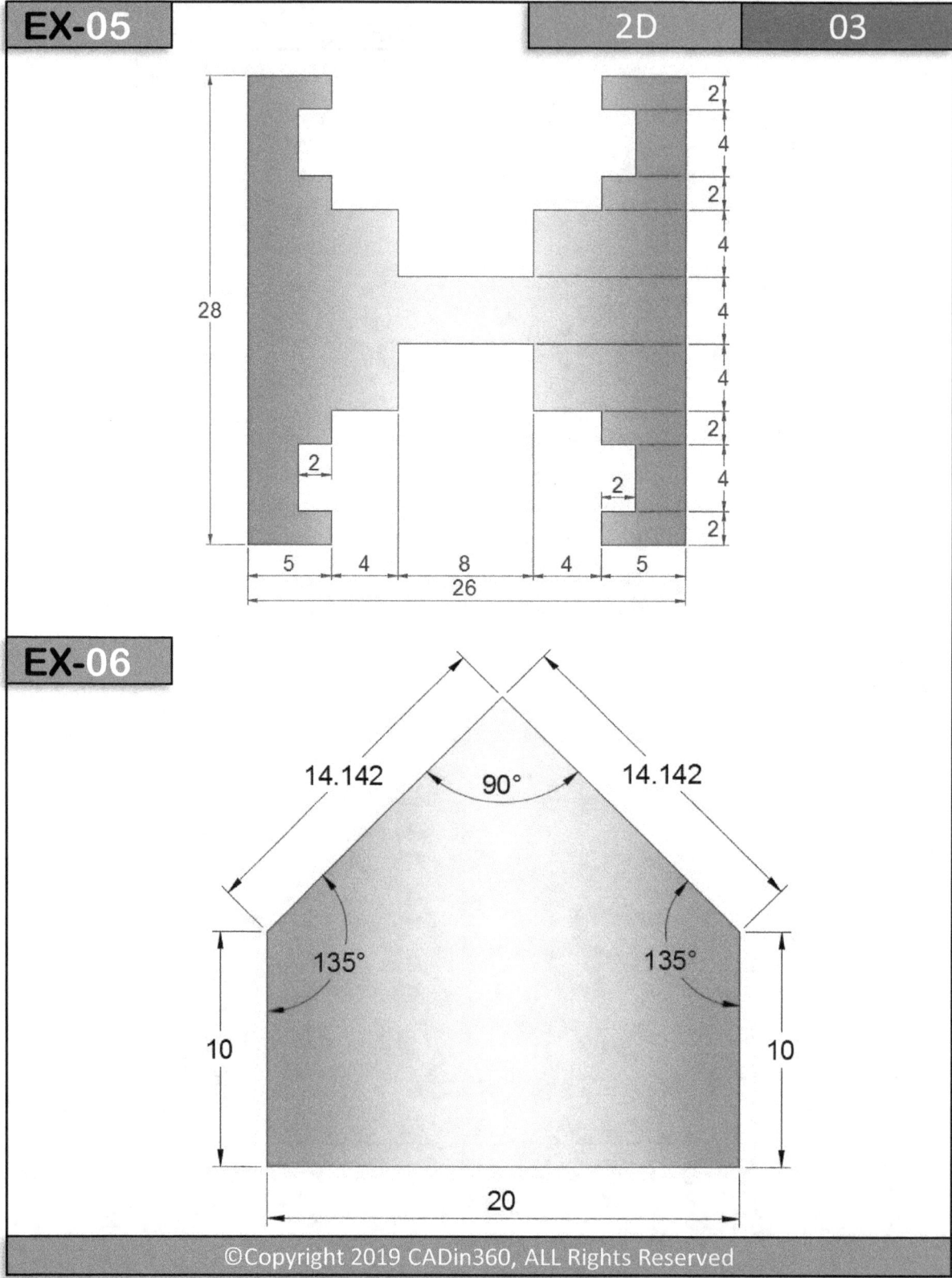

EX-05

2D   03

EX-06

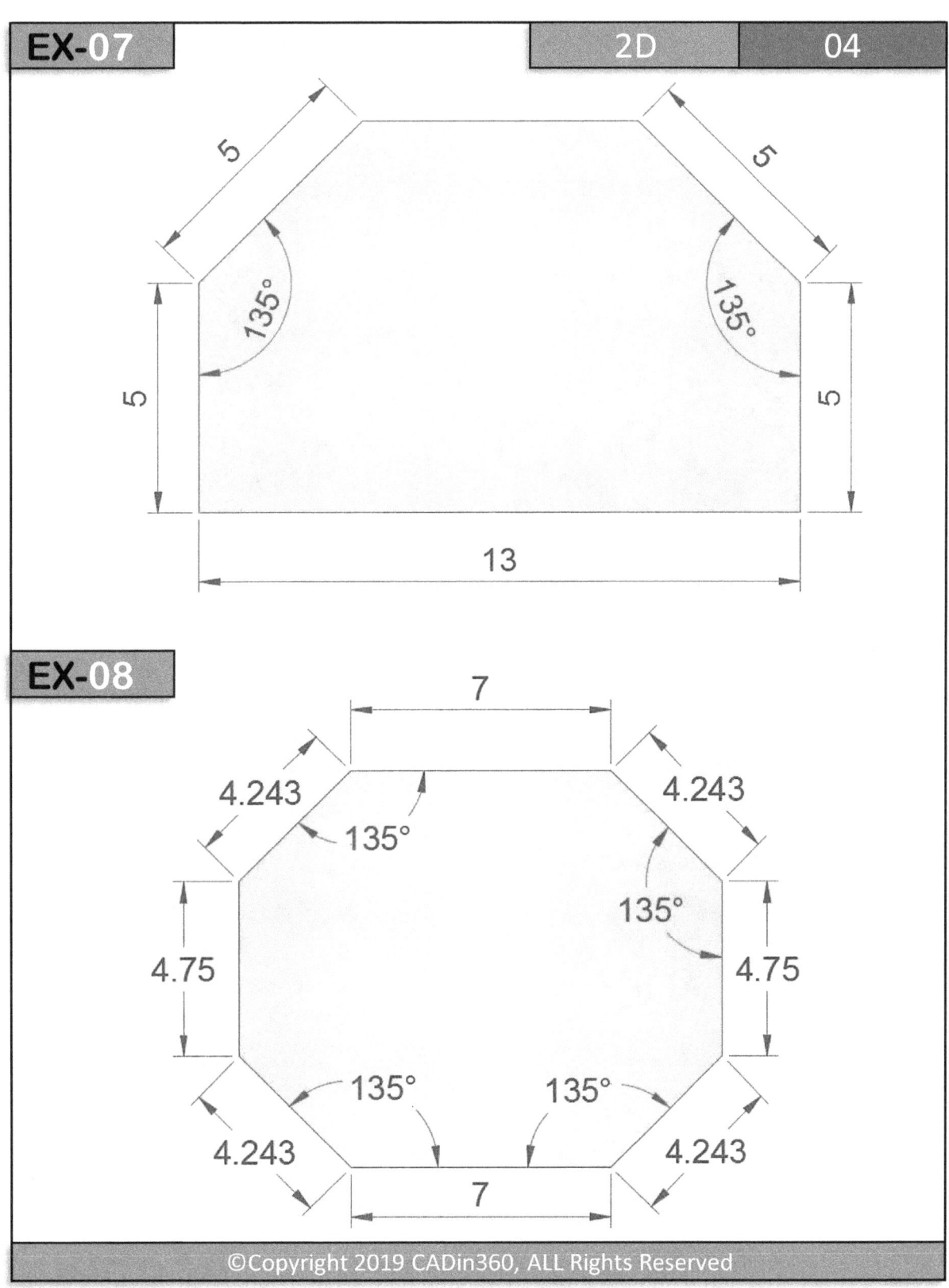

EX-08

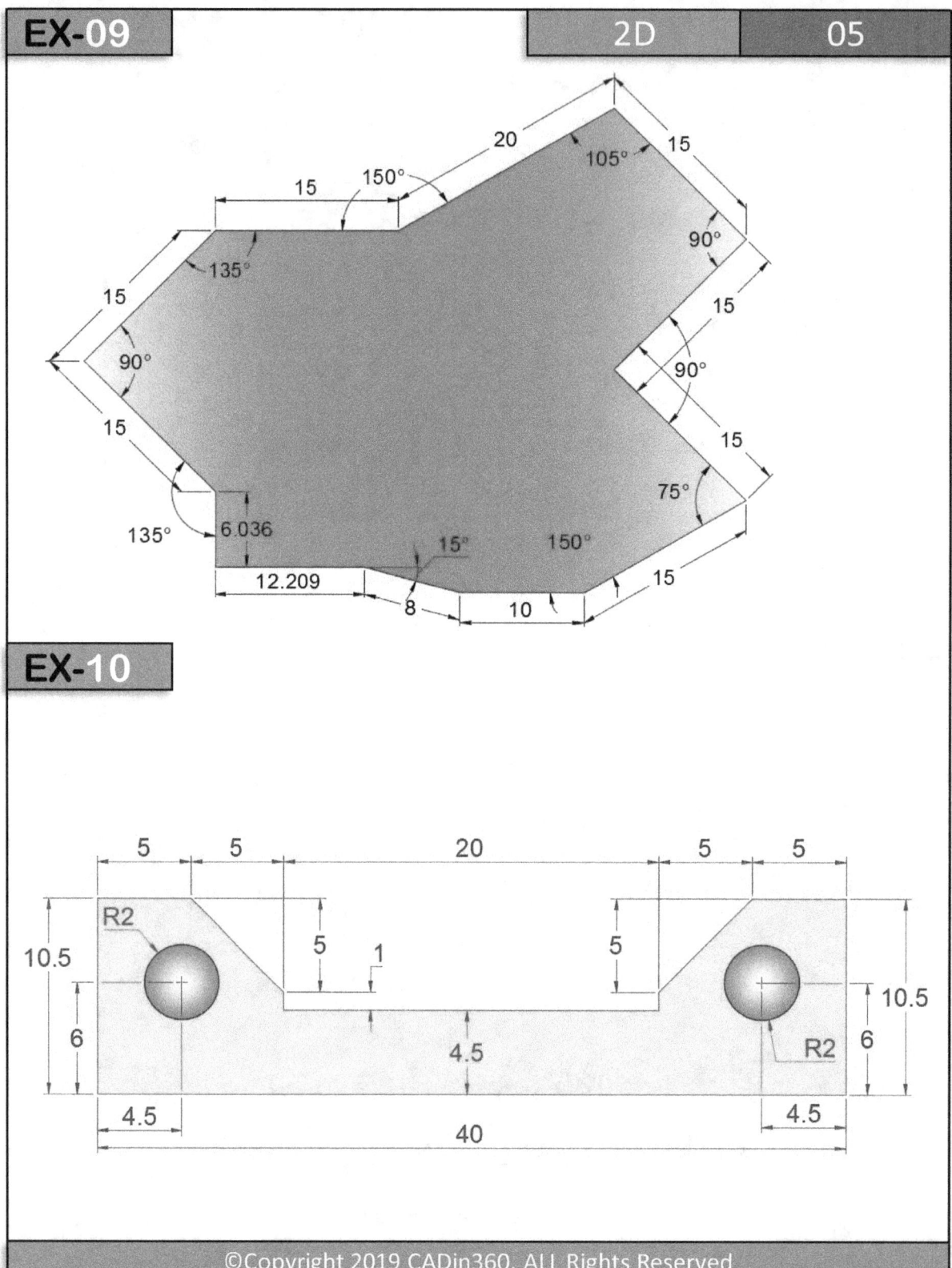

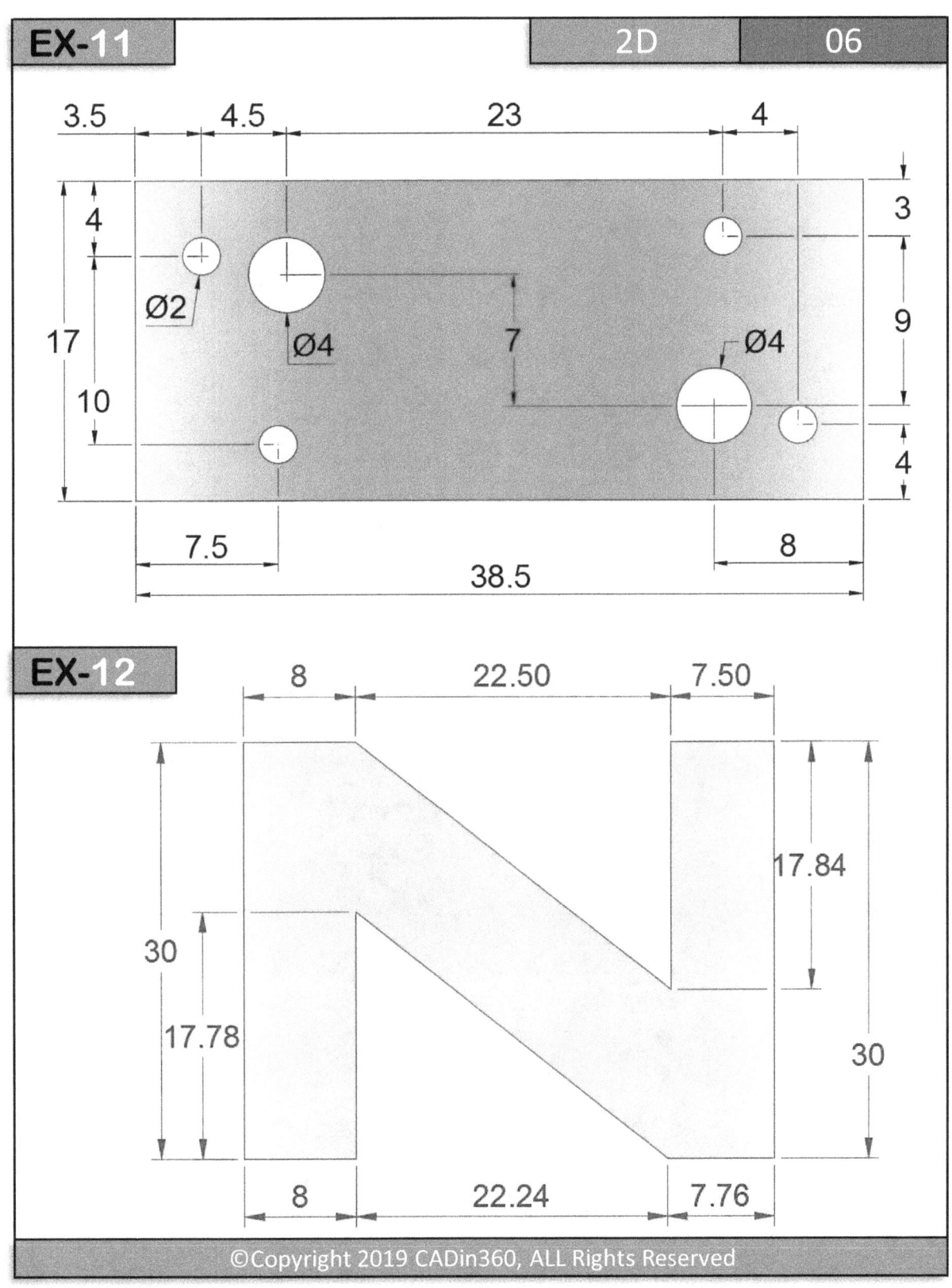

EX-11   2D   06

3.5   4.5   23   4

4

3

Ø2

Ø4

17   7   9

10   Ø4

4

7.5

8

38.5

EX-12

8   22.50   7.50

17.84

30

17.78

30

8   22.24   7.76

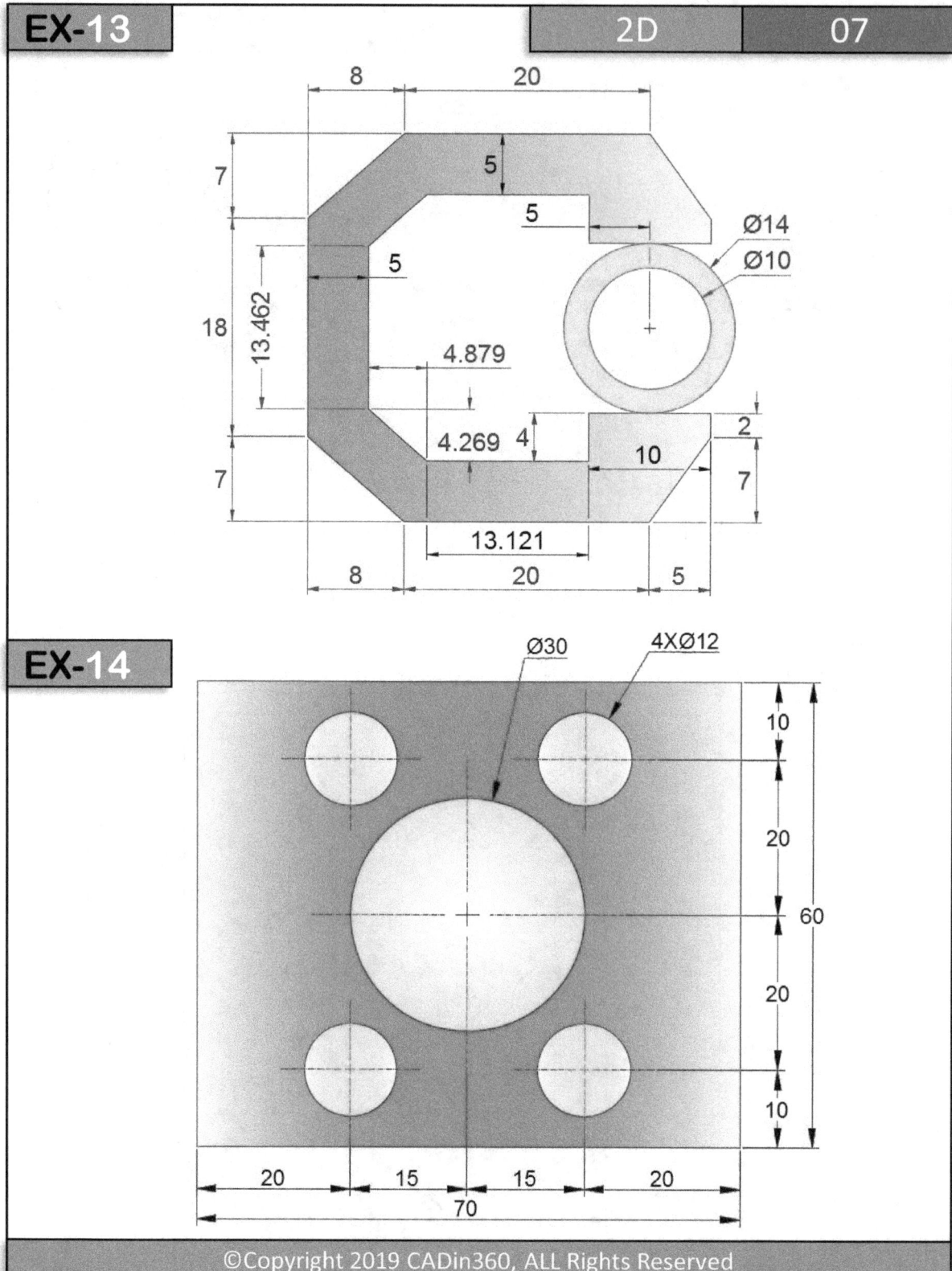

EX-13

2D   07

Ø14
Ø10

EX-14

Ø30   4XØ12

©Copyright 2019 CADin360, ALL Rights Reserved

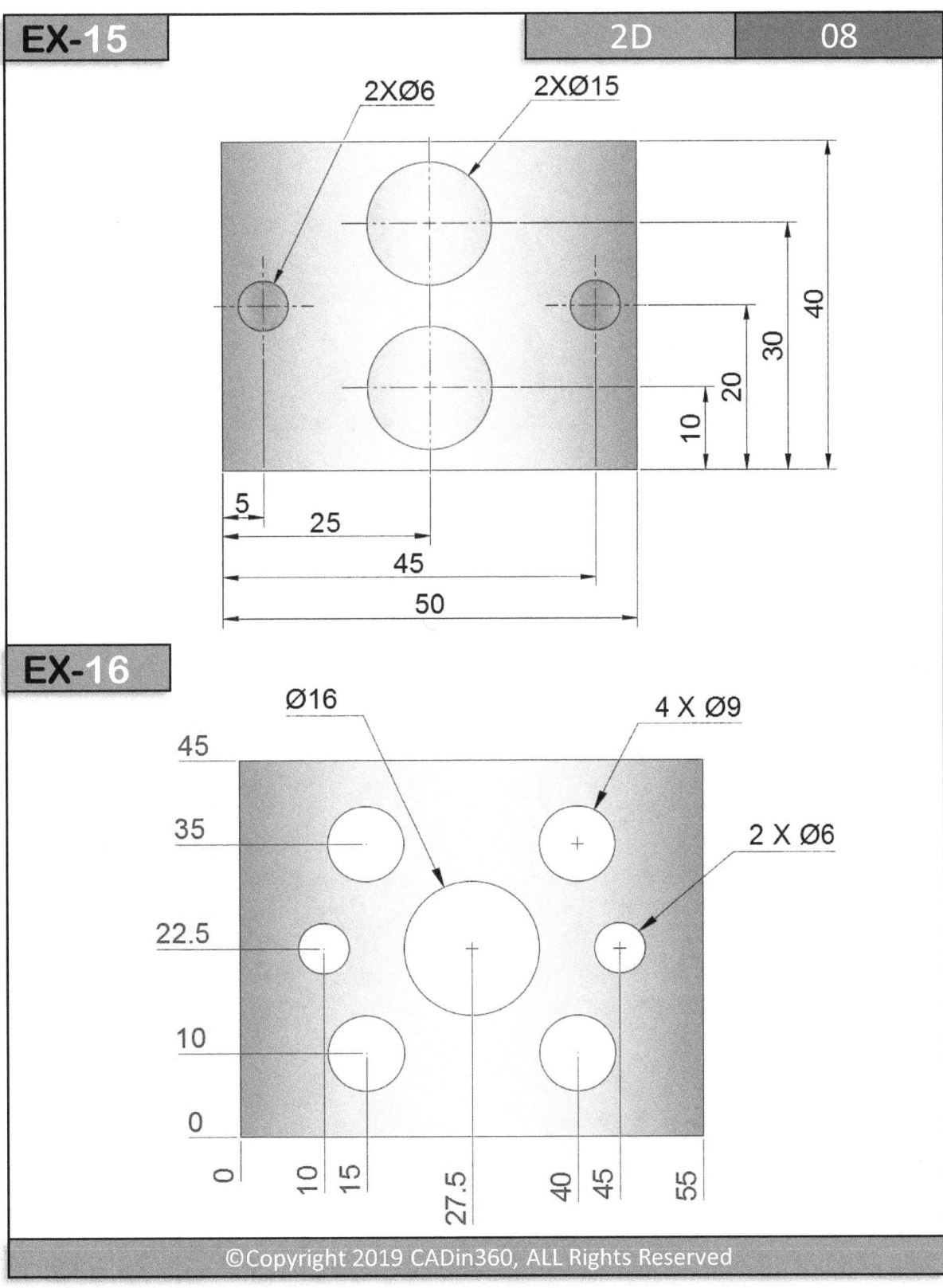

EX-15

2D 08

2XØ6 2XØ15

40
30
20
10
5
25
45
50

EX-16

Ø16 4 X Ø9

2 X Ø6

45
35
22.5
10
0

0 10 15 27.5 40 45 55

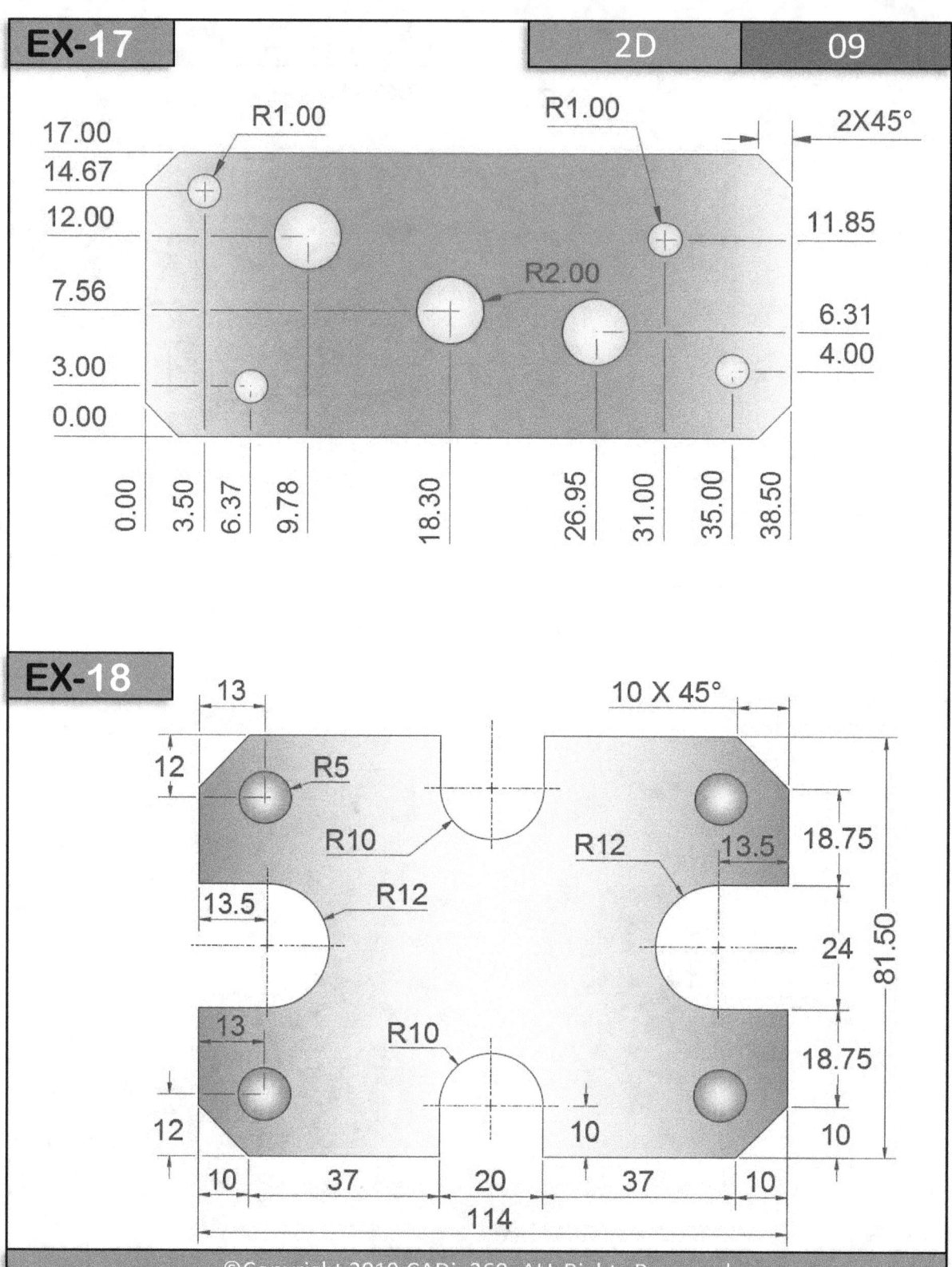

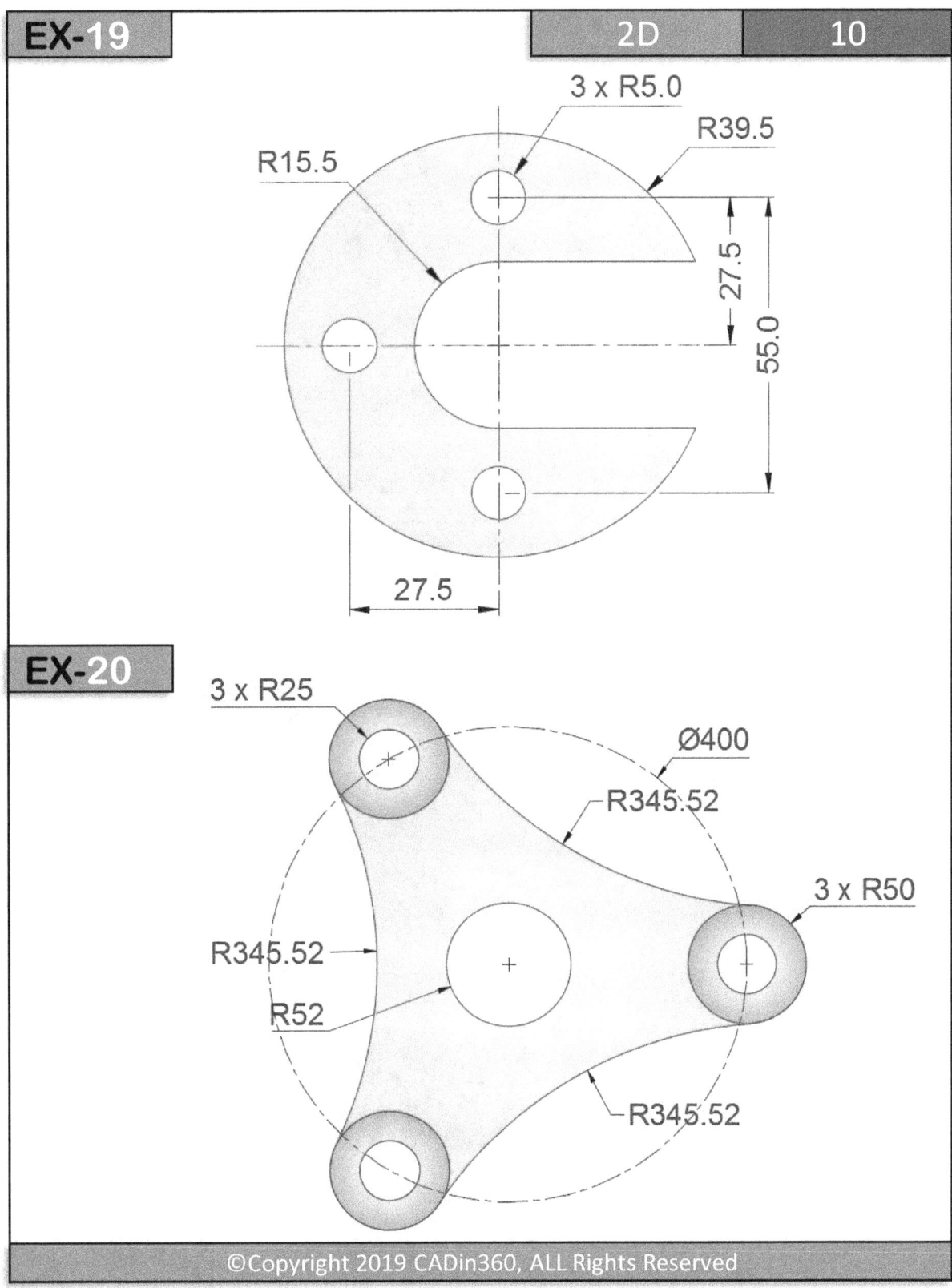

EX-19

3 x R5.0
R39.5
R15.5
27.5
55.0
27.5

EX-20

3 x R25
Ø400
R345.52
R345.52
3 x R50
R52
R345.52

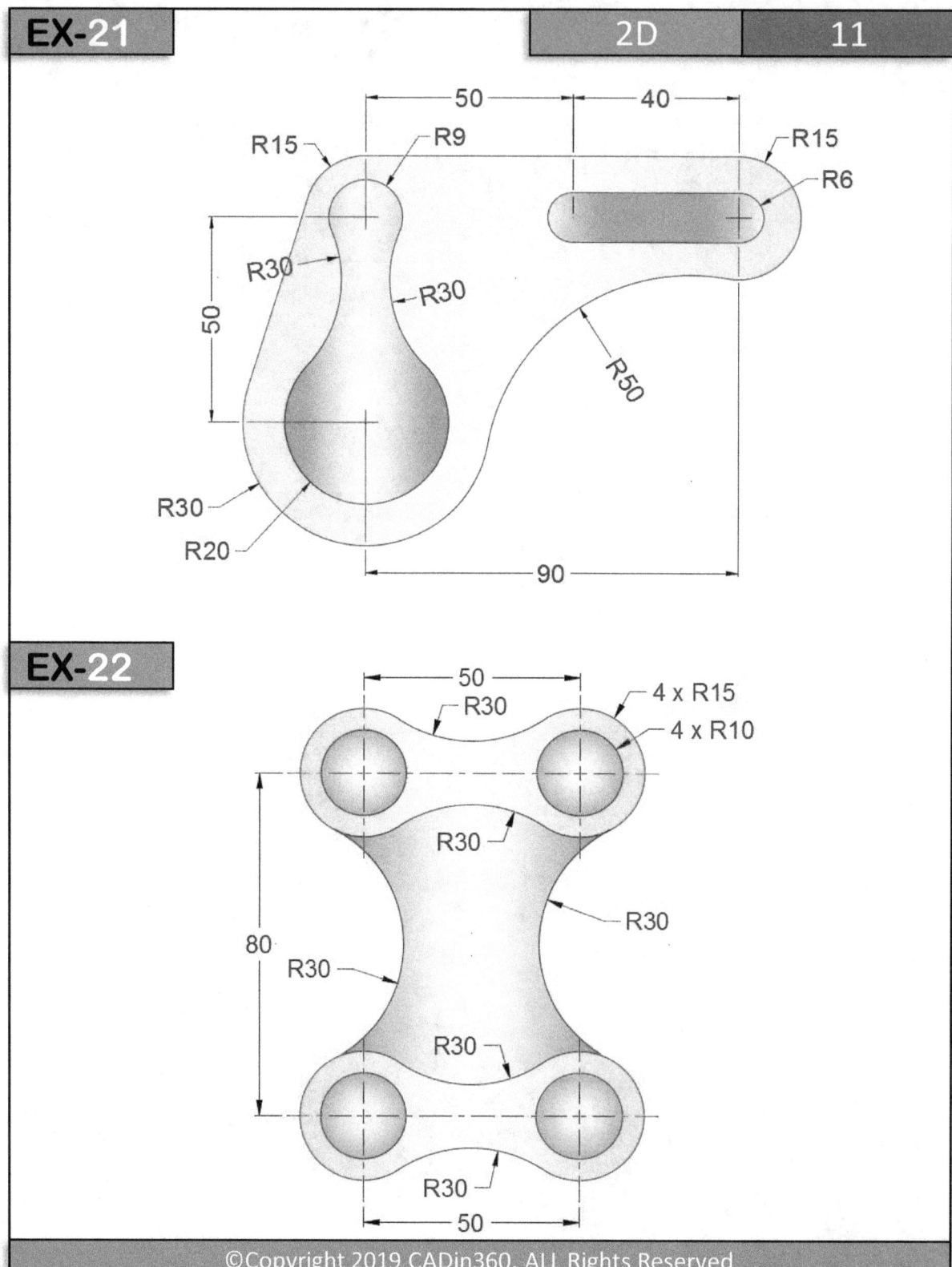

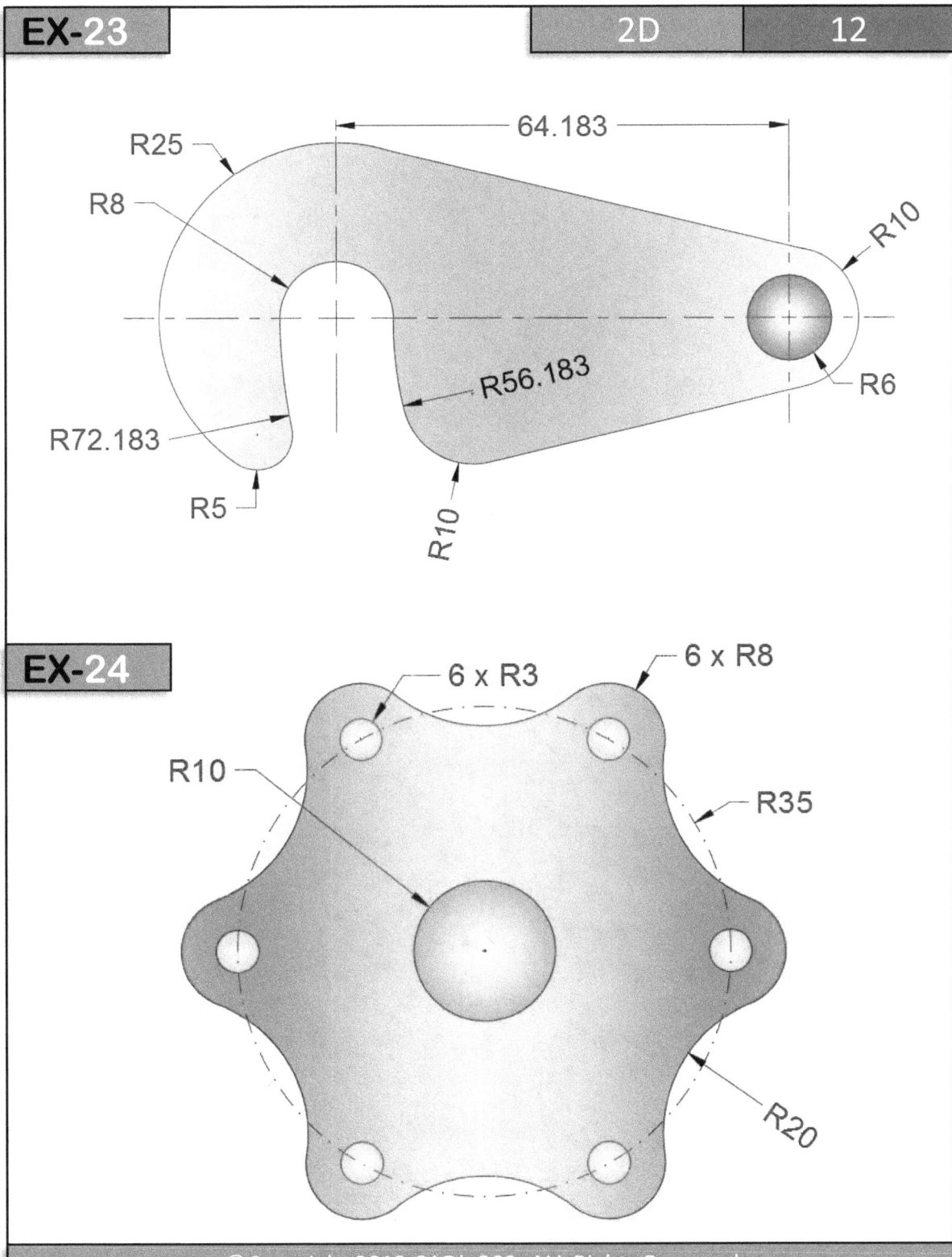

EX-23

2D 12

R25
R8
64.183
R10
R56.183
R72.183
R5
R10
R6

EX-24

6 x R3
6 x R8
R10
R35
R20

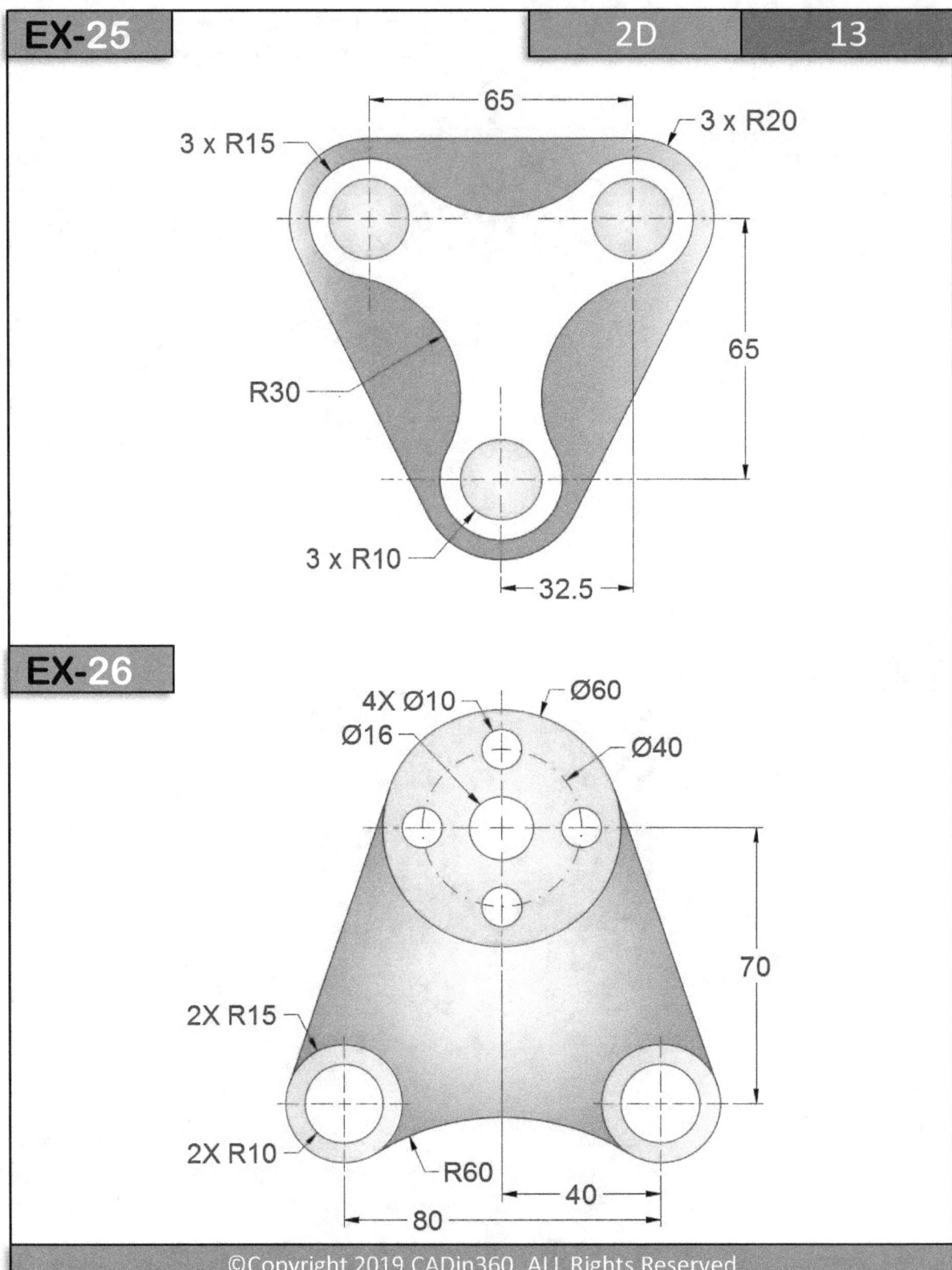

EX-25

2D 13

65

3 x R15

3 x R20

65

R30

3 x R10

32.5

EX-26

4X Ø10

Ø60

Ø16

Ø40

70

2X R15

2X R10

R60

40

80

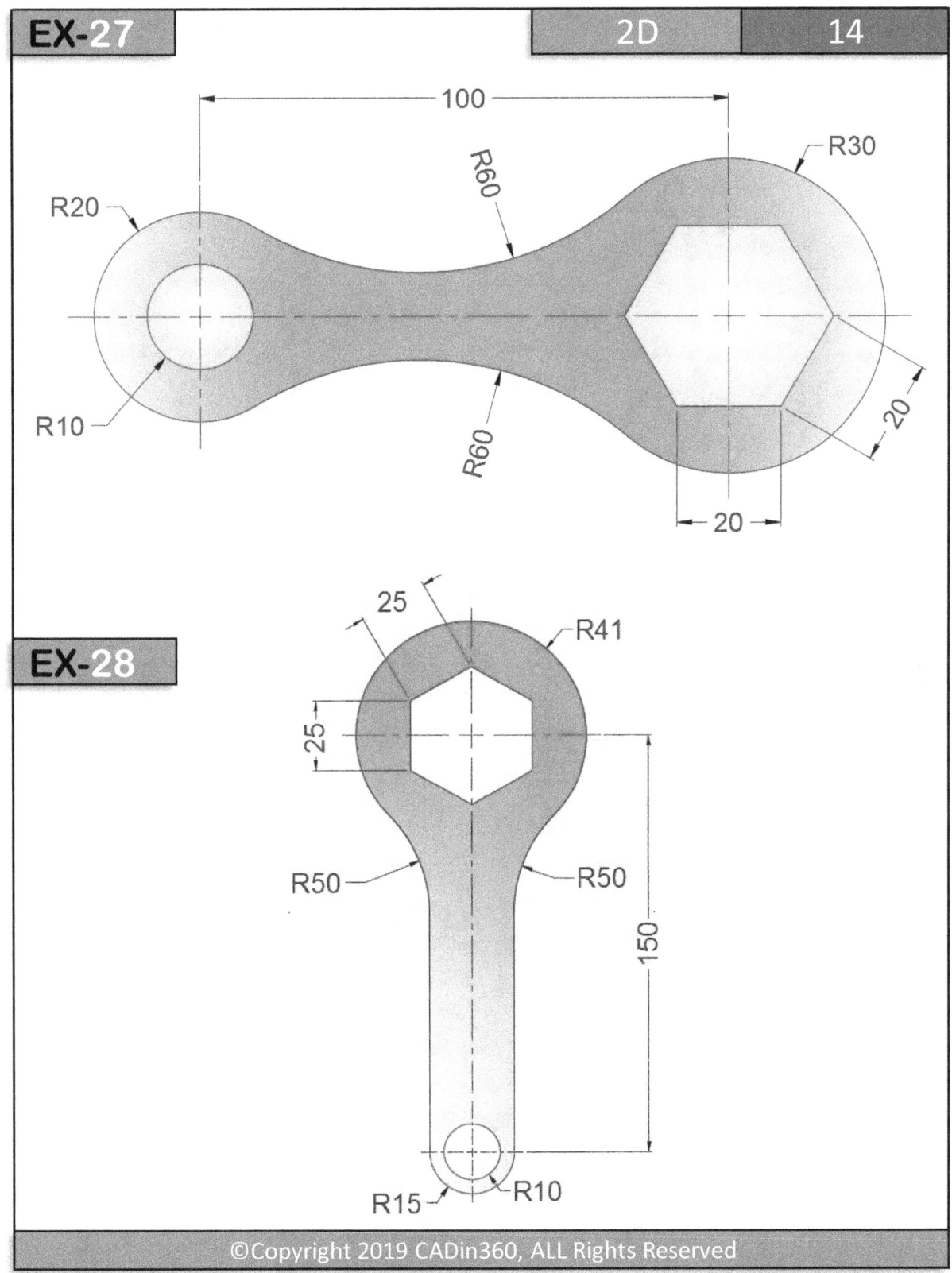

EX-28

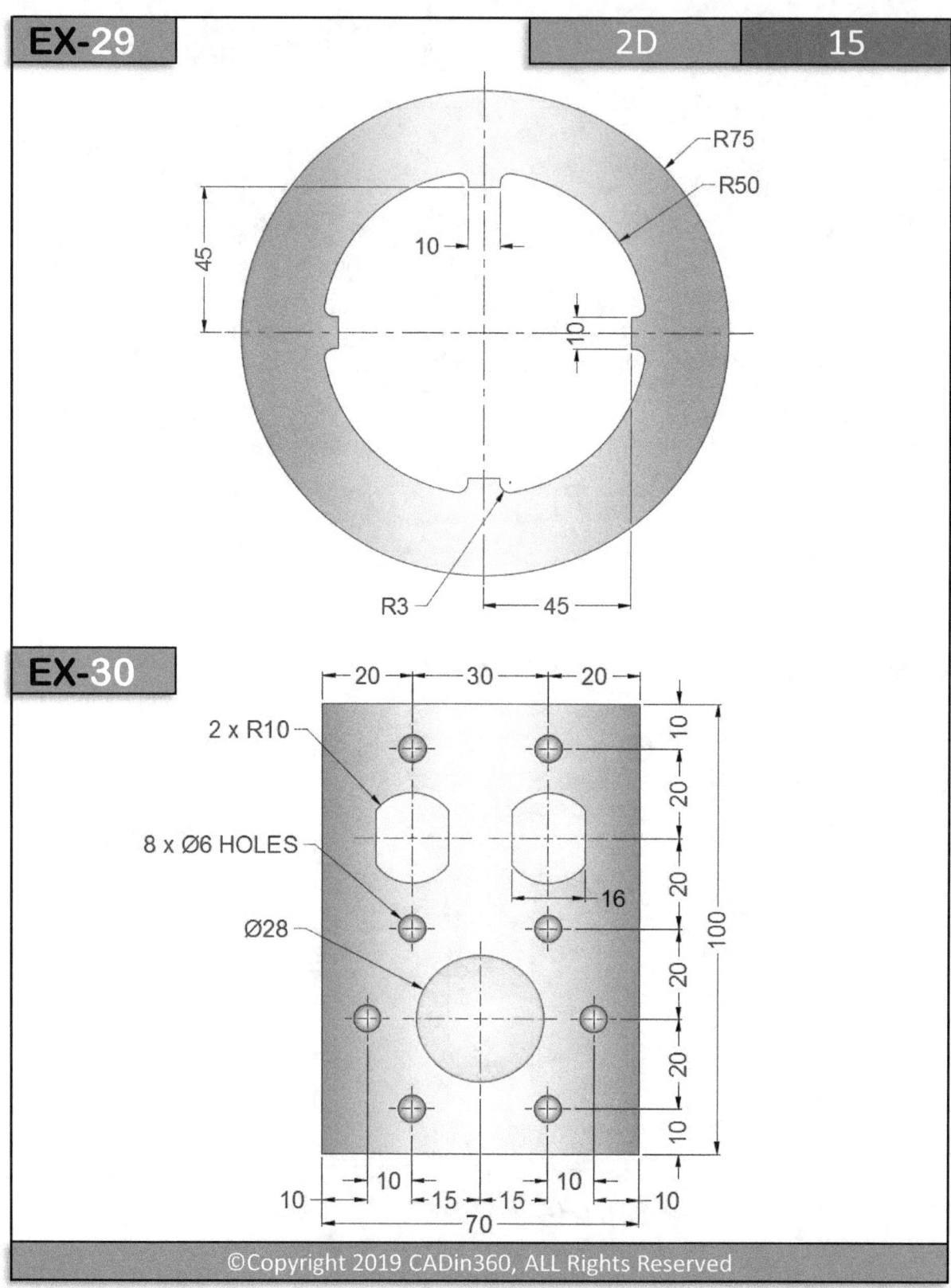

EX-29

R75
R50
45
10
10
R3
45

EX-30

20 30 20
2 x R10
10
20
8 x Ø6 HOLES
16
20
Ø28
100
20
20
10
10 10
10 15 15 10
70

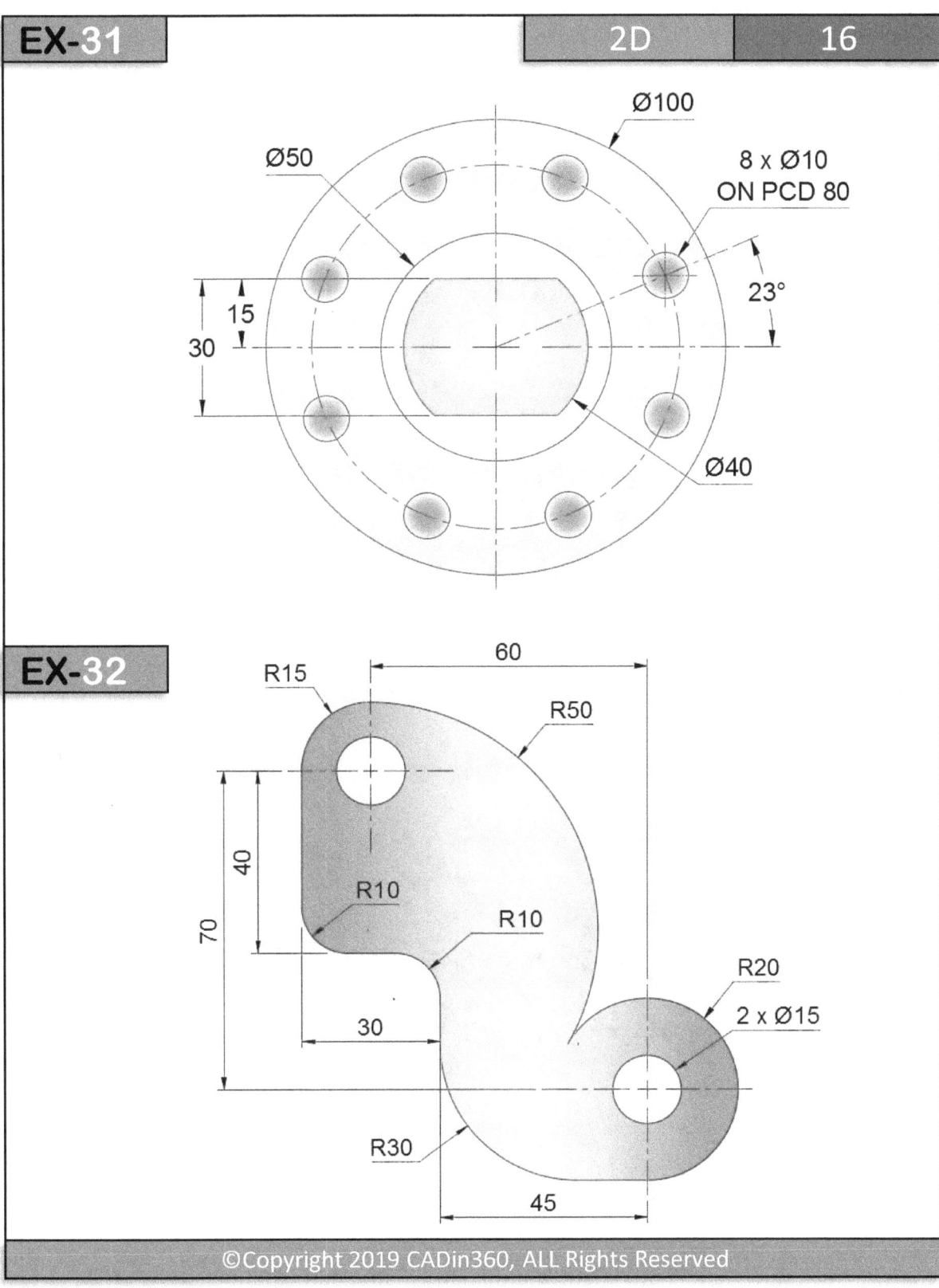

EX-31   2D   16

Ø100

Ø50

8 x Ø10
ON PCD 80

23°

15

30

Ø40

EX-32

R15

60

R50

R10

R10

40

70

R20

2 x Ø15

30

R30

45

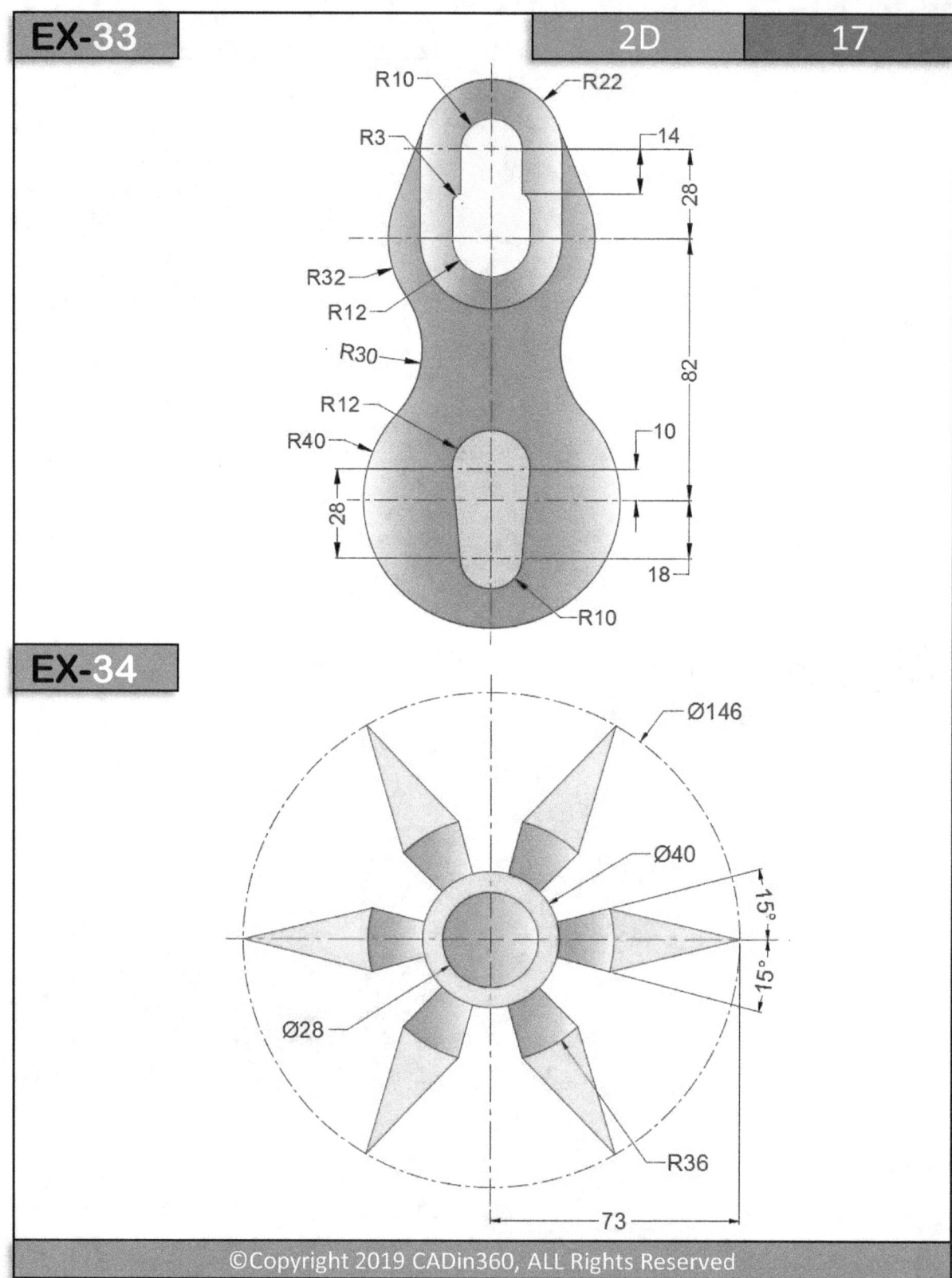

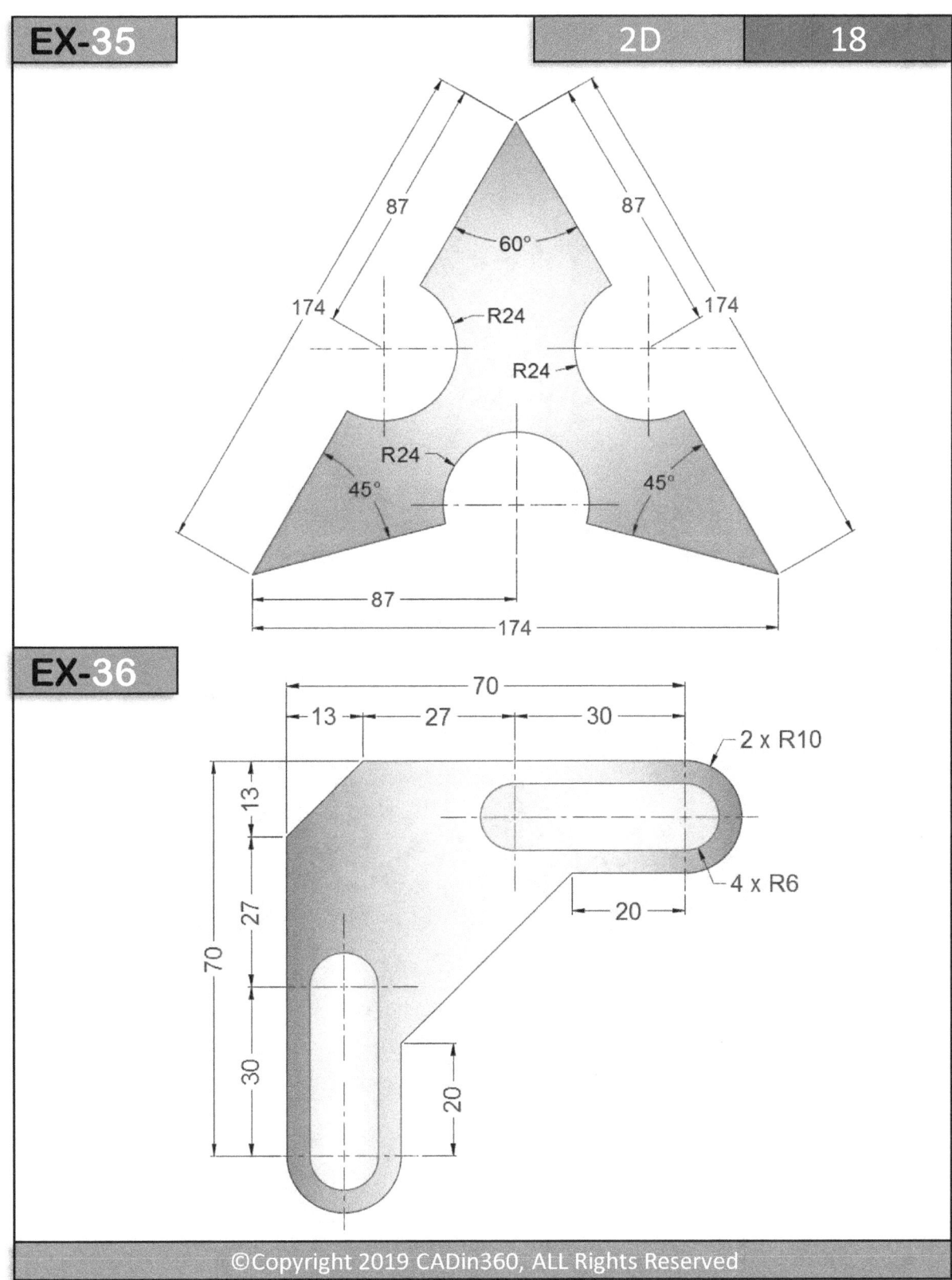

EX-35

2D   18

87   60°   87
174   R24   174
      R24
R24   45°   45°
87
174

EX-36

70
13   27   30
2 x R10
13
27   4 x R6
70   20
30   20

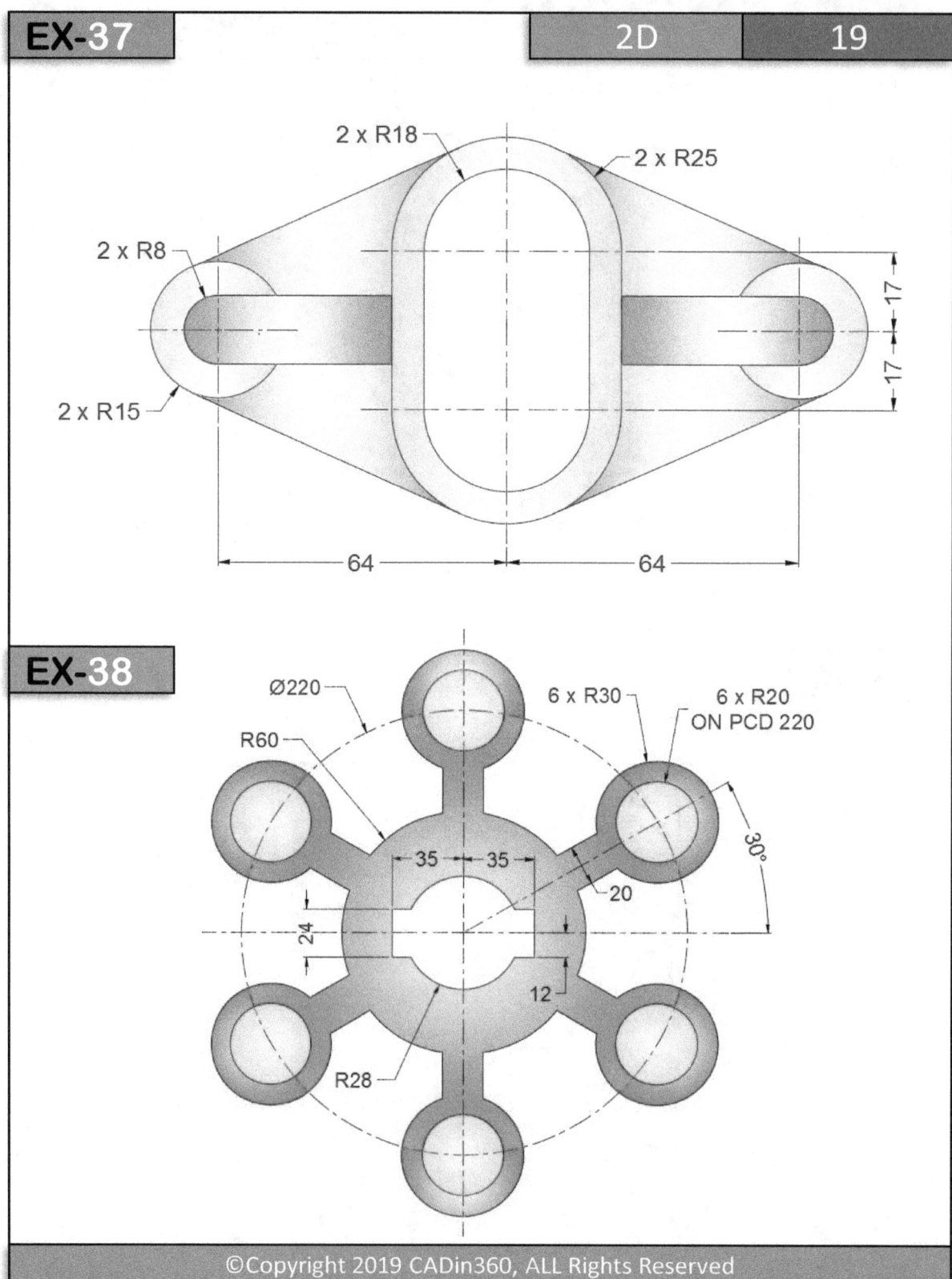

EX-38

2 x R18

2 x R25

2 x R8

2 x R15

17

17

64

64

Ø220

6 x R30

6 x R20
ON PCD 220

R60

35

35

30°

20

24

12

R28

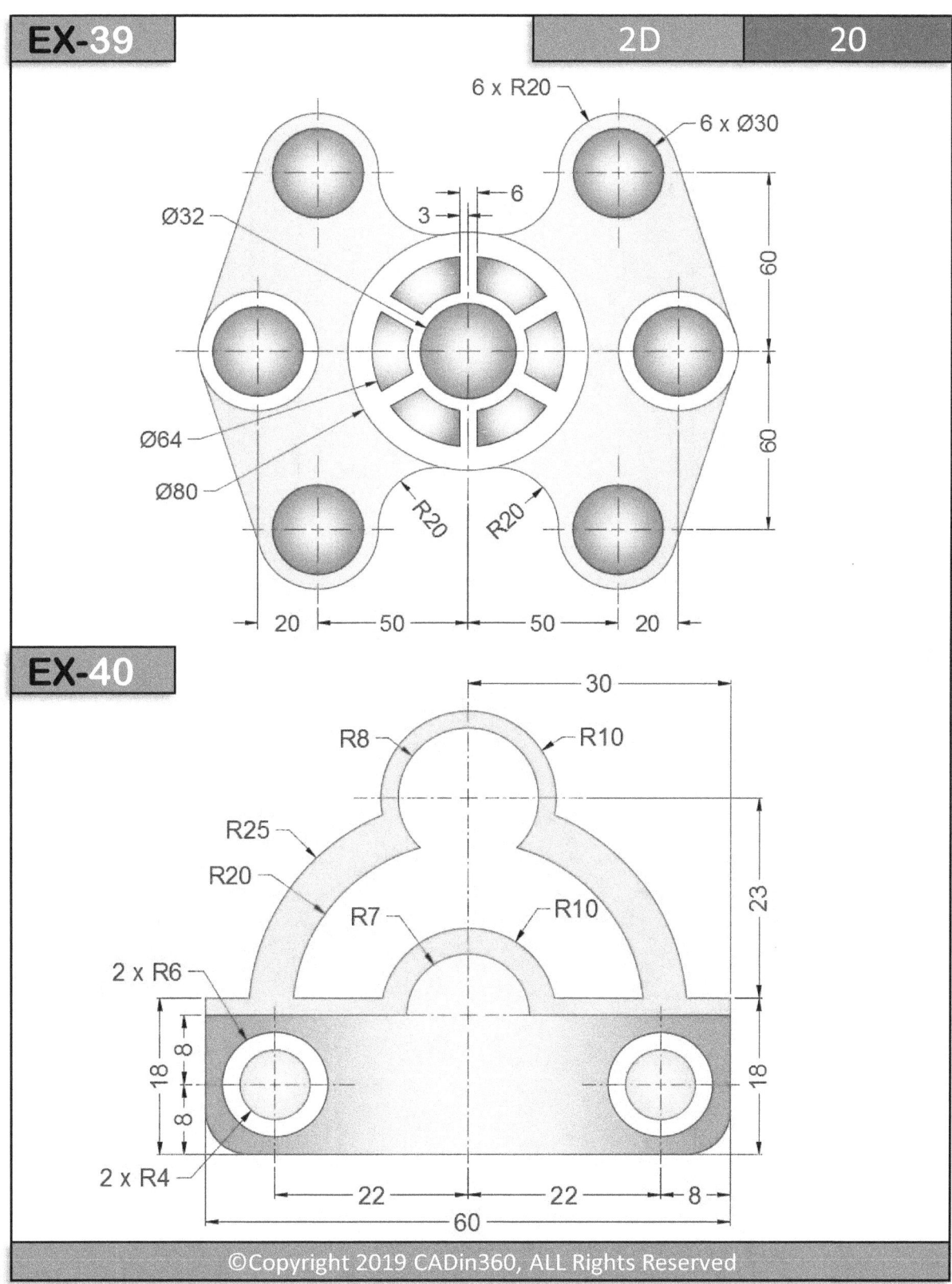

EX-39 2D 20

6 x R20
6 x Ø30
Ø32
3 6
60
60
Ø64
Ø80
R20
R20
20 50 50 20

EX-40

30
R8 R10
R25
R20
R7 R10
23
2 x R6
18 8
18
8
2 x R4
22 22 8
60

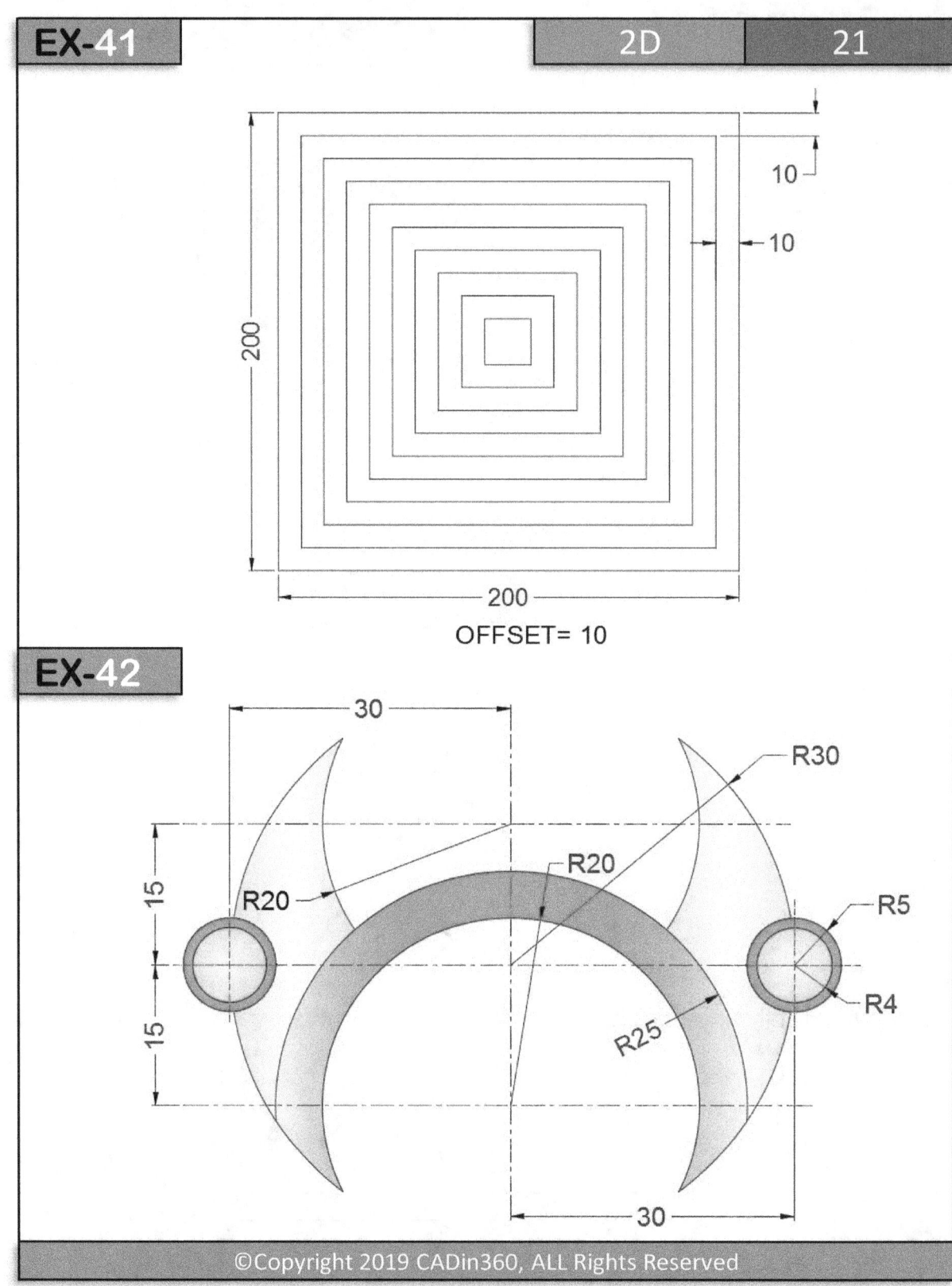

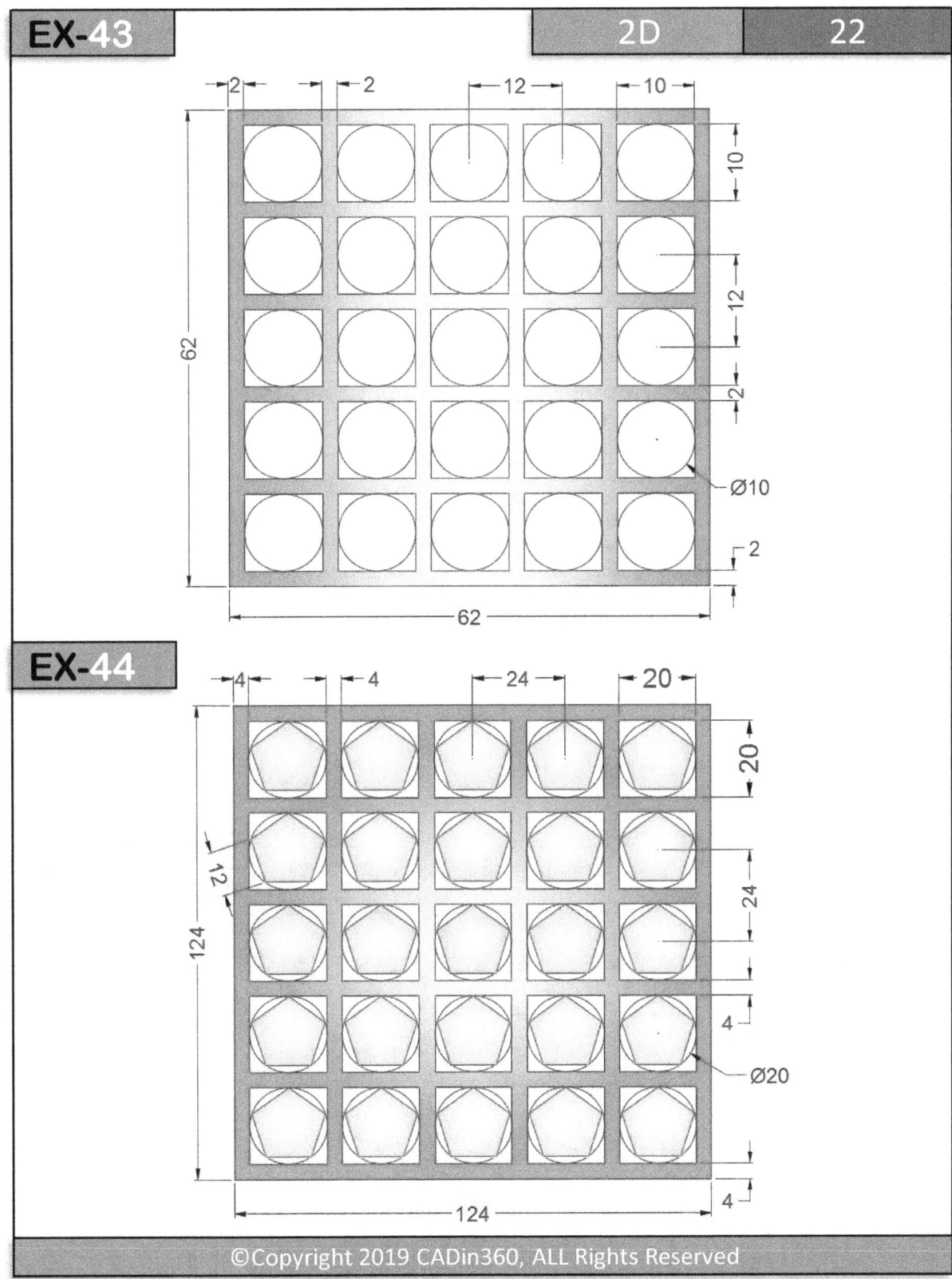

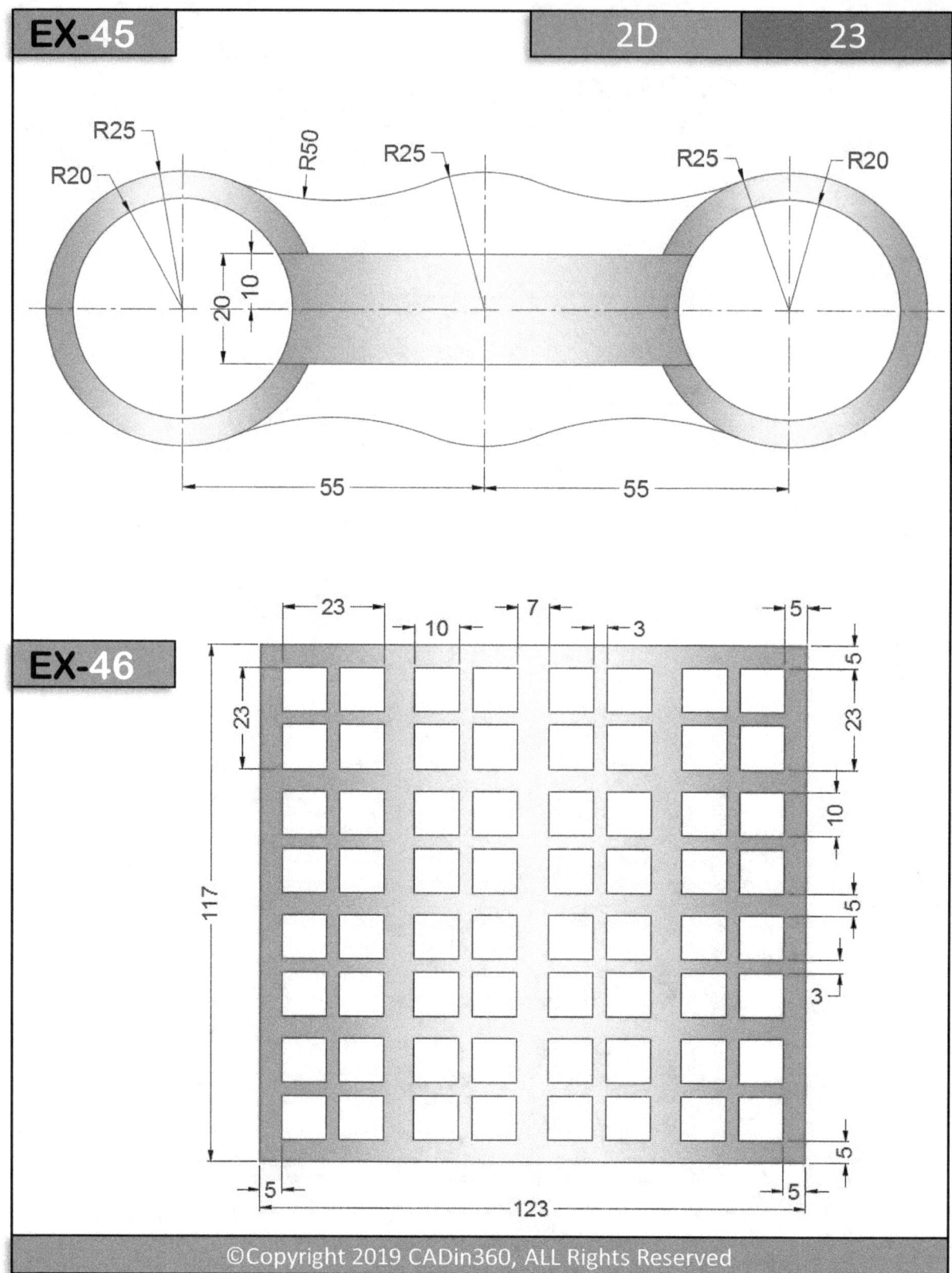

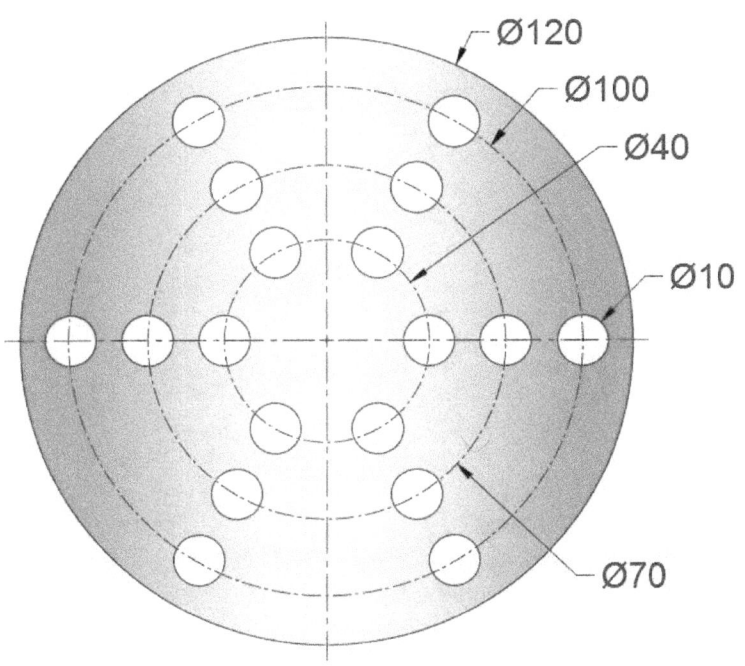

Ø120
Ø100
Ø40
Ø10
Ø70

EX-48

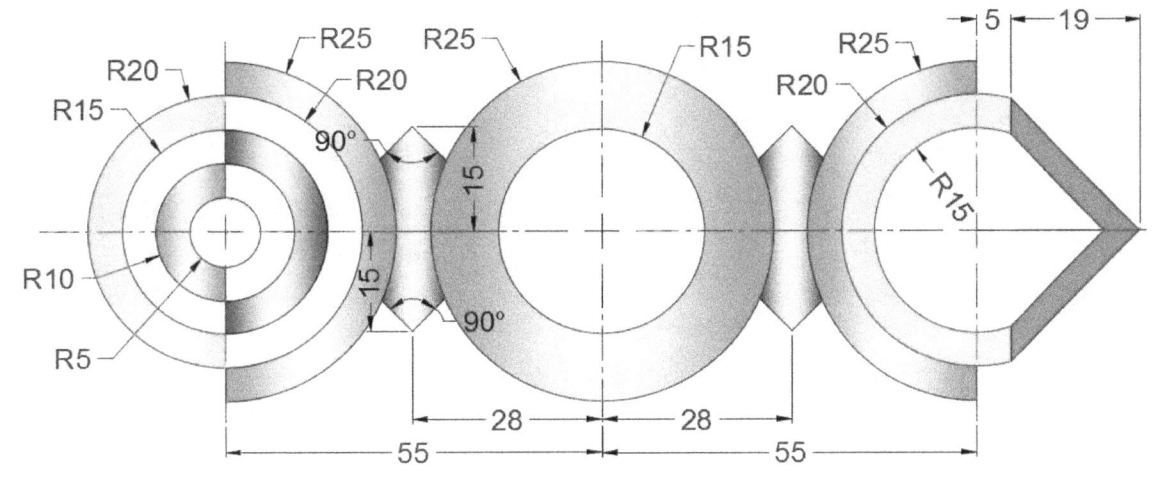

R25   R25   R15   R25   5   19
R20   R20   R15
R15   90°   R20
R10   15   15   R15
R5   90°
28   28
55   55

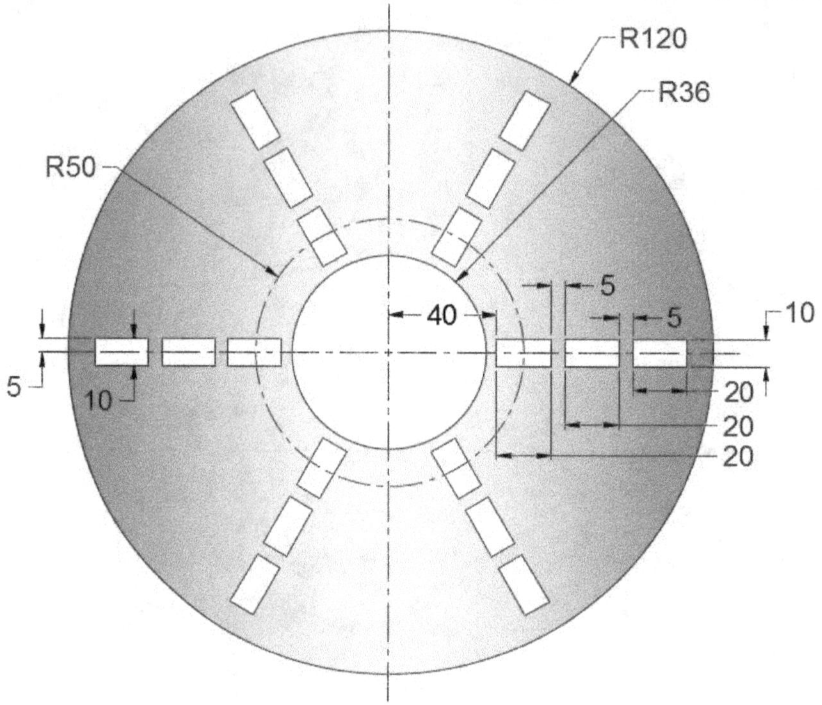

R120

R36

R50

40

5

5

10

5

10

20

20

20

EX-50

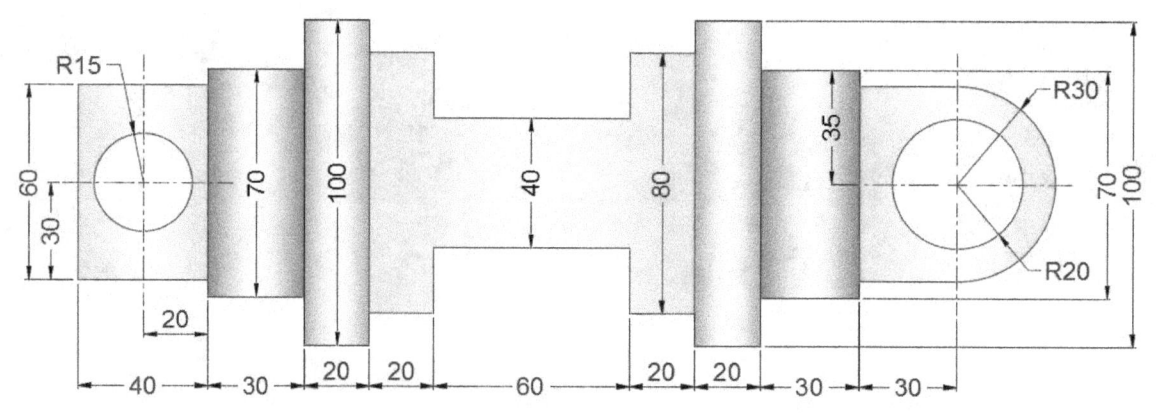

R15

R30

R20

60

30

70

100

40

80

35

70

100

20

40

30

20

20

60

20

20

30

30

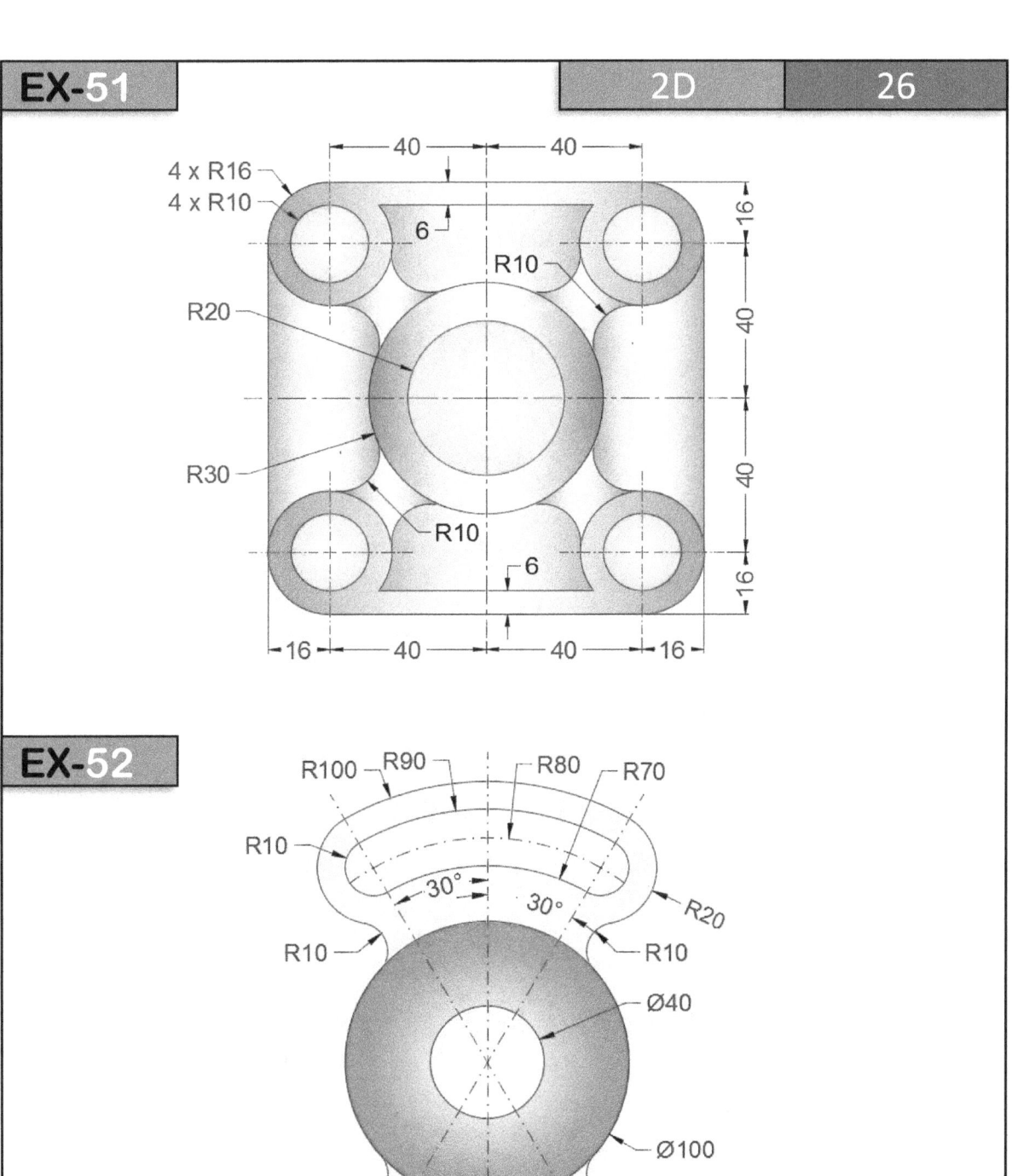

EX-52

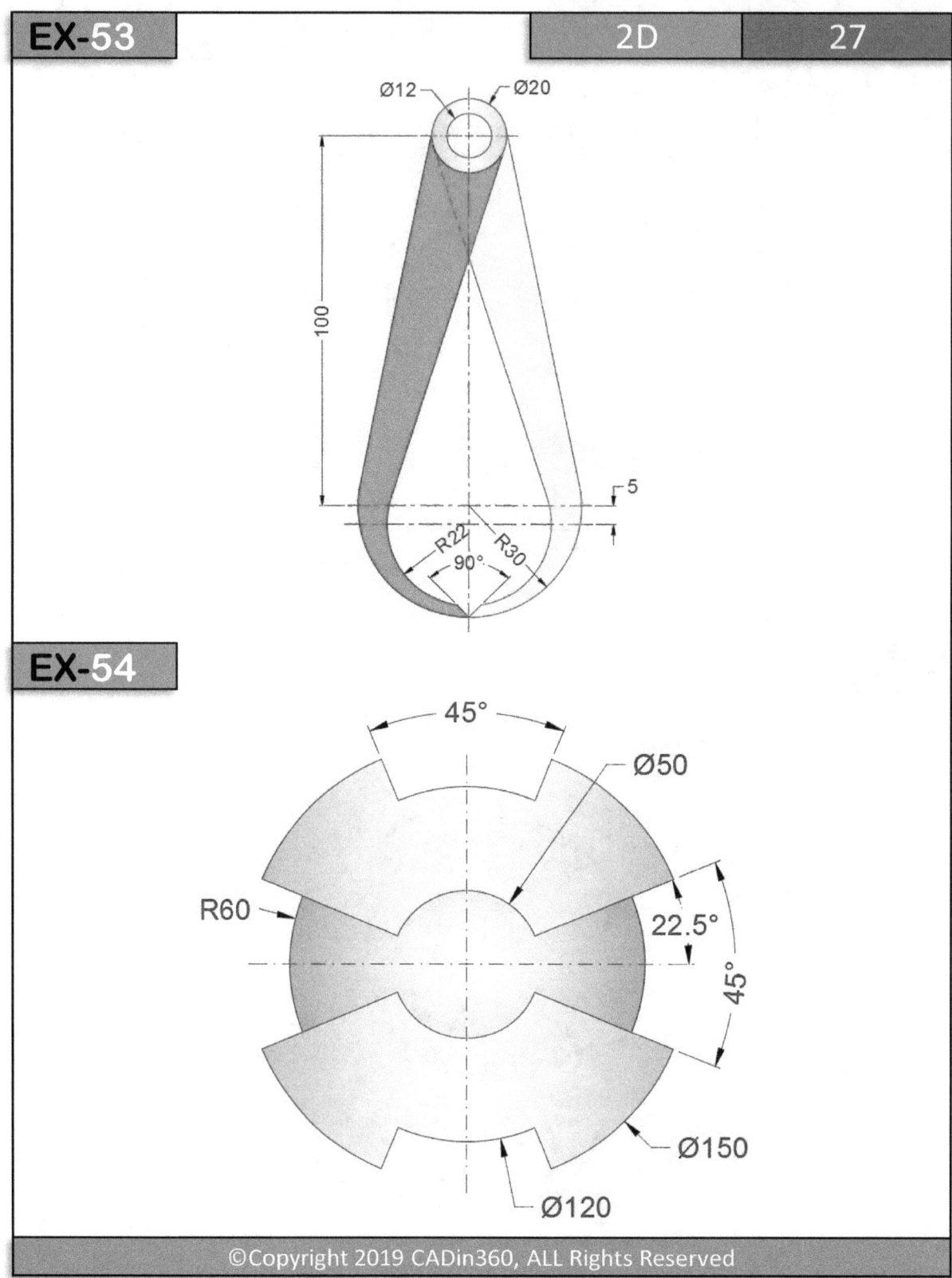

EX-53

2D 27

Ø12 Ø20

100

5

R22 R30
90°

EX-54

45°

Ø50

R60

22.5°

45°

Ø150

Ø120

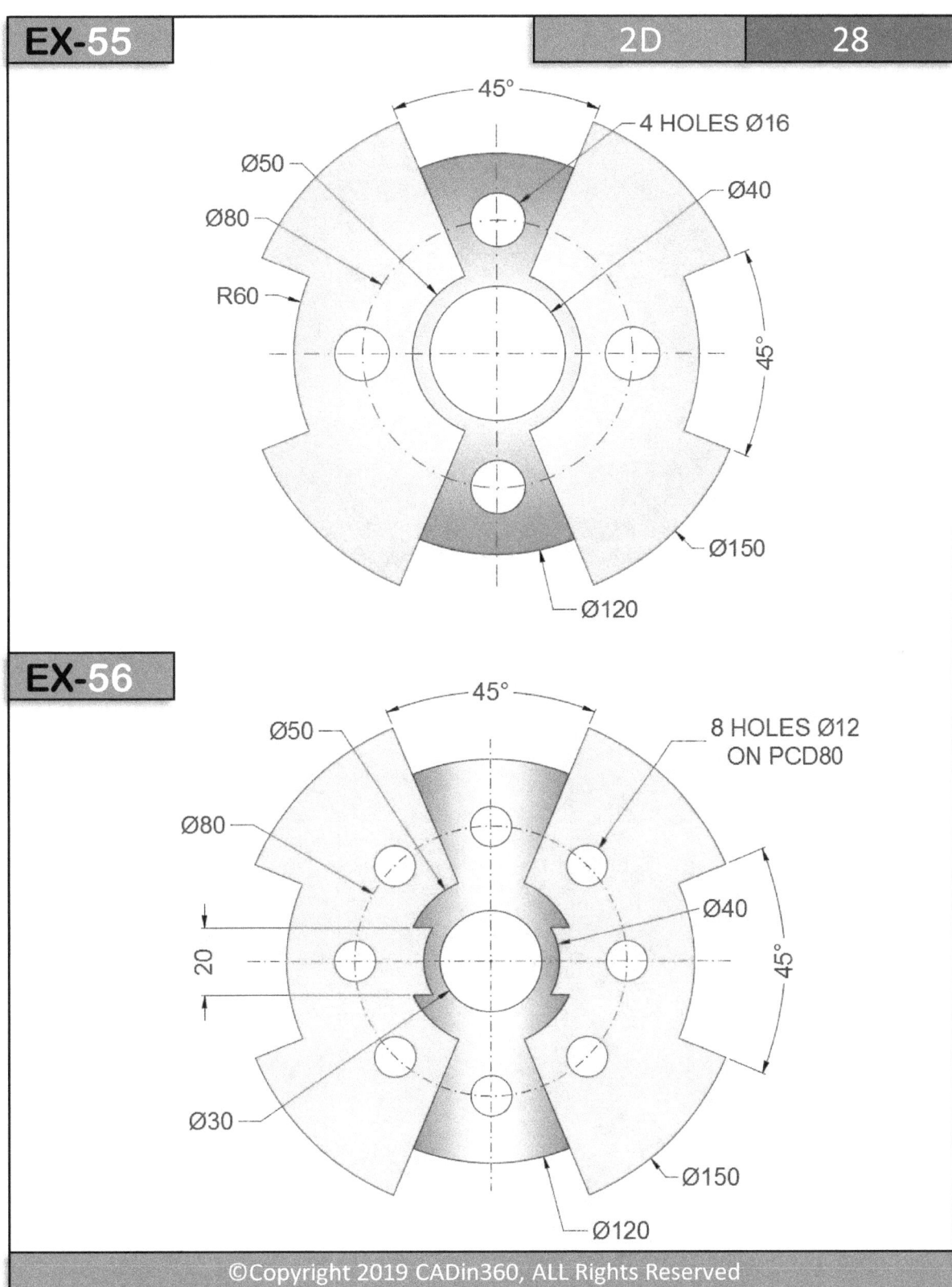

EX-55 | 2D | 28

45°
4 HOLES Ø16
Ø50
Ø80
Ø40
R60
45°
Ø150
Ø120

EX-56

45°
Ø50
8 HOLES Ø12
ON PCD80
Ø80
Ø40
20
45°
Ø30
Ø150
Ø120

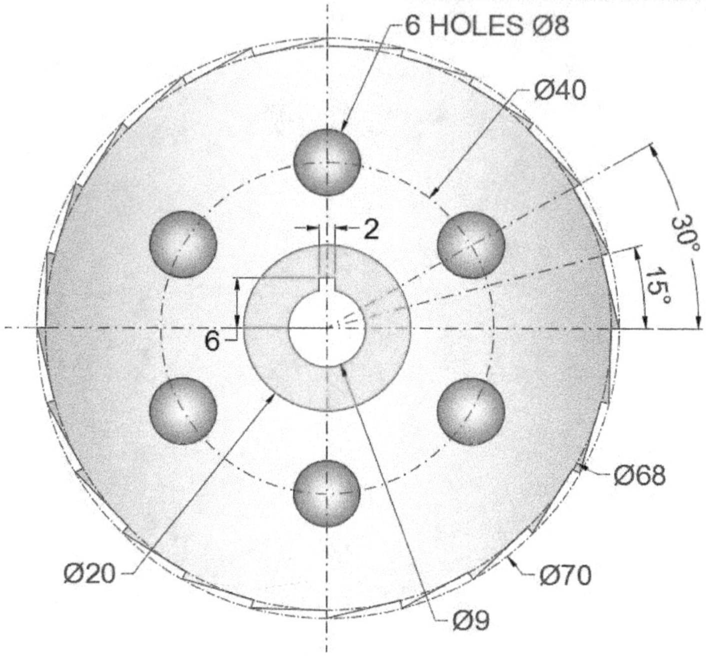

EX-57

2D   29

6 HOLES Ø8
Ø40
2
6
30°
15°
Ø68
Ø20
Ø70
Ø9

EX-58

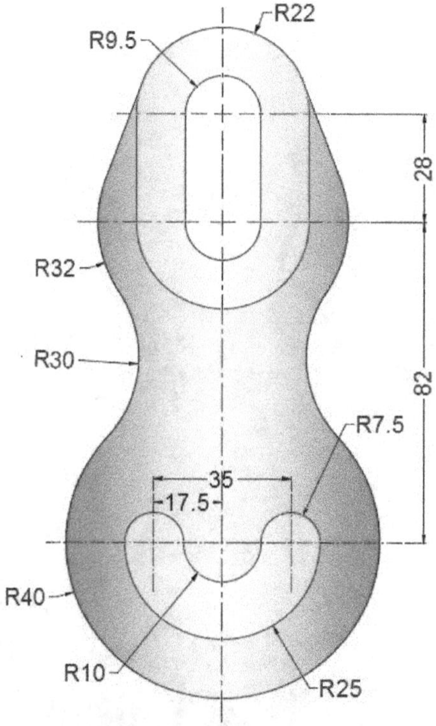

R22
R9.5
28
R32
R30
82
R7.5
35
17.5
R40
R10
R25

R12 — R22

28

R22

70

2 x R20

70 | 70

2 x R30

86.60

20 | 20

R50

2 x R15

2 x R25

100

3 x R30

3 x R20

R80

98

70 | 70

86

R80

2 x R15

2 x R25

R80

50

100

R80

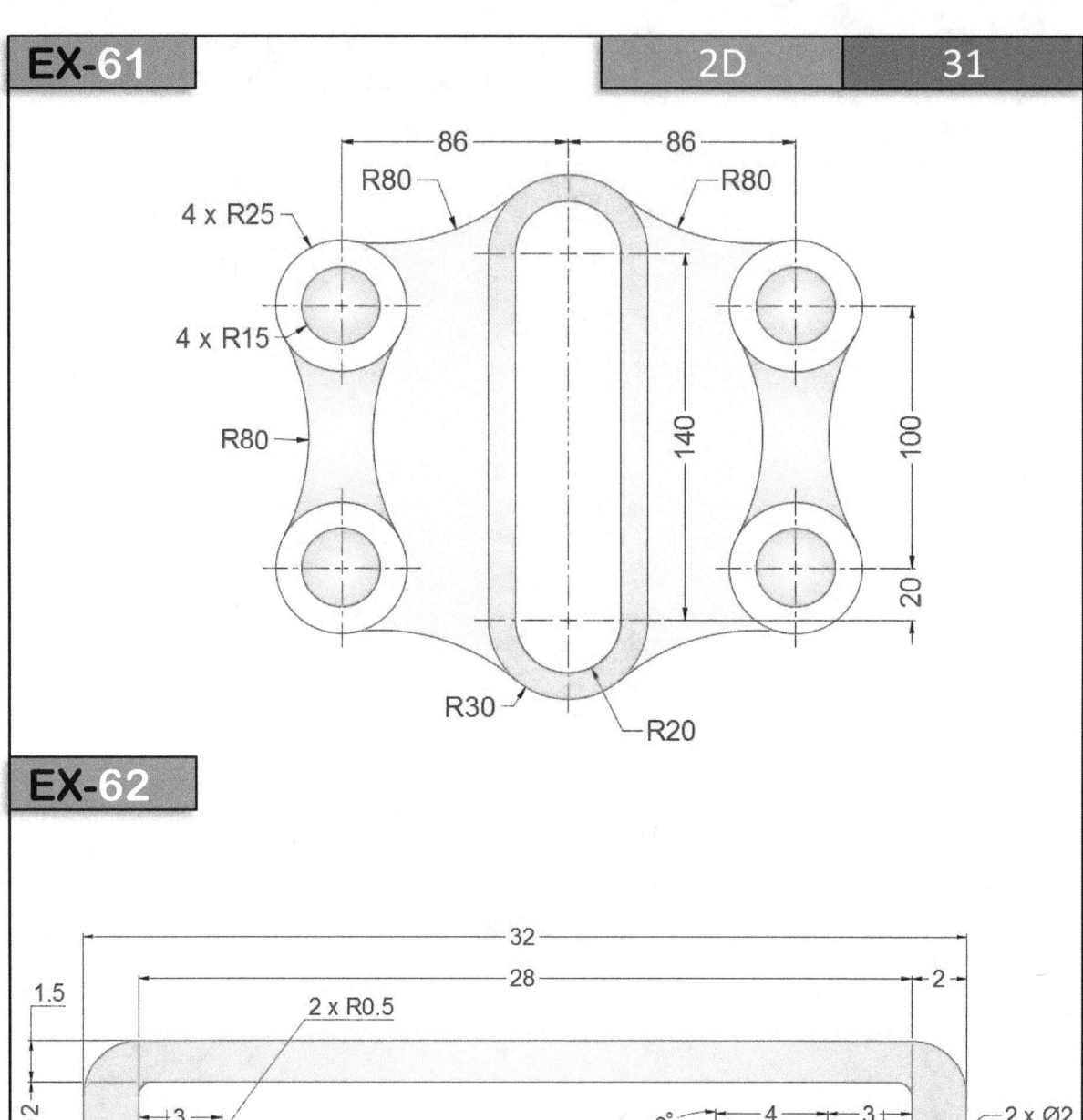

EX-62

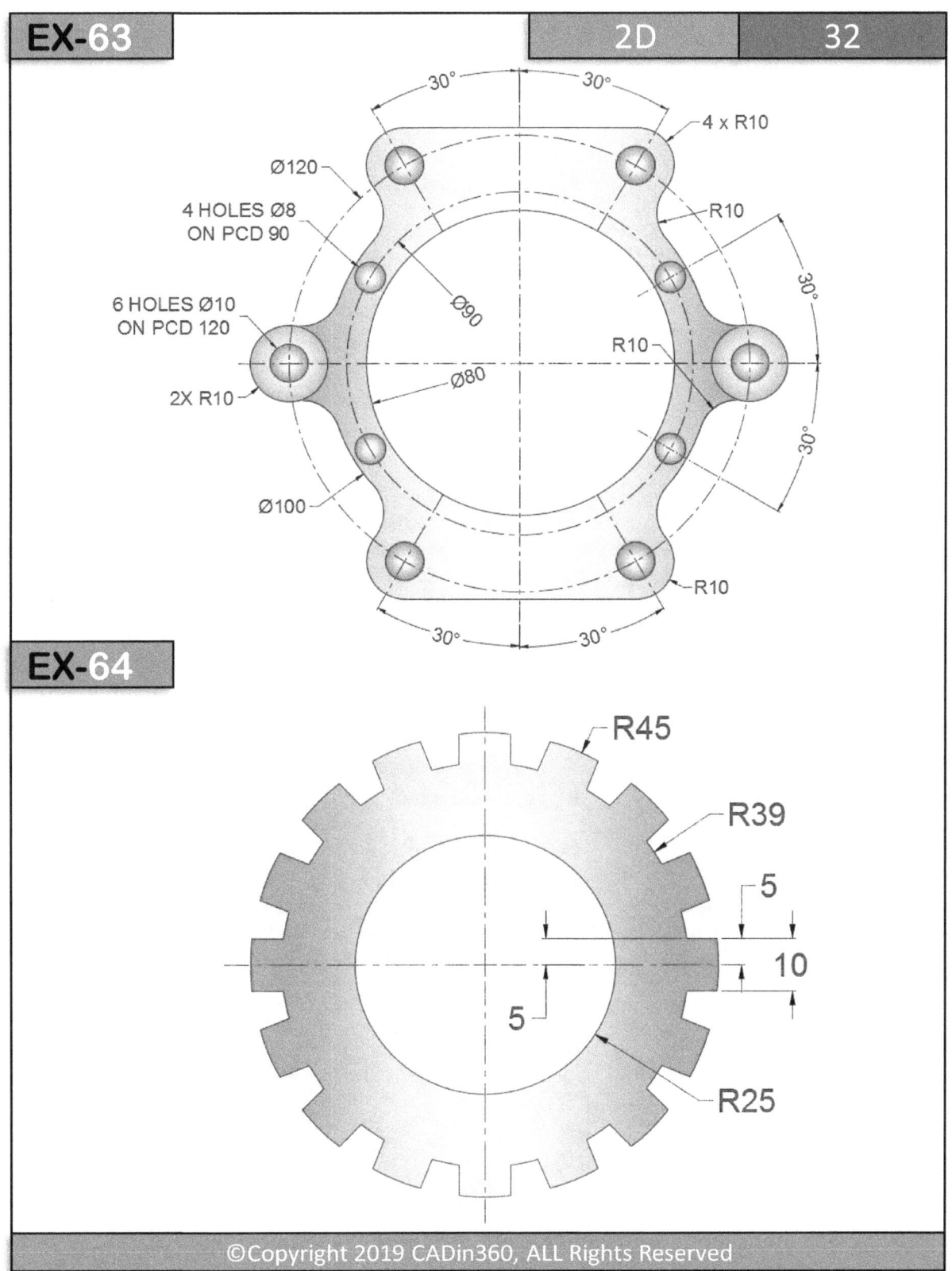

EX-64

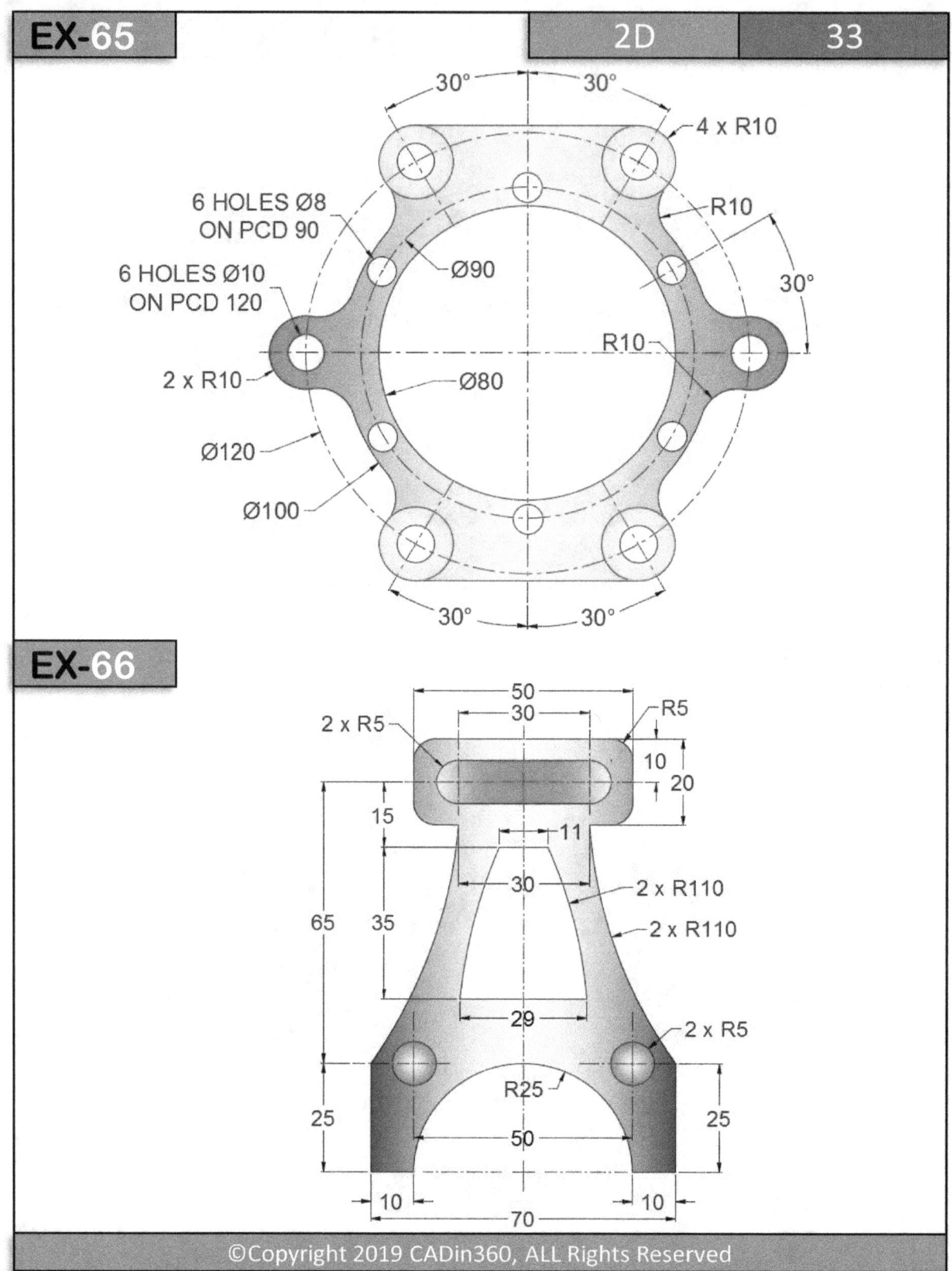

EX-66

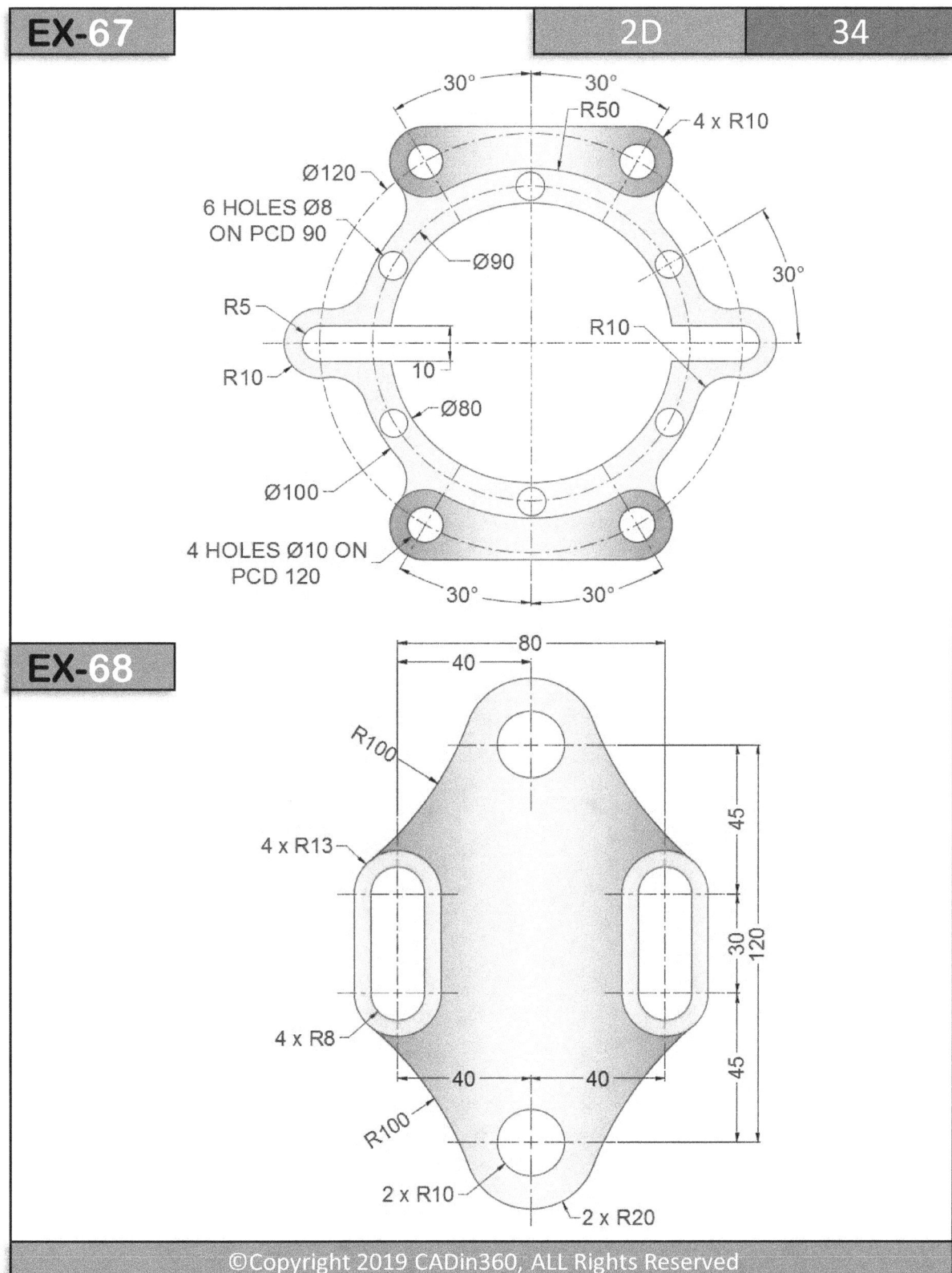

EX-67

2D

34

30° 30°
R50
4 x R10
Ø120
6 HOLES Ø8
ON PCD 90
Ø90
30°
R5
R10
R10
10
Ø80
Ø100
4 HOLES Ø10 ON
PCD 120
30° 30°

EX-68

80
40
R100
45
4 x R13
30
120
4 x R8
40
40
45
R100
2 x R10
2 x R20

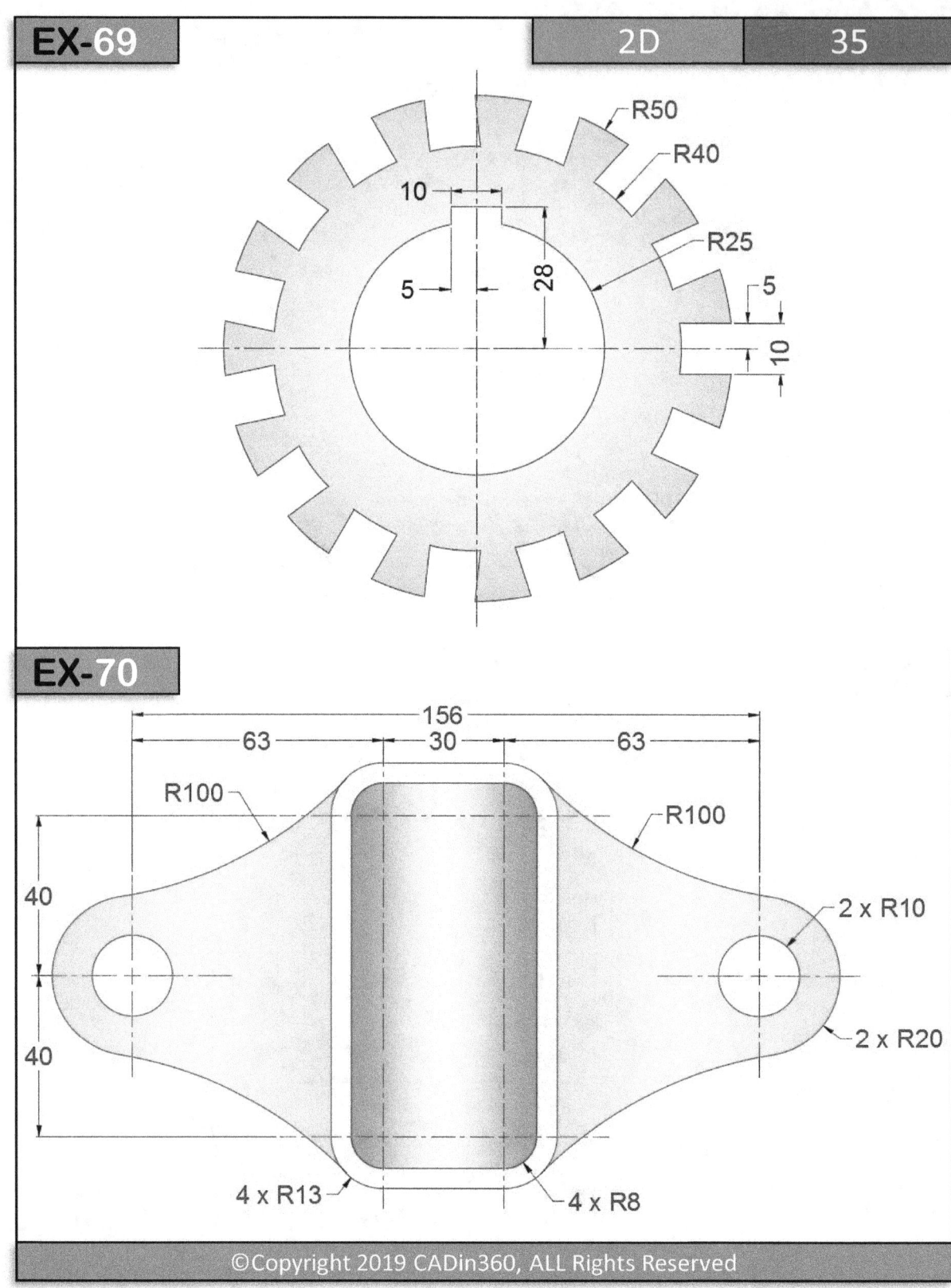

EX-70

EX-71

25   65   20   65   25

4 x R5

35   15   15

4 x R5

10

4 x R3

35

50

29

11

30

40

50

70

R25

R110

R110

30

40

4 x R5

10

10

EX-72

38   98   98   38

53   23   30   23

R165

4X R5

4X R8

15

2X Ø45

44

R165

17

45

60

45

75

52.5

105

4X R8

R165

45

2X Ø75

15

83   60   53   60

300

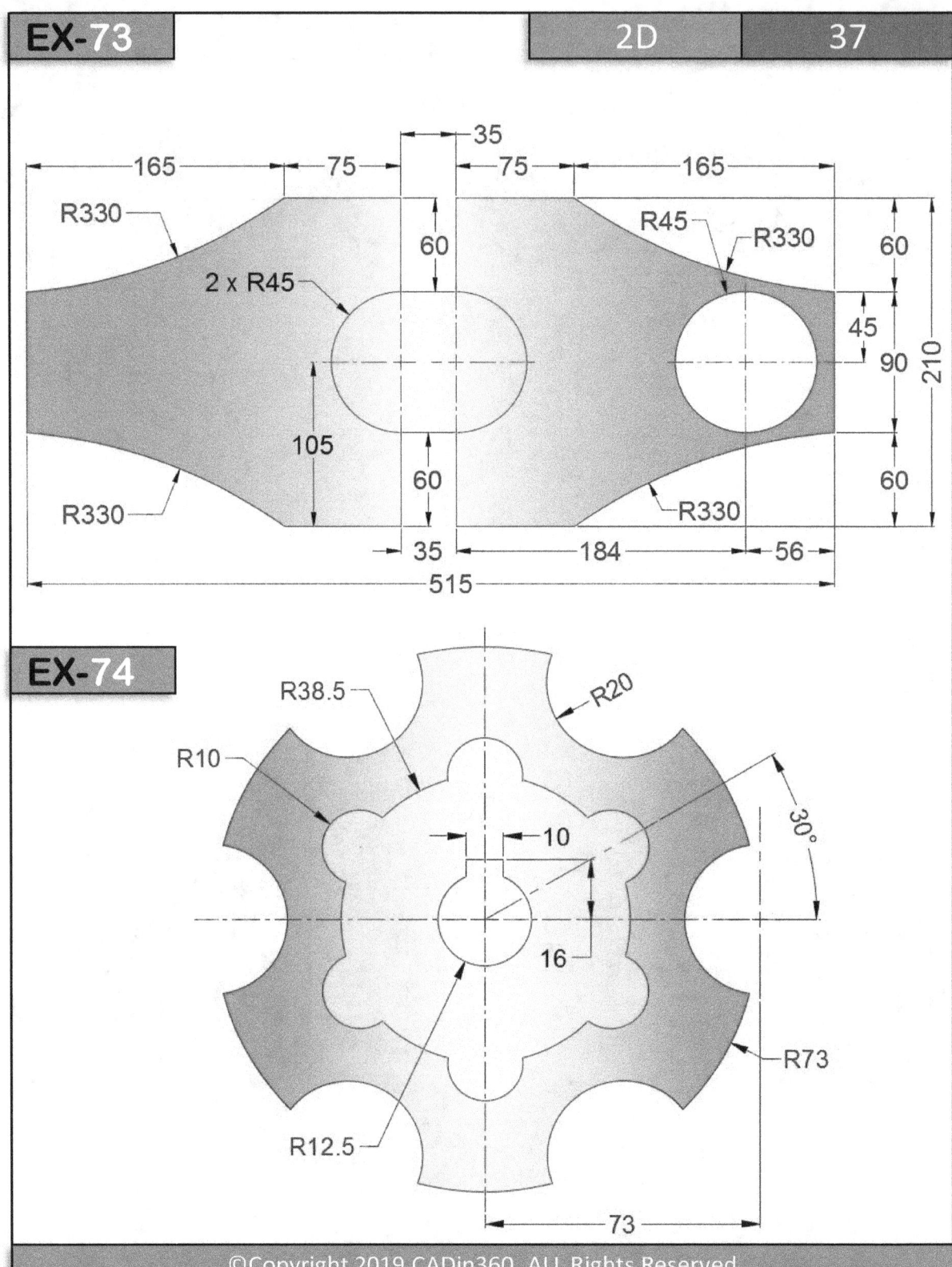

EX-74

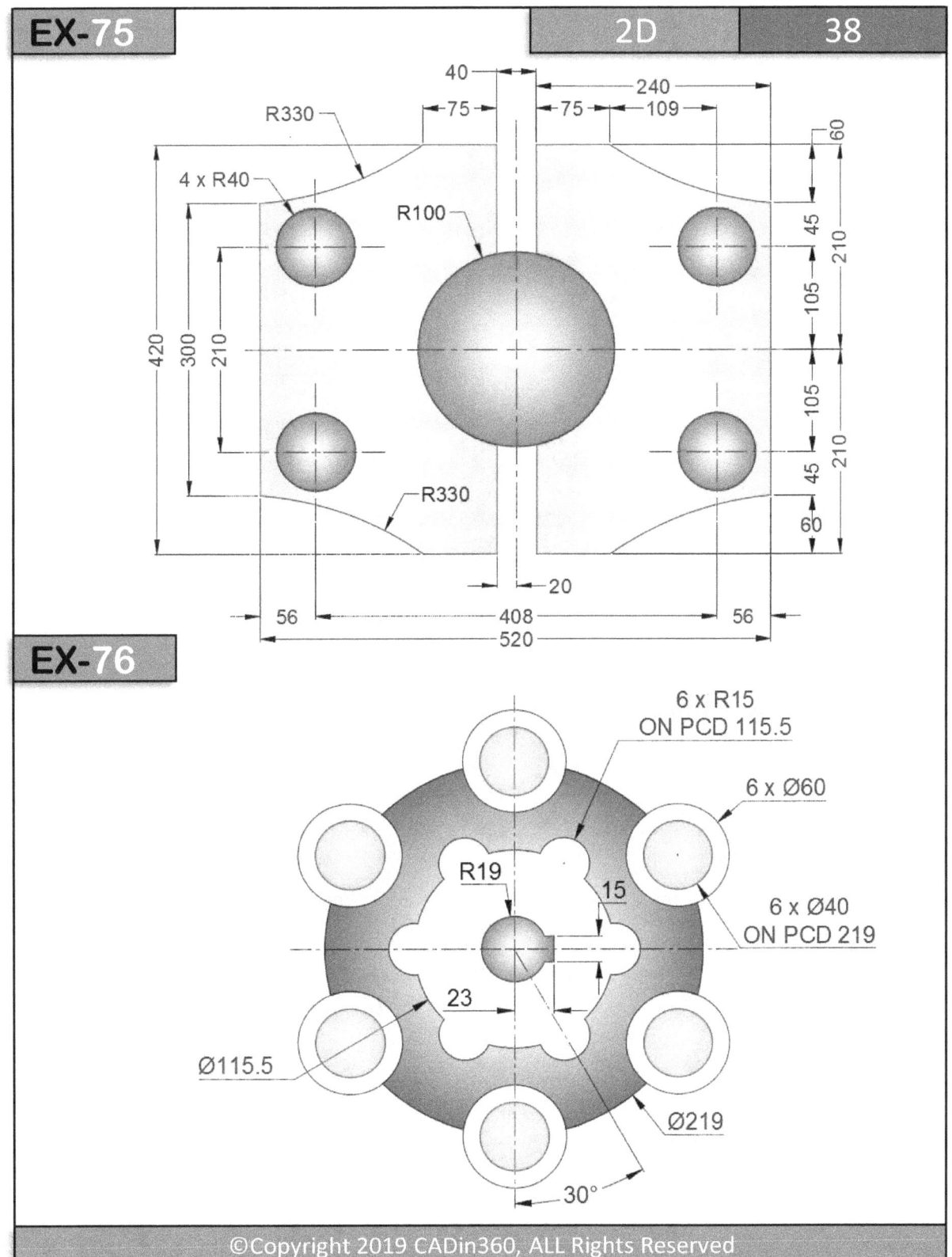

EX-78

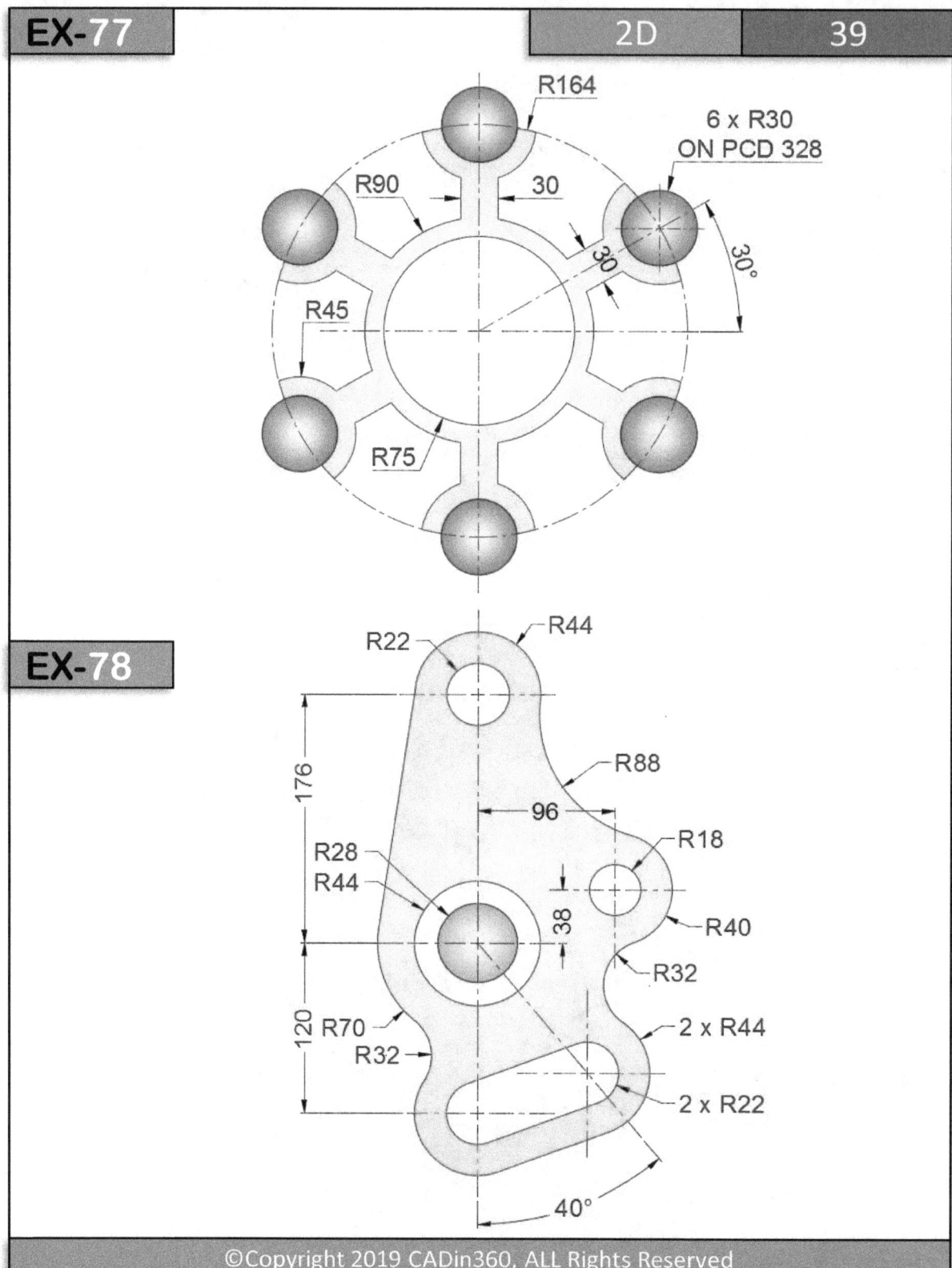

R164

6 x R30
ON PCD 328

R90

30

30

30°

R45

R75

R22

R44

R88

96

R28
R44

R18

38

R40

R32

176

R70

R32

120

2 x R44

2 x R22

40°

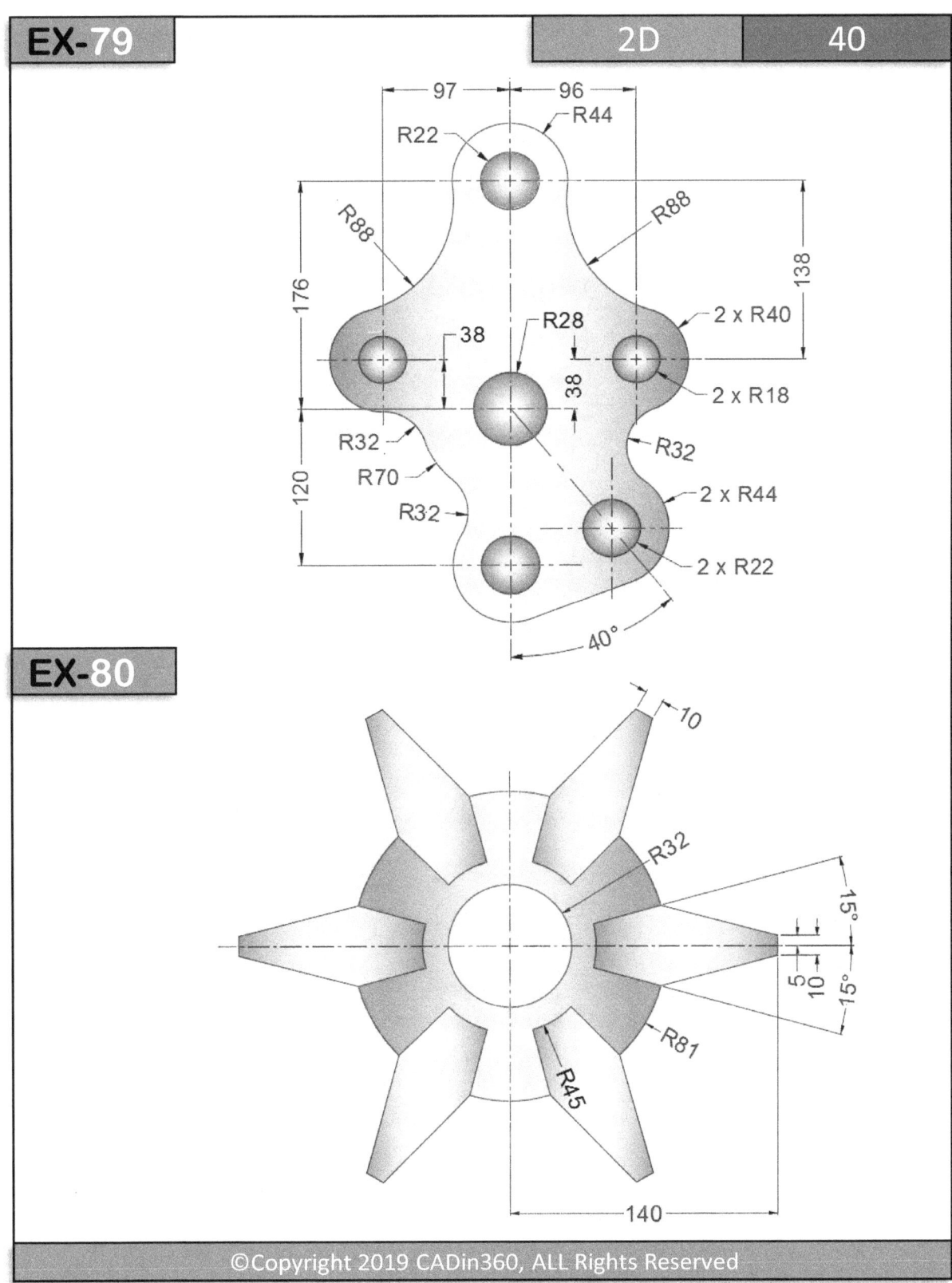

EX-80

EX-82

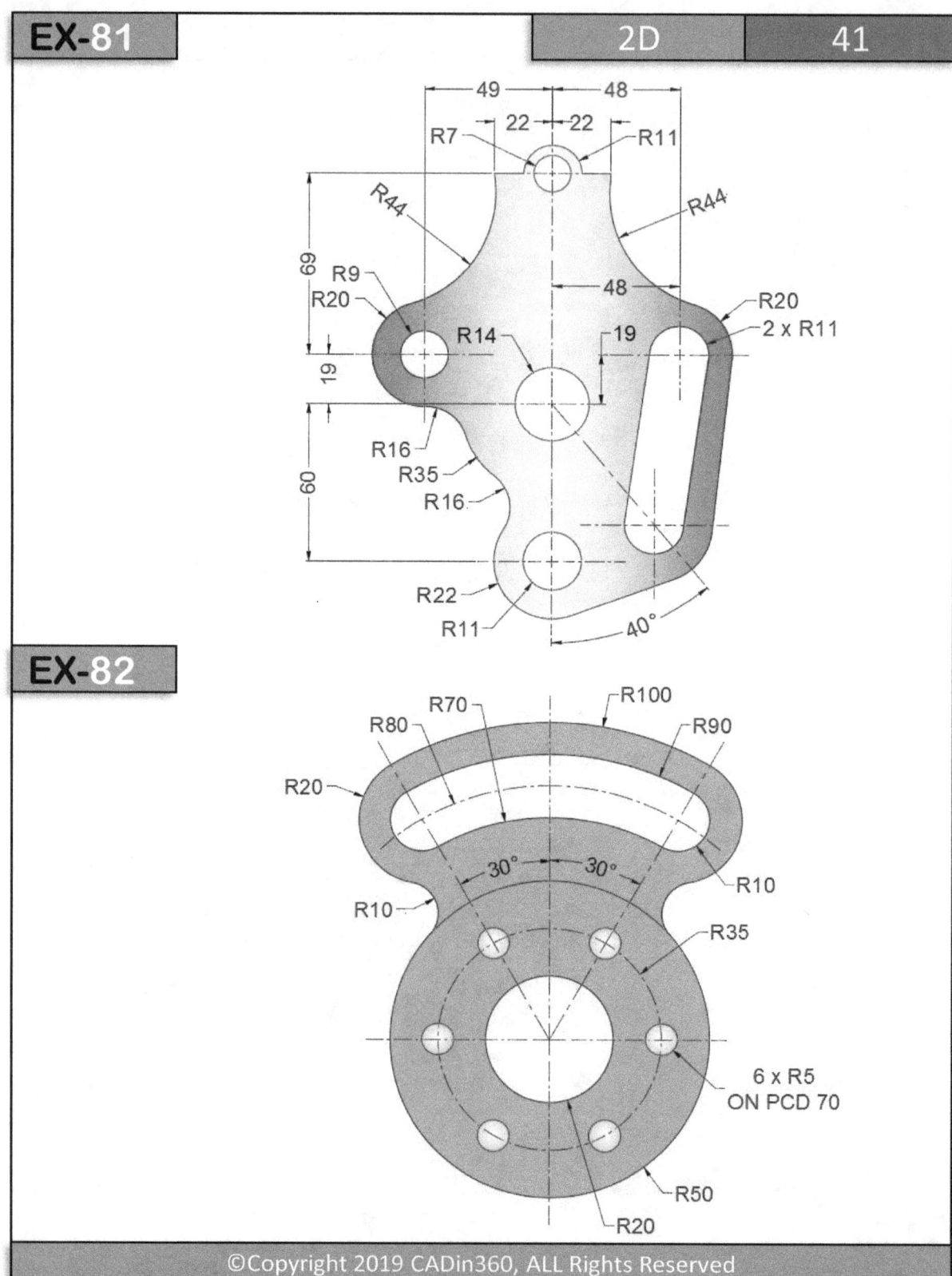

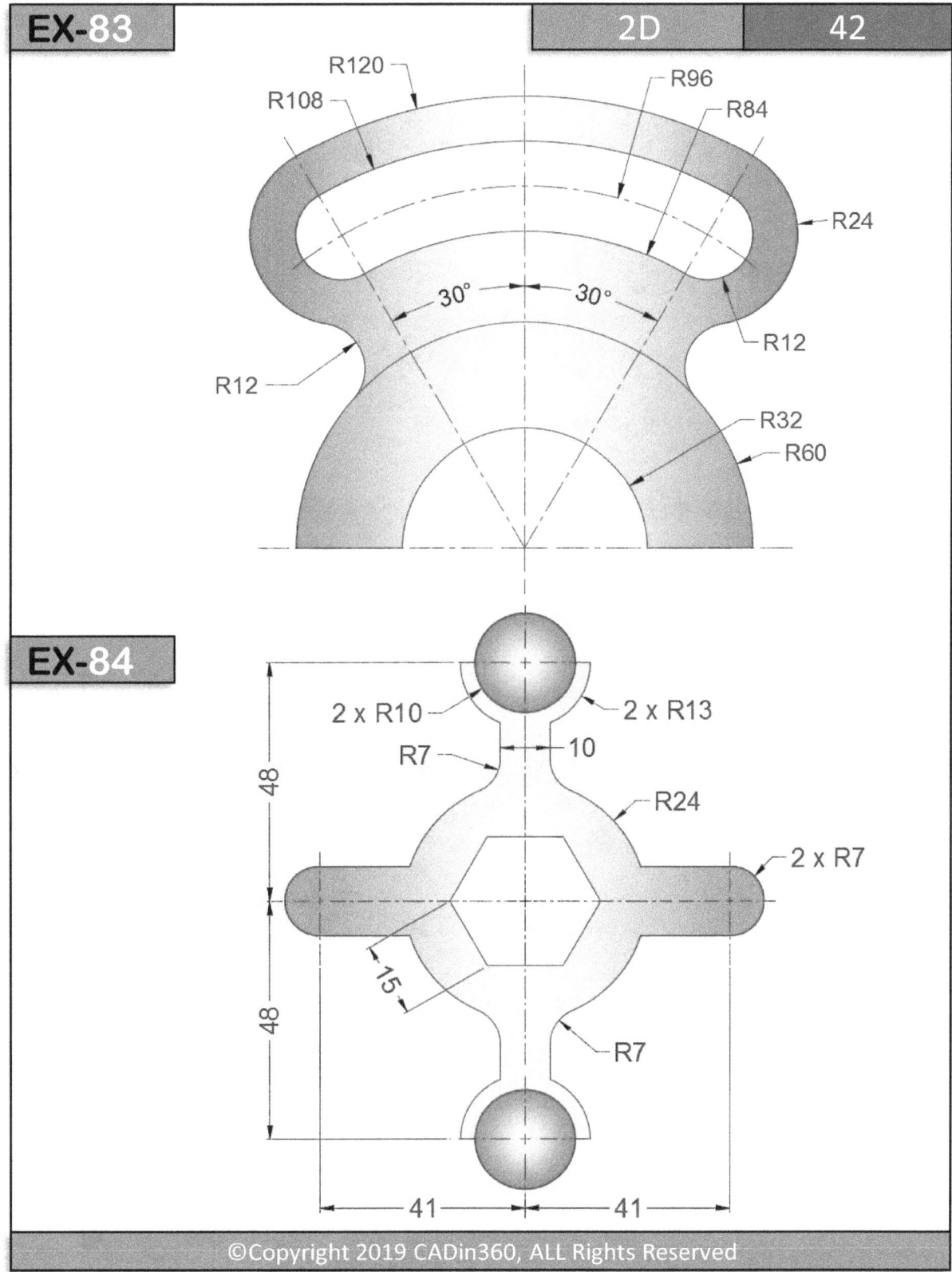

EX-83   2D   42

R120
R108
R96
R84
R24
30° 30°
R12
R12
R32
R60

EX-84

2 x R10
2 x R13
R7
10
R24
2 x R7
48
15
48
R7
41
41

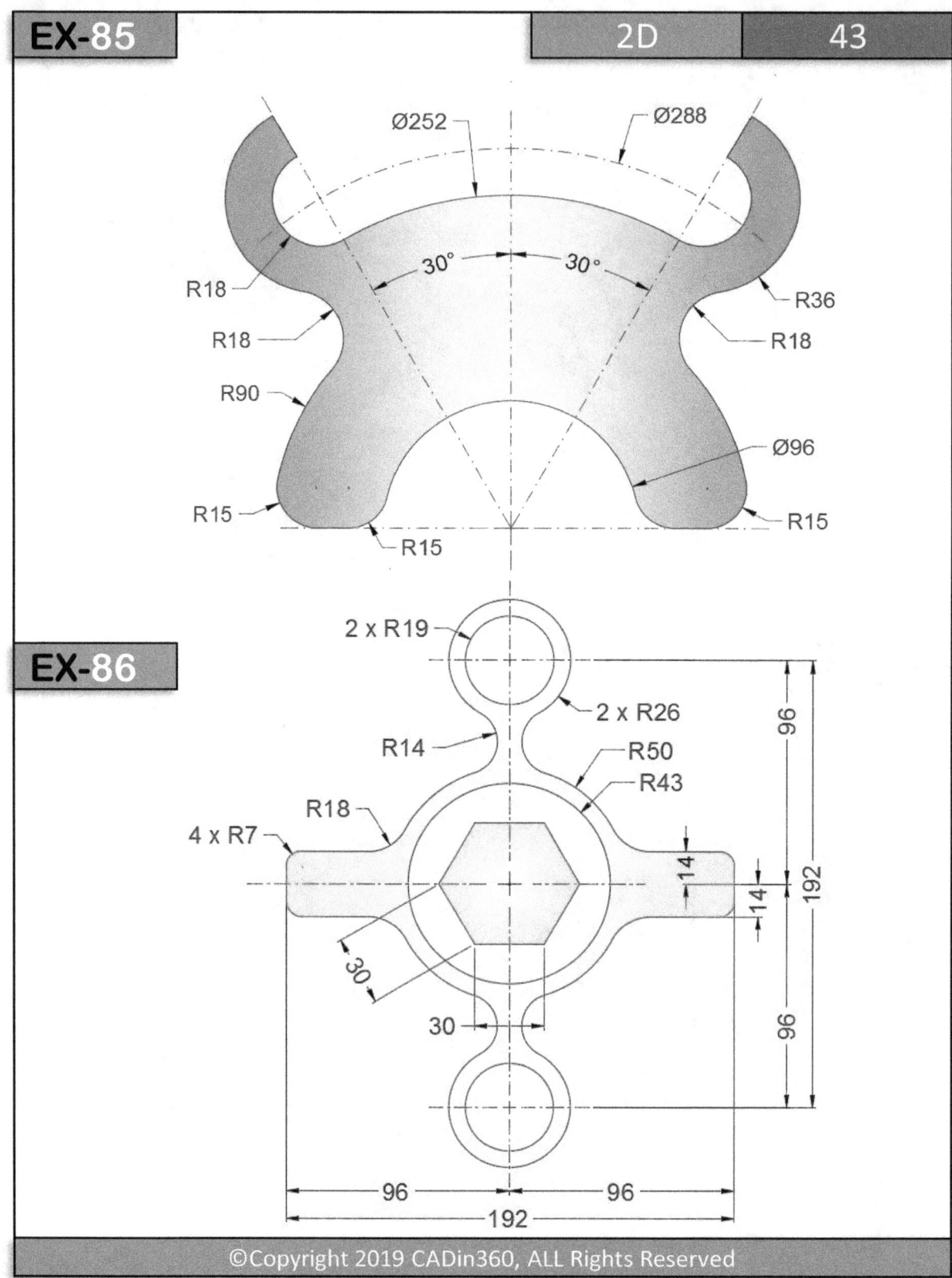

Ø252   Ø288

R18

R18

30°   30°

R36

R18

R90

Ø96

R15

R15

R15

EX-86

2 x R19

2 x R26

R14

R50

R43

R18

4 x R7

30

96

14

14

192

30

96

96

96

192

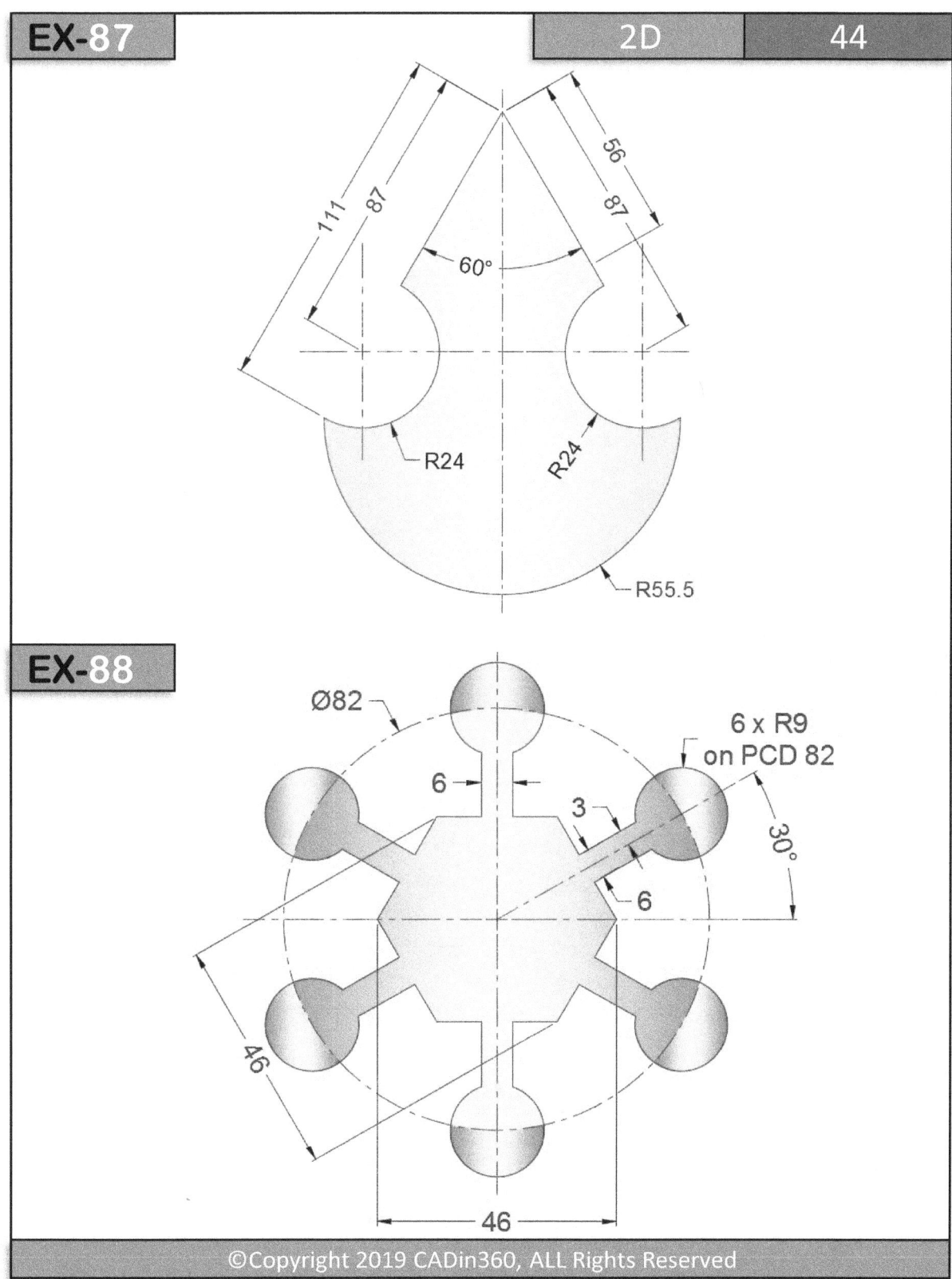

EX-88

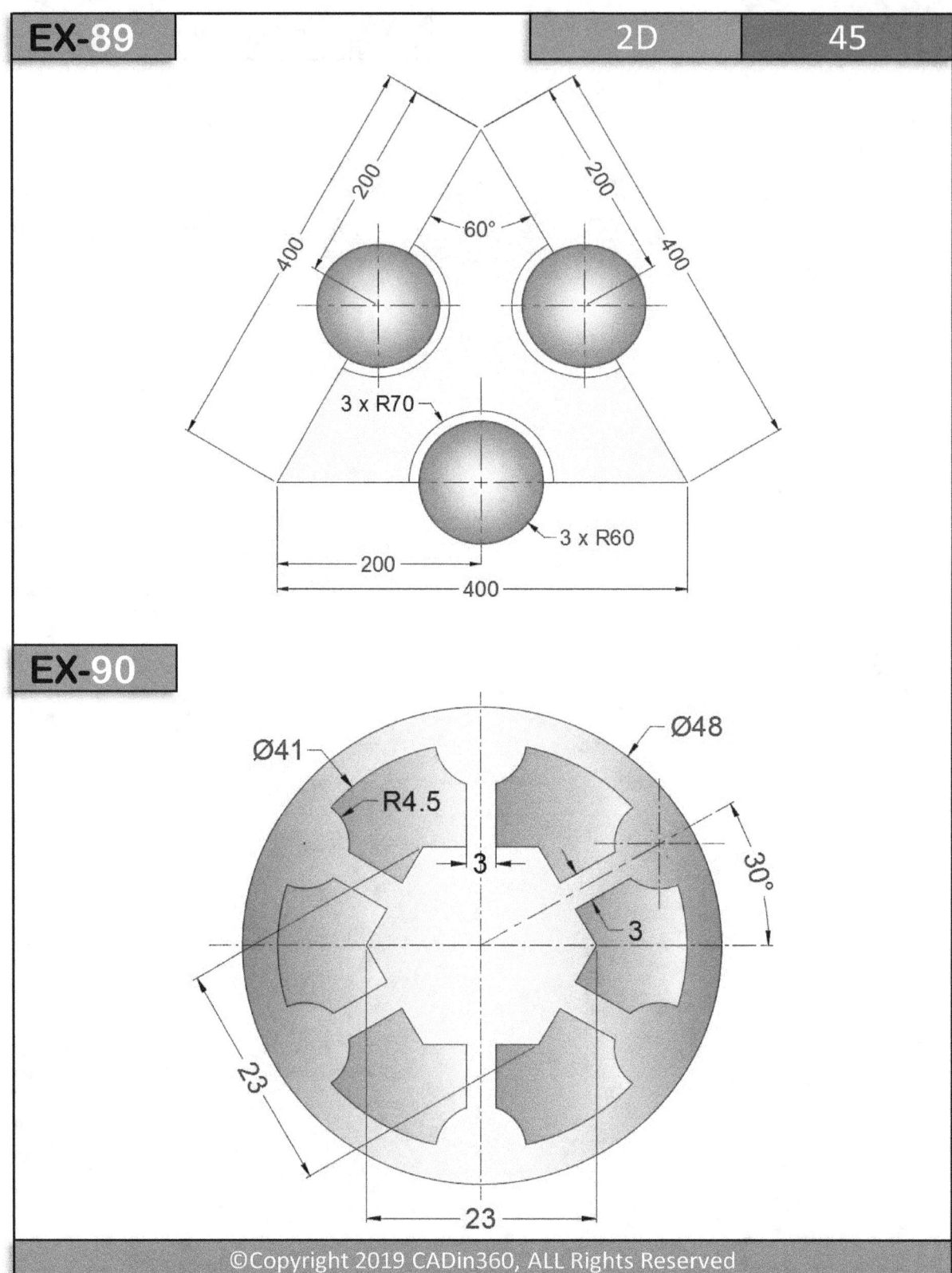

EX-89   2D   45

200
400
60°
200
400
3 x R70
3 x R60
200
400

EX-90

Ø41
R4.5
Ø48
3
30°
3
23
23

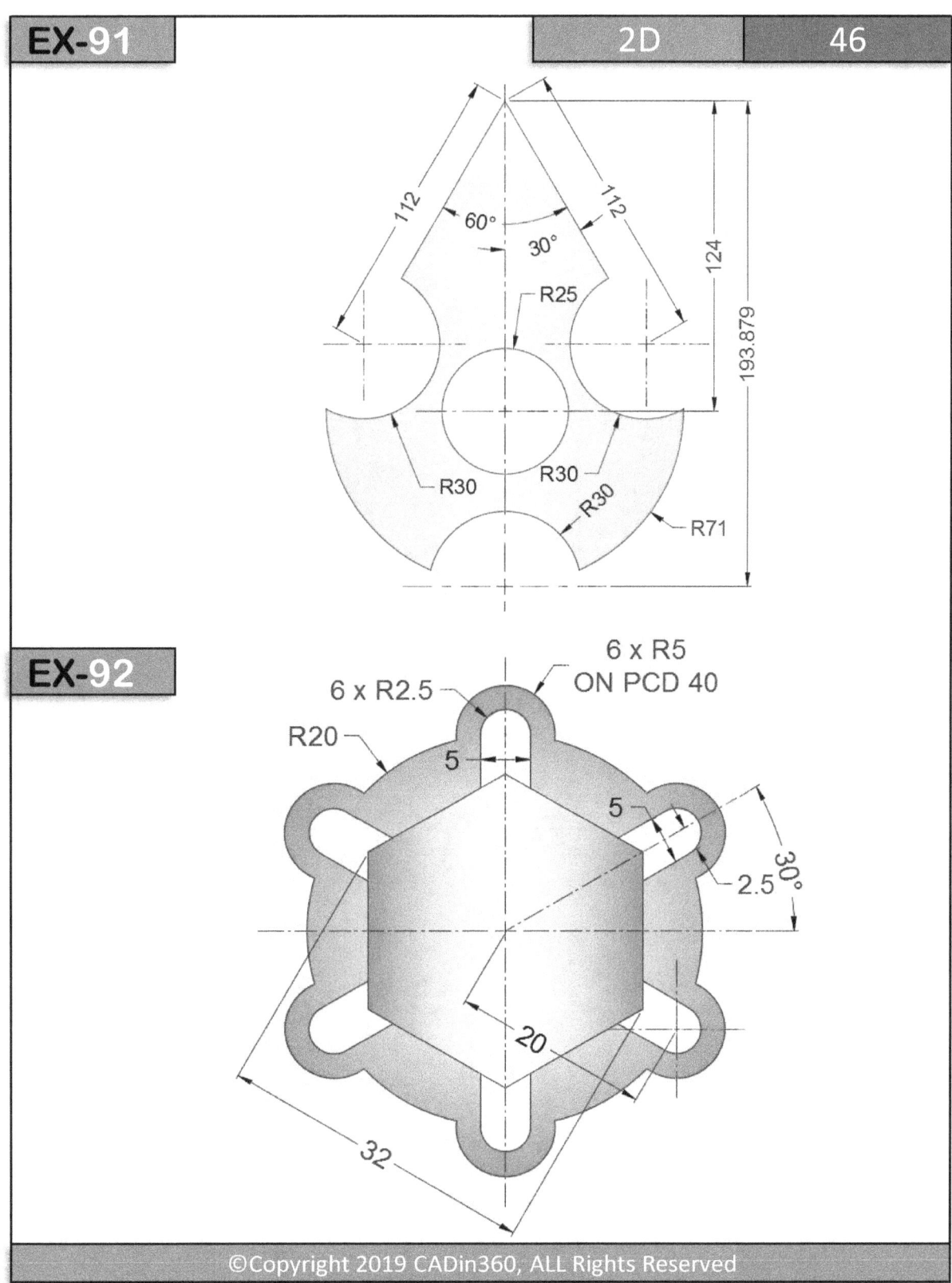

EX-92

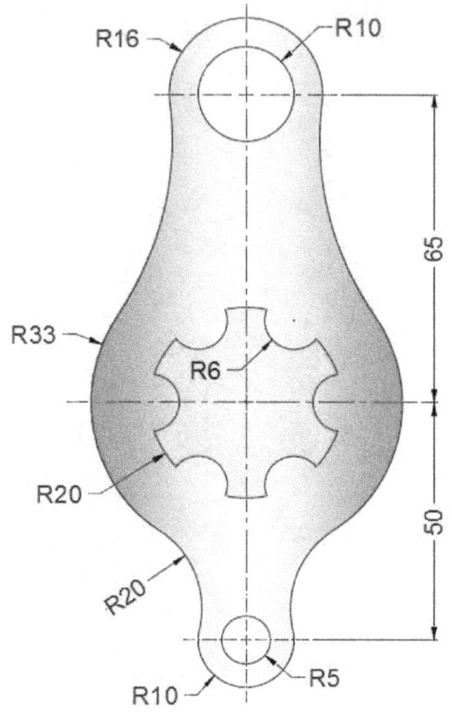

R16　R10

65

R33

R6

R20

50

R20

R10　R5

EX-94

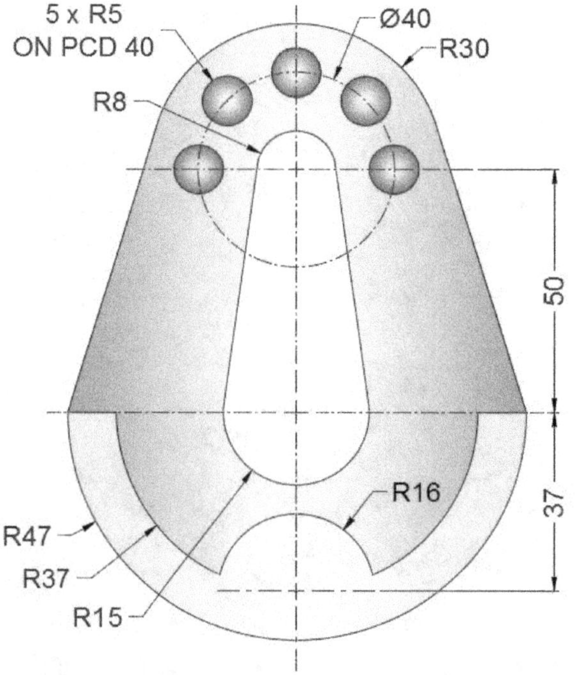

5 x R5
ON PCD 40

Ø40
R30

R8

50

R16

37

R47

R37

R15

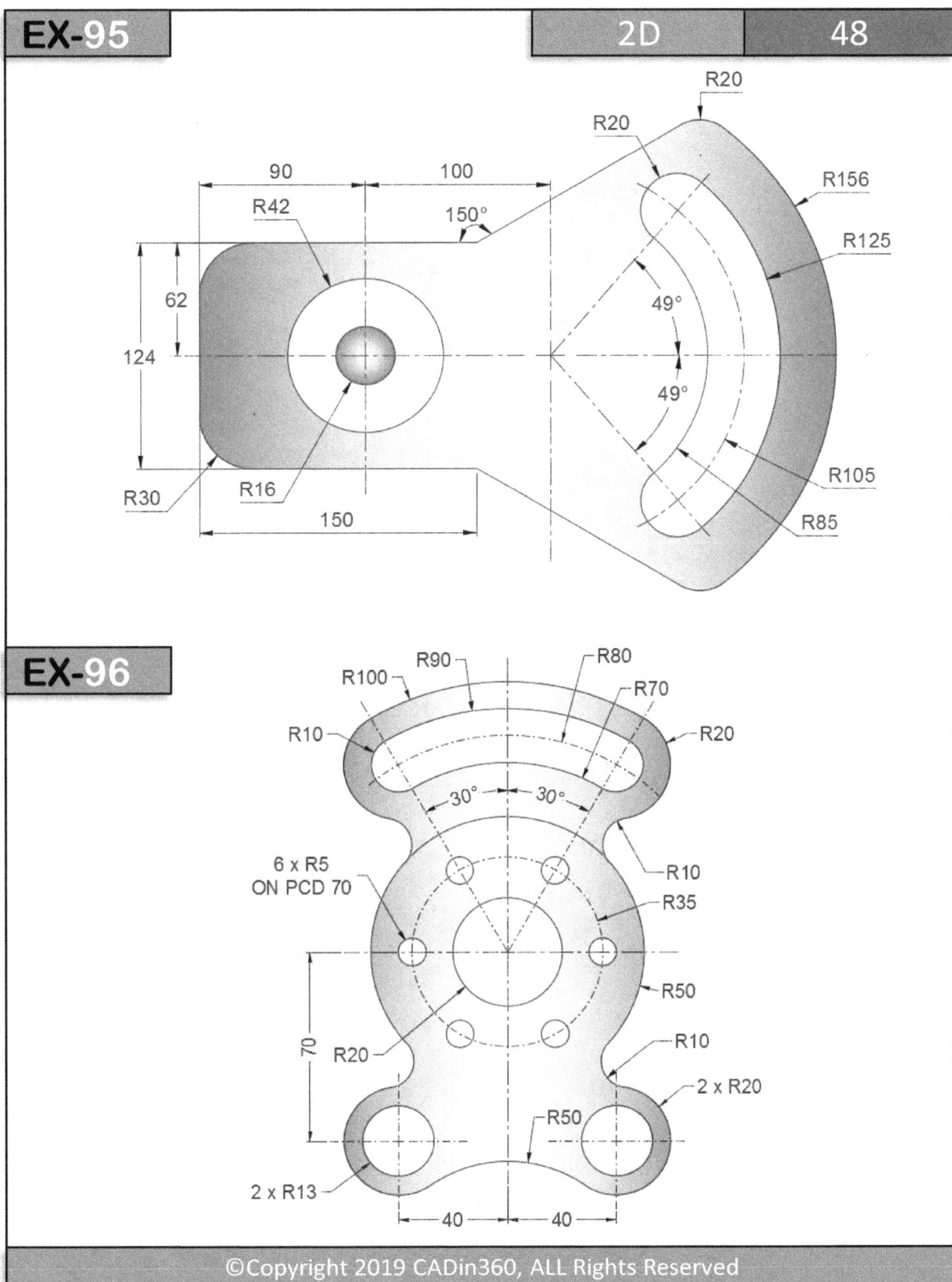

EX-96

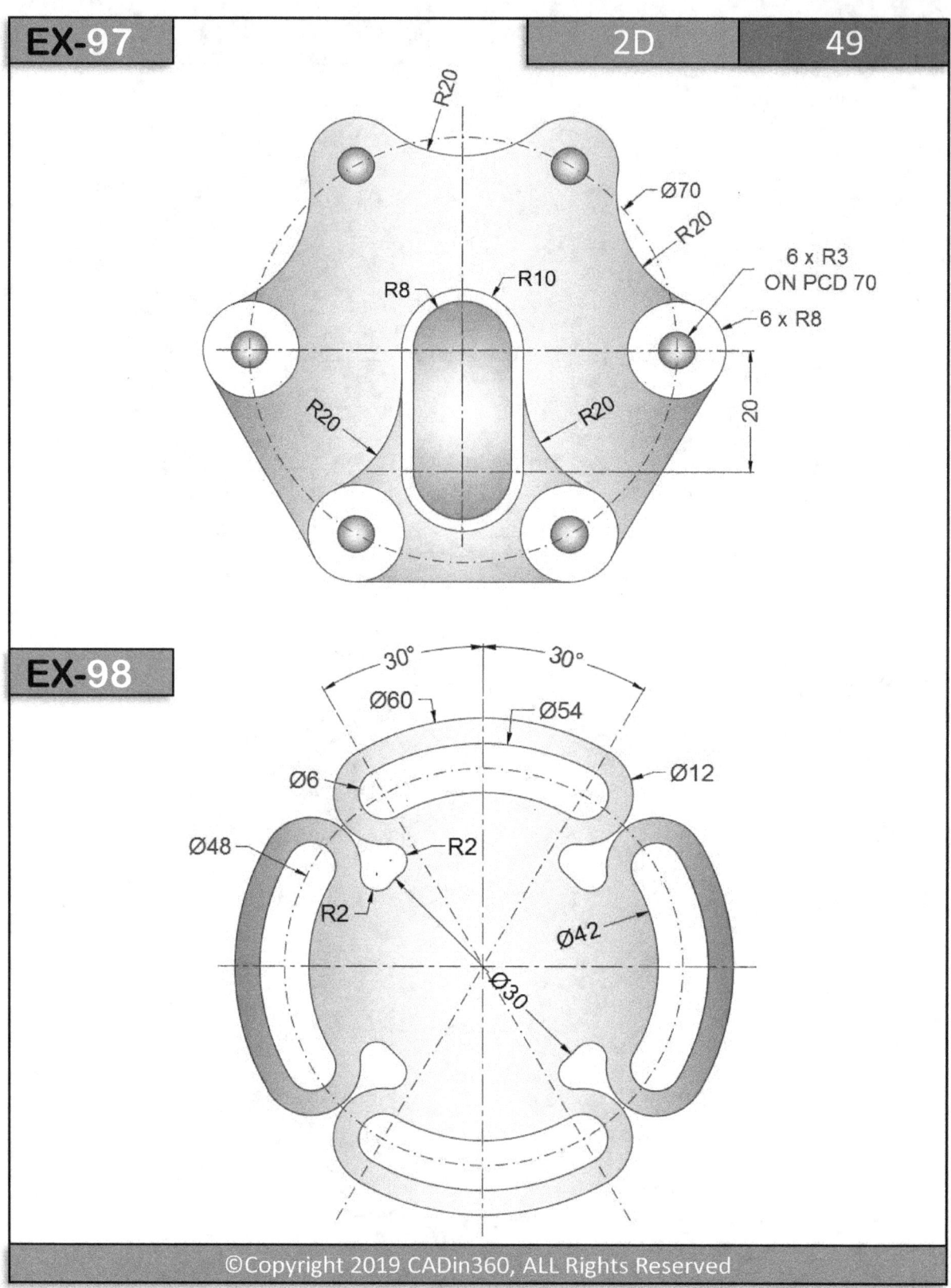

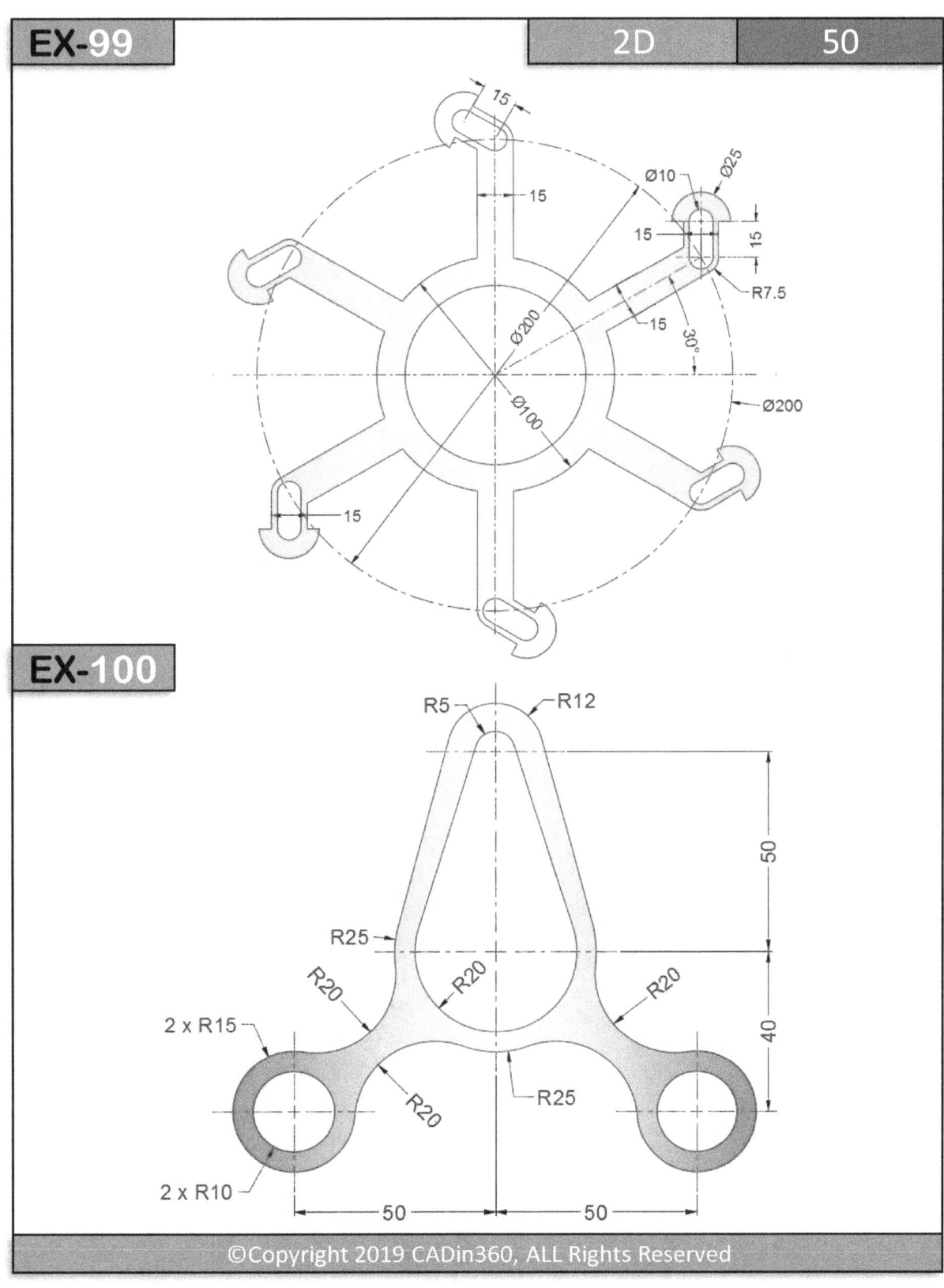

EX-100

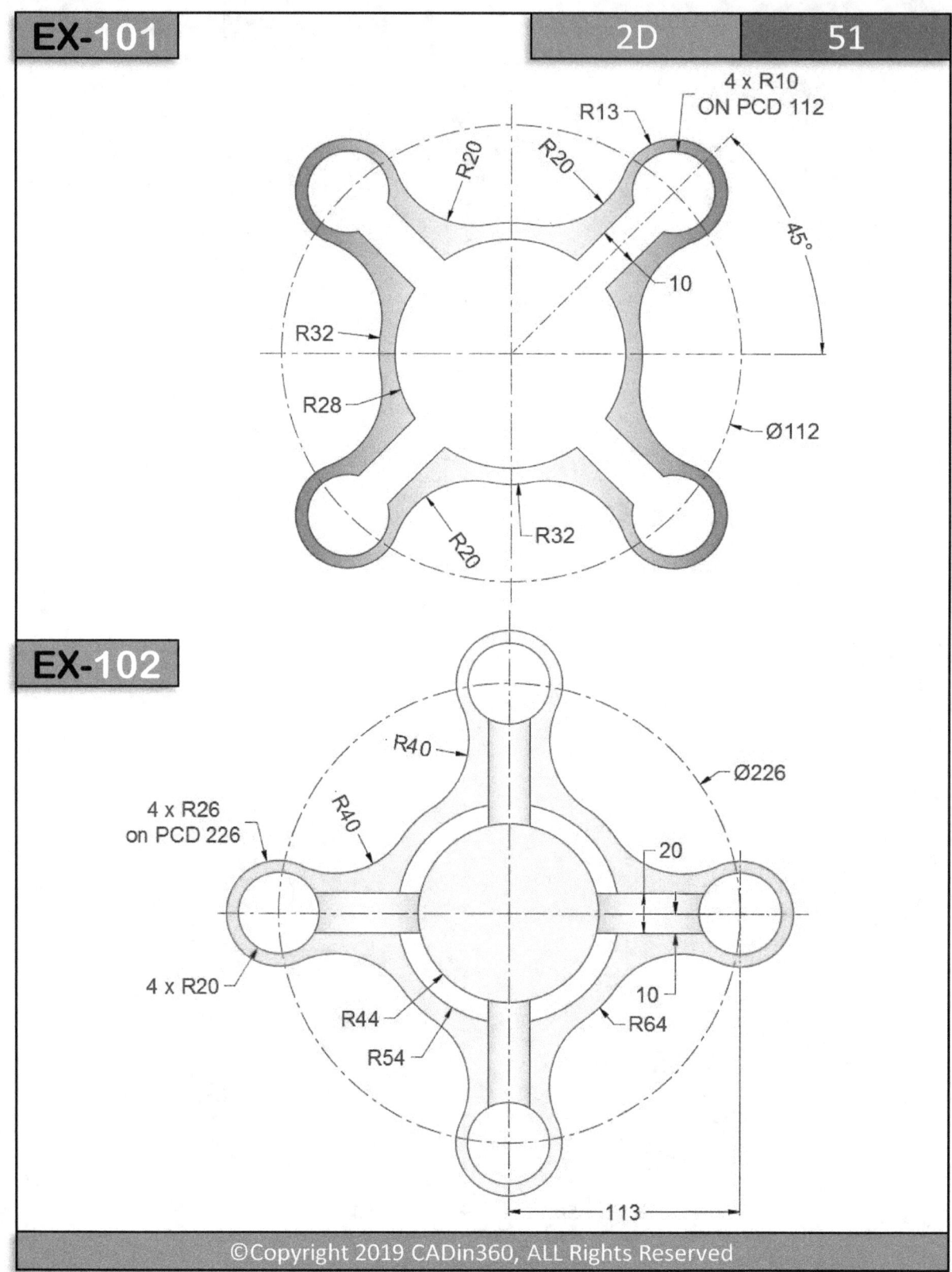

EX-102

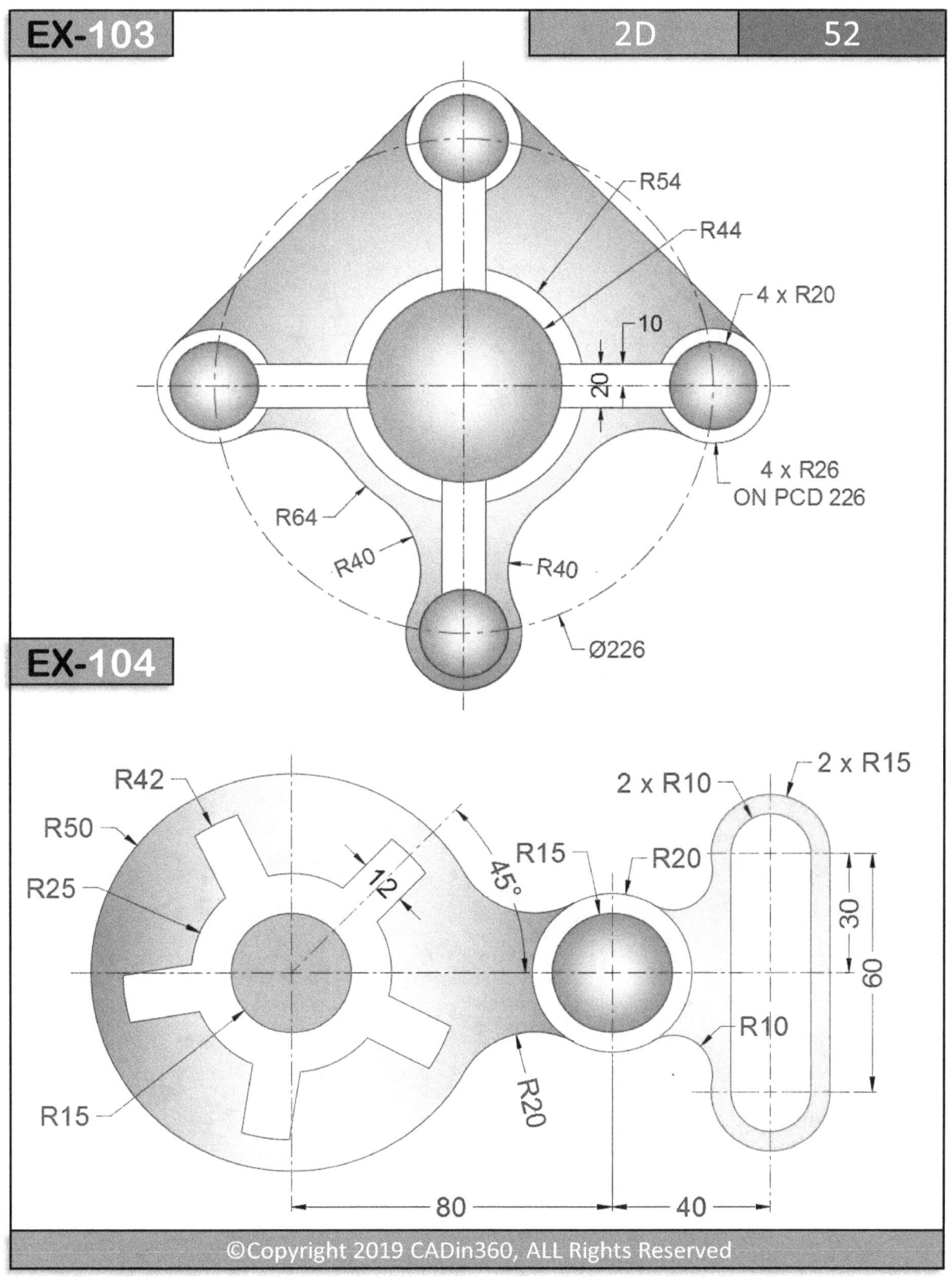

EX-104

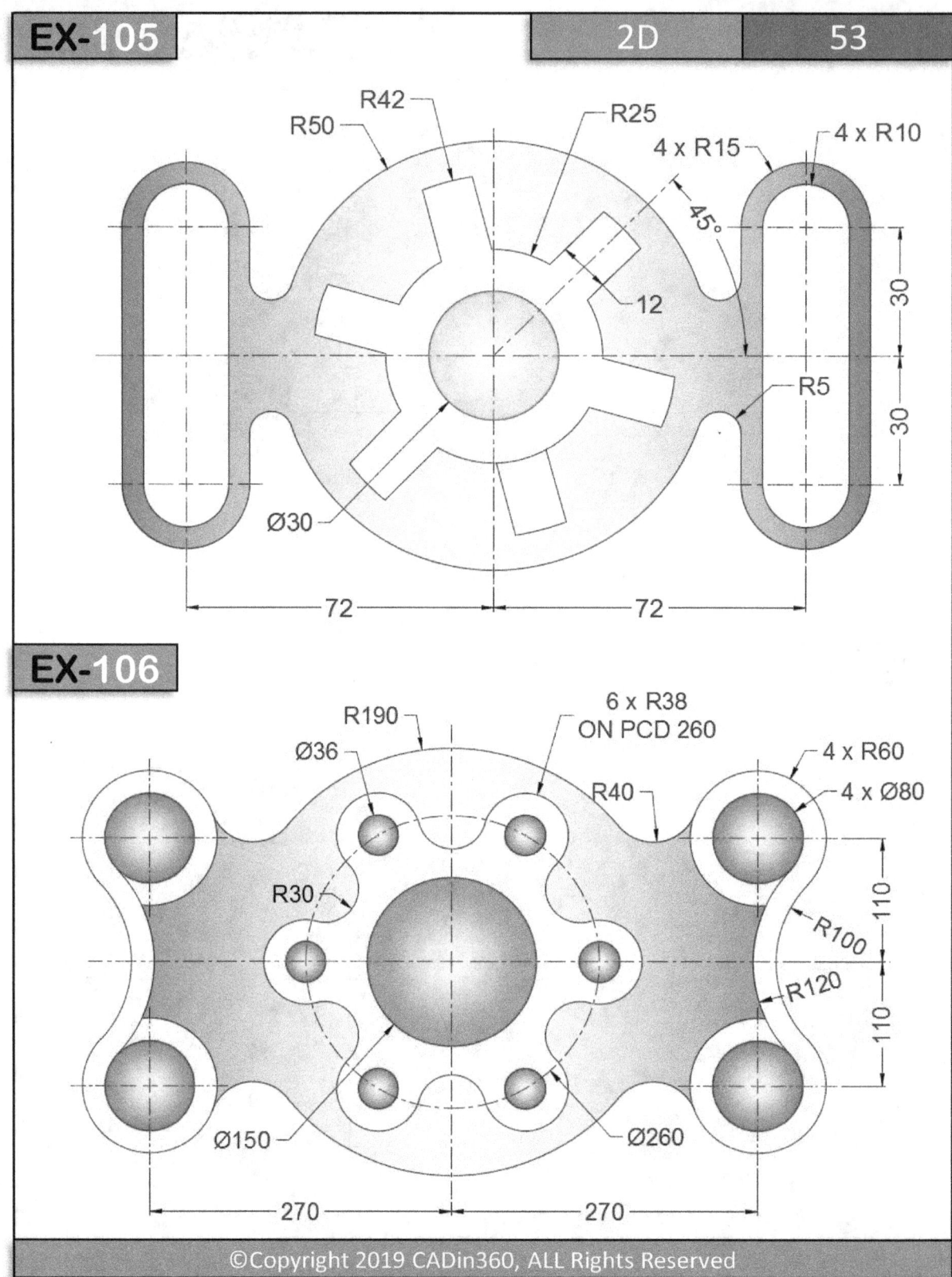

EX-105 2D 53

R50 R42 R25 4 x R15 4 x R10

45°

12

R5

30

30

Ø30

72 72

EX-106

R190 6 x R38 ON PCD 260 4 x R60

Ø36 R40 4 x Ø80

R30 110

R100

R120

110

Ø150 Ø260

270 270

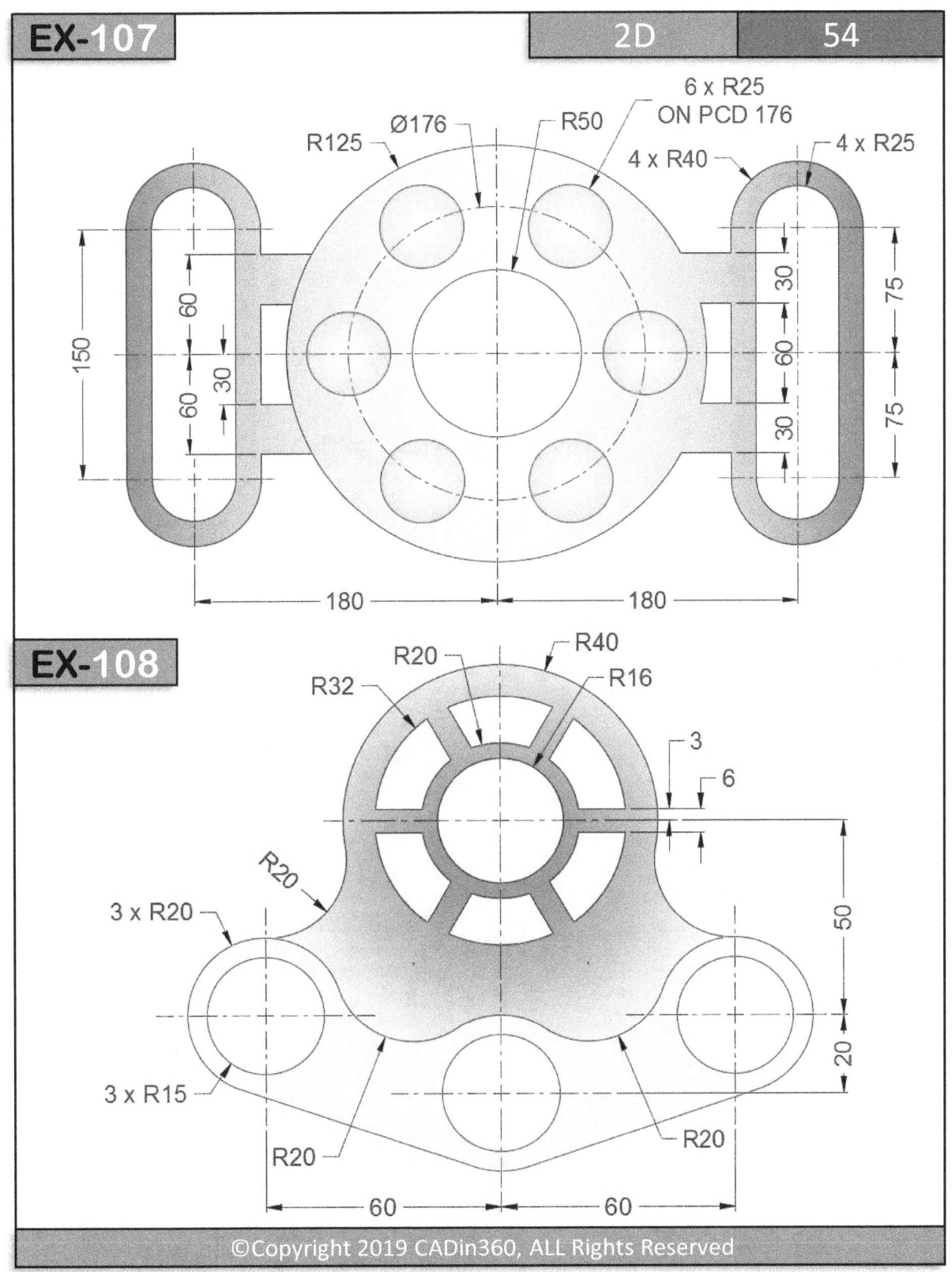

EX-107 | 2D | 54

R125
Ø176
R50
6 x R25
ON PCD 176
4 x R40
4 x R25

150
60
60
30
30
60
30
75
75

180
180

EX-108

R20
R40
R16
R32
R20

3
6

R20
3 x R20
50
20

3 x R15
R20
R20

60
60

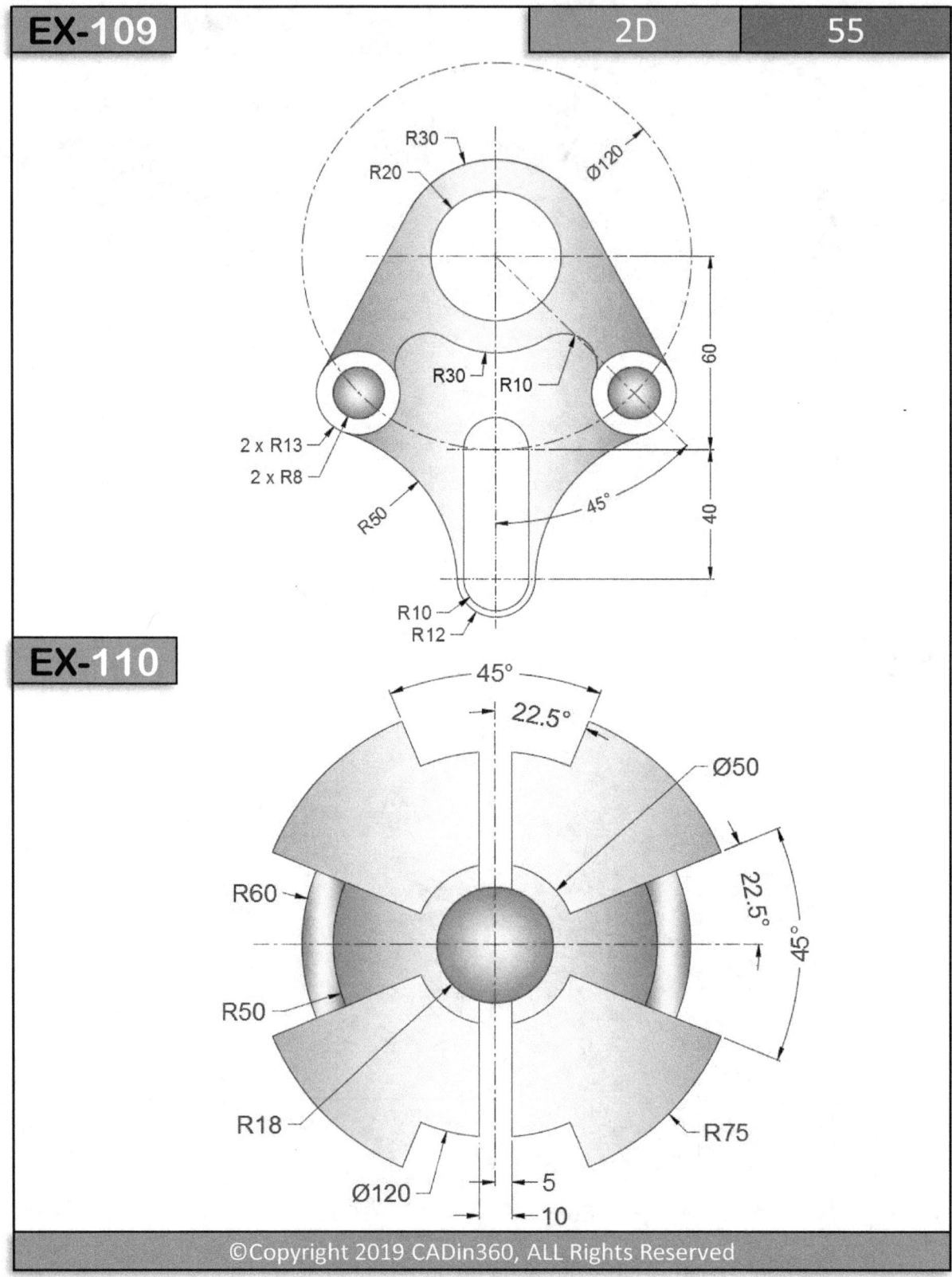

EX-110

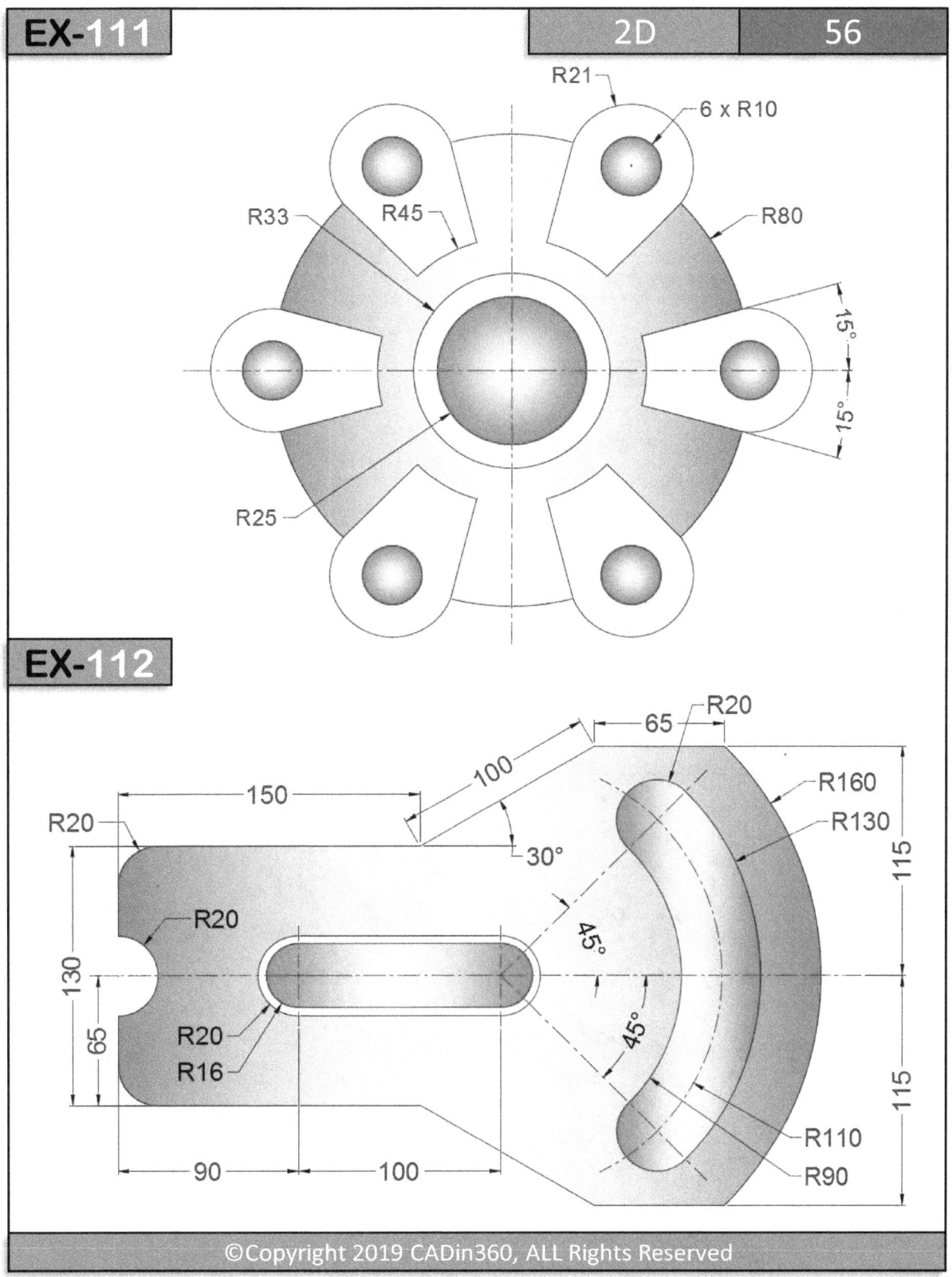

EX-112

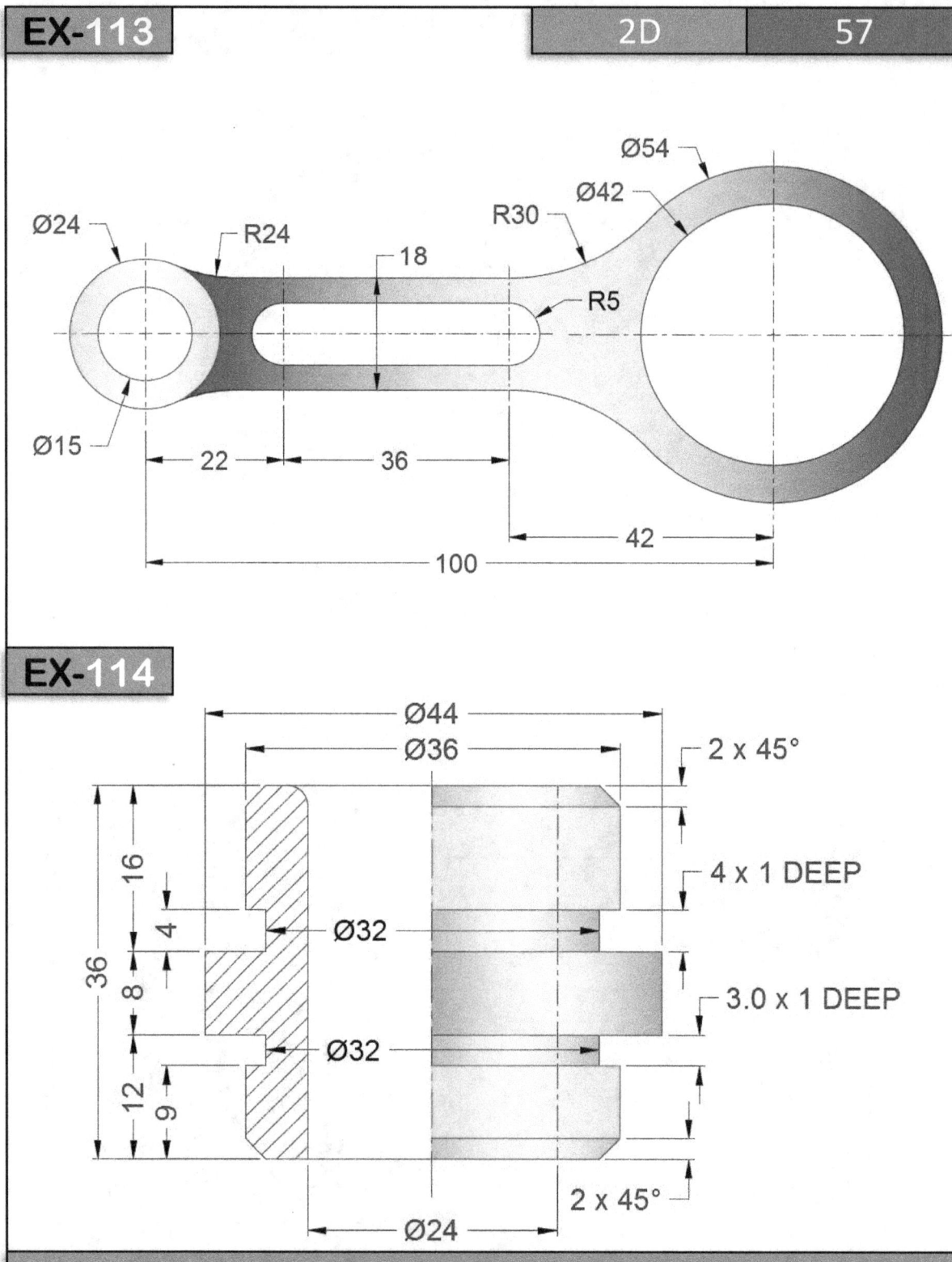

EX-113   2D   57

Ø24   R24   18   R30   Ø54   Ø42
Ø15   R5
22   36
42
100

EX-114

Ø44
Ø36
2 x 45°
16
4
Ø32
4 x 1 DEEP
36
8
3.0 x 1 DEEP
Ø32
12
9
2 x 45°
Ø24

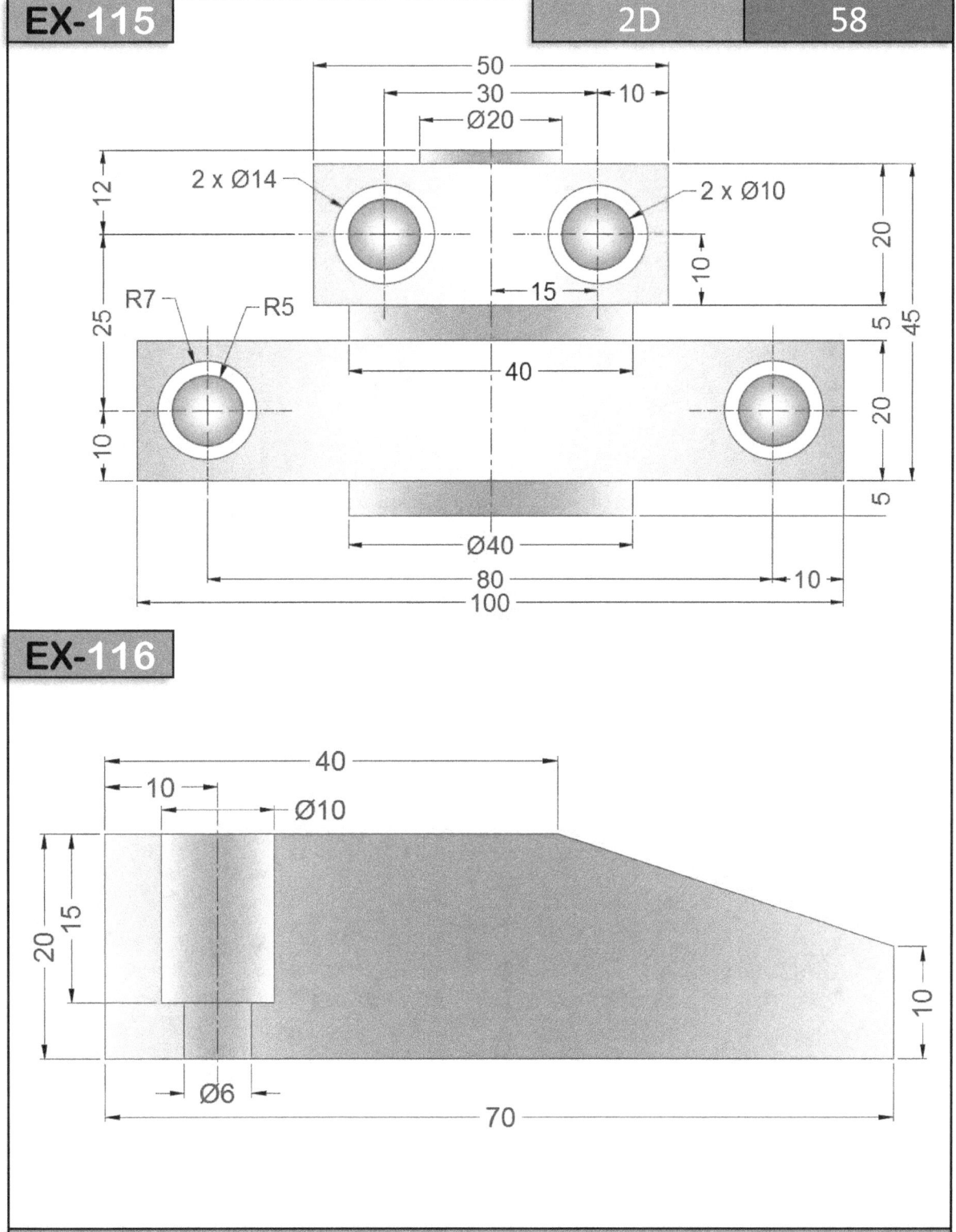

EX-115

2D

58

EX-116

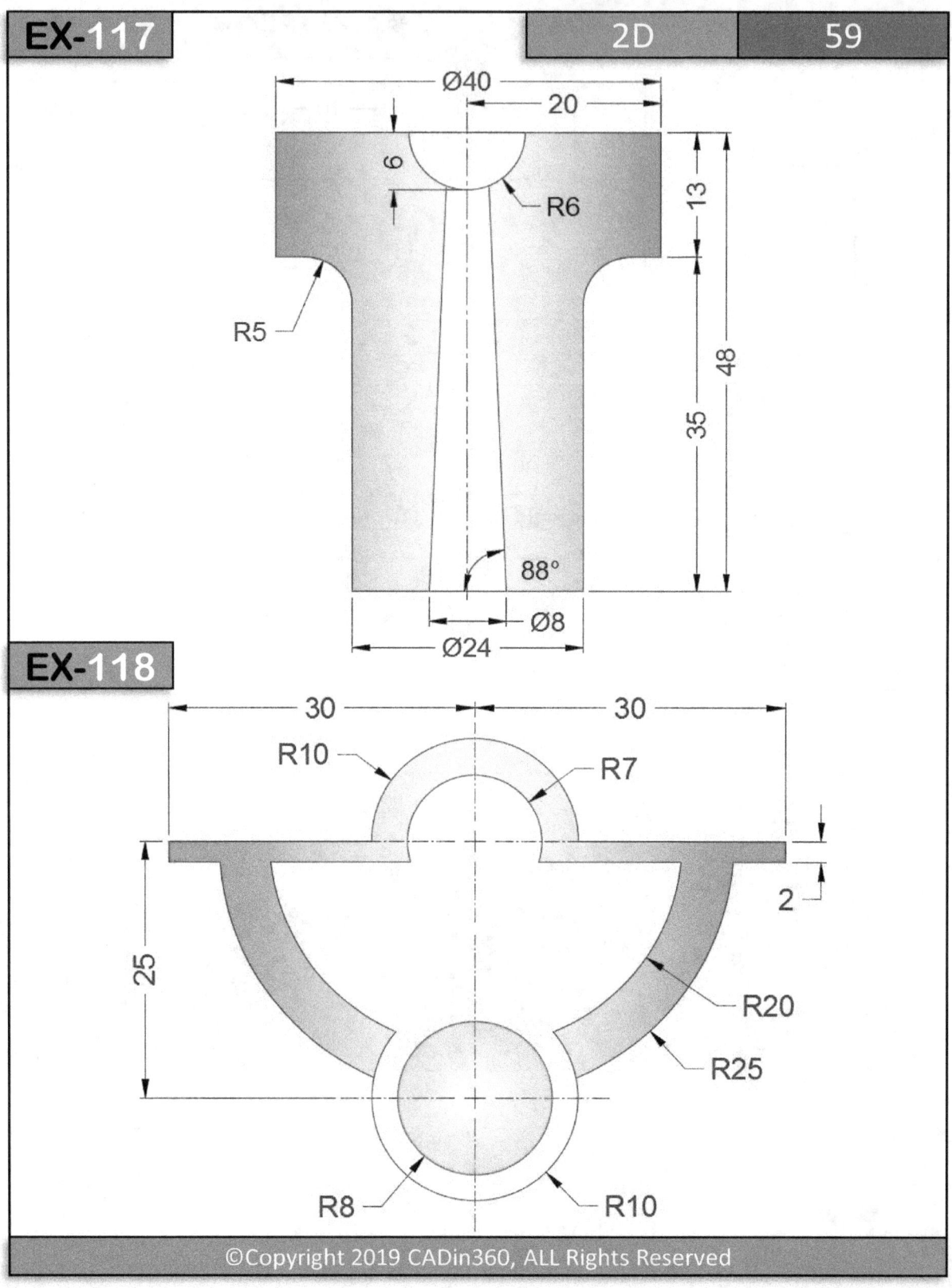

EX-118

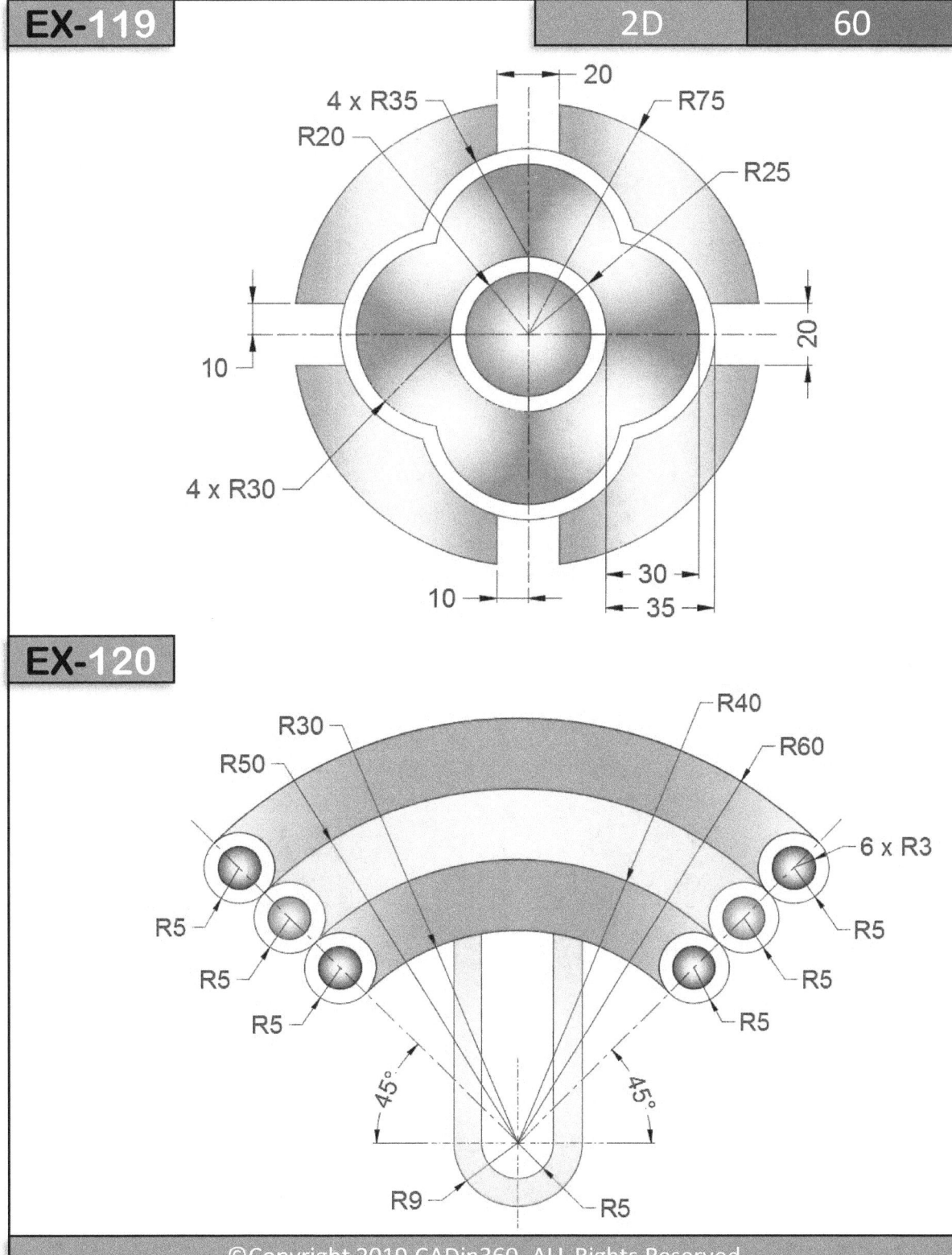

EX-119

2D   60

4 x R35
R20
20
R75
R25
10
20
4 x R30
10
30
35

EX-120

R30
R50
R40
R60
6 x R3
R5
R5
R5
R5
R5
R5
45°
45°
R9
R5

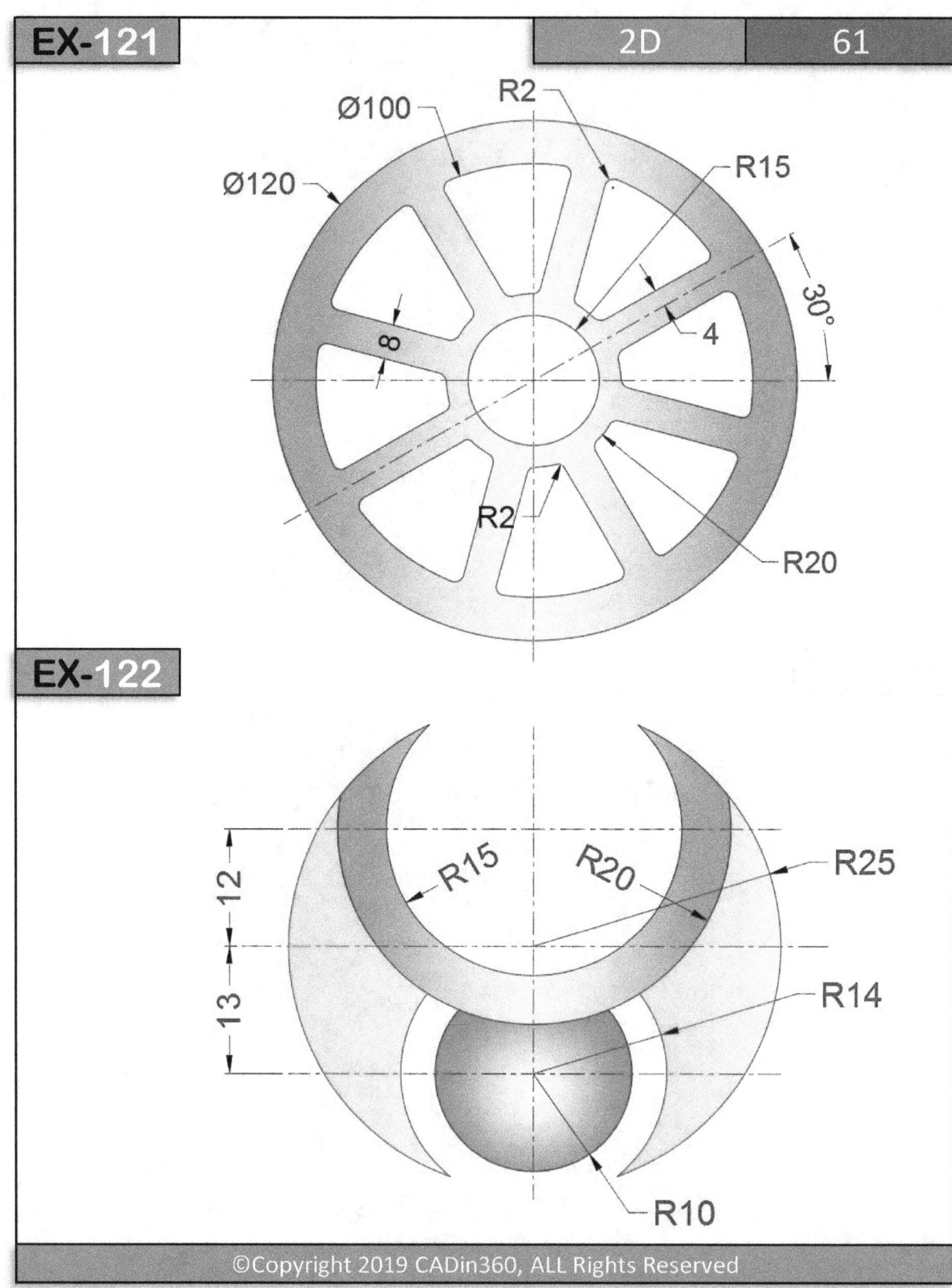

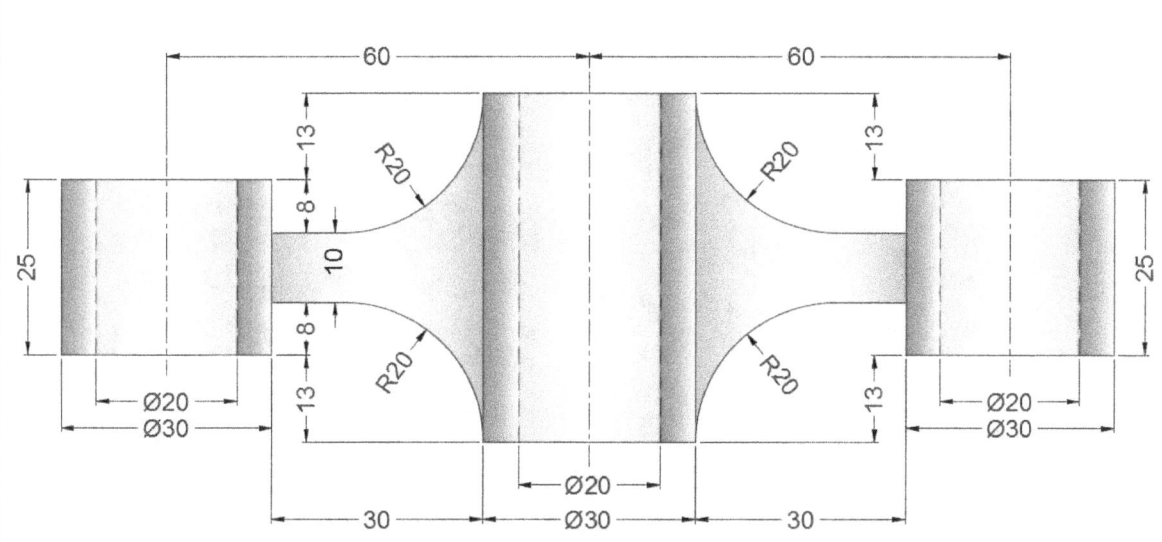

EX-124

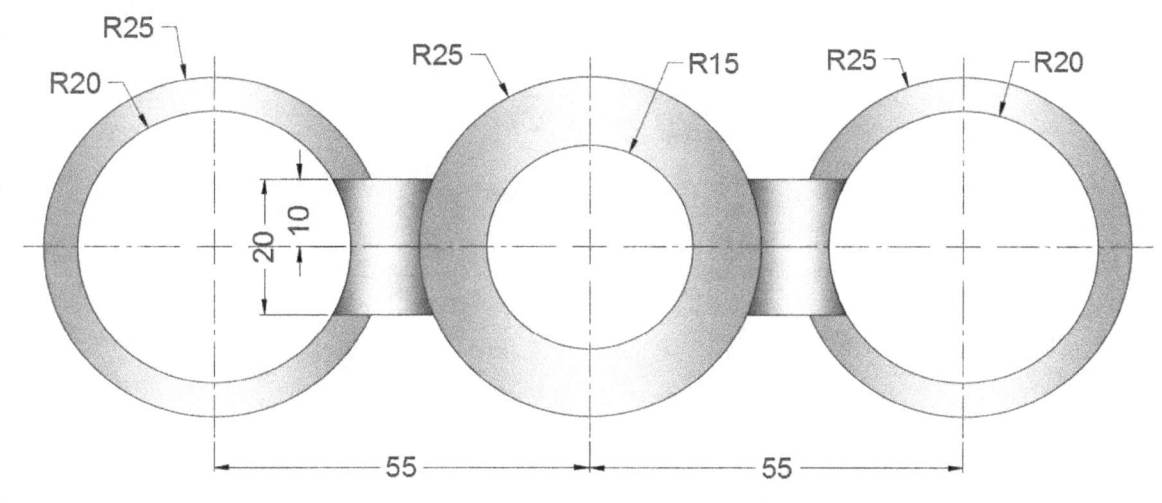

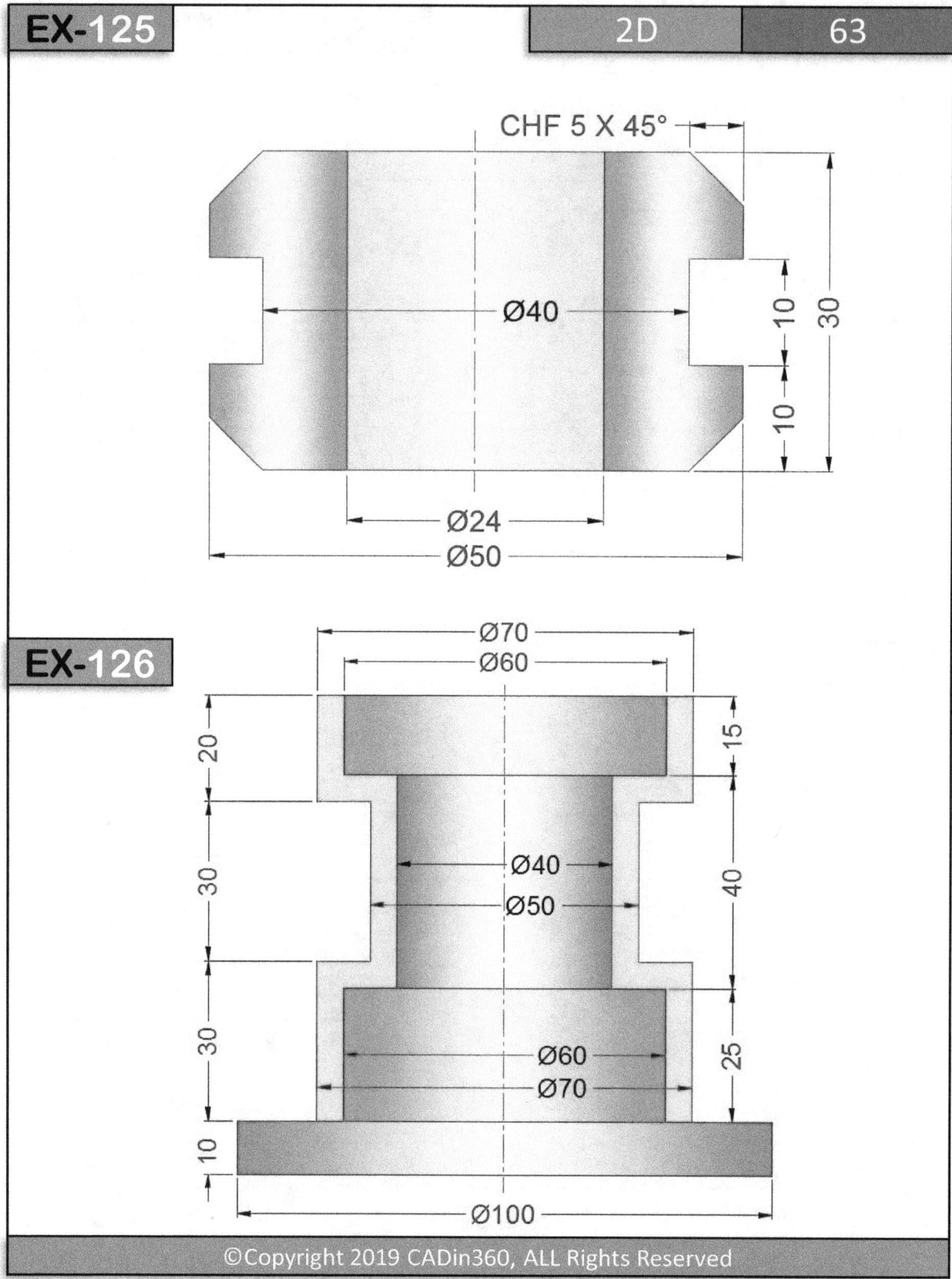

EX-125 2D 63

CHF 5 X 45°

Ø40

10
10
30
10

Ø24
Ø50

EX-126

Ø70
Ø60

20

15

30

Ø40
Ø50

40

30

Ø60
Ø70

25

10

Ø100

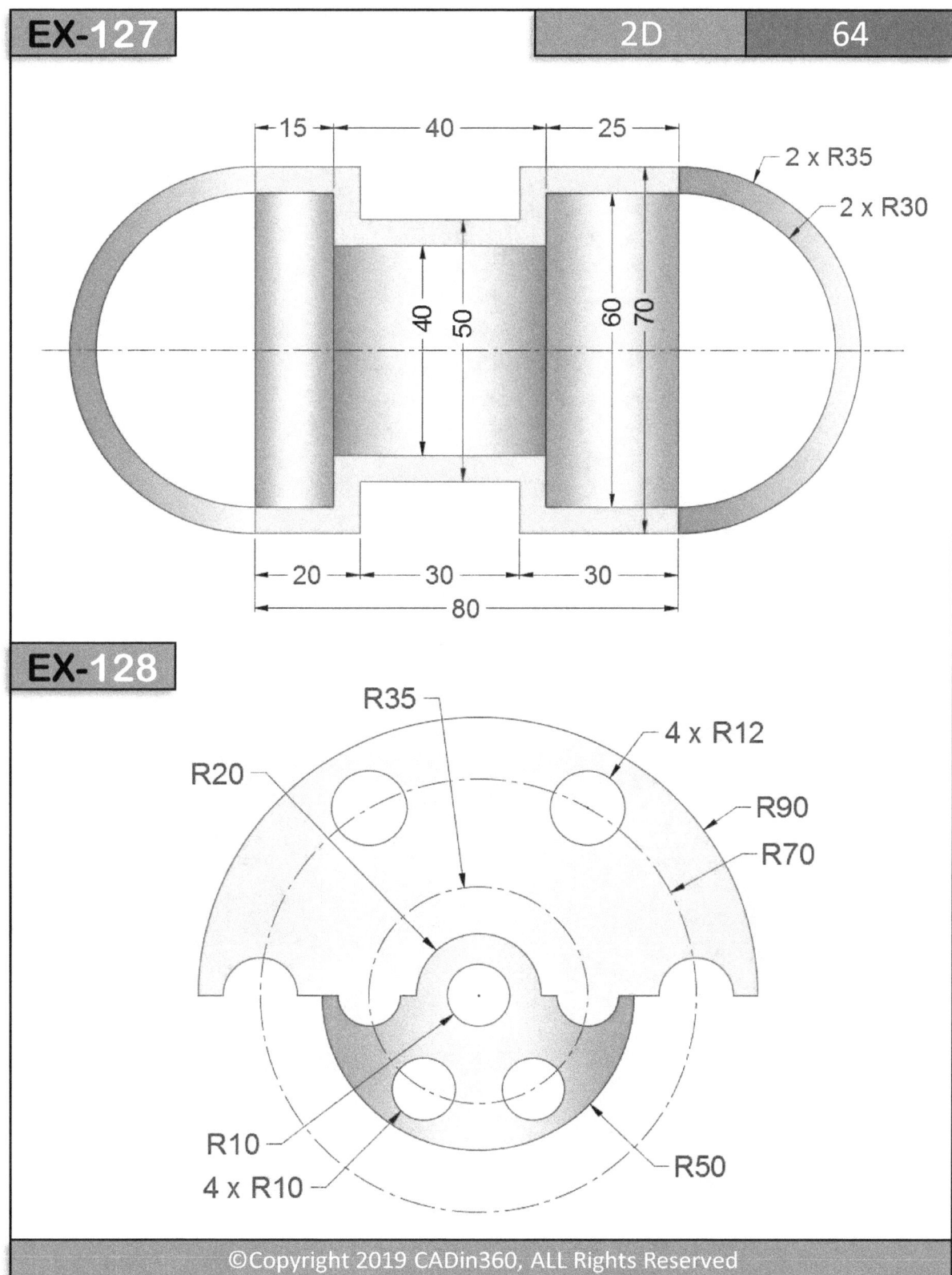

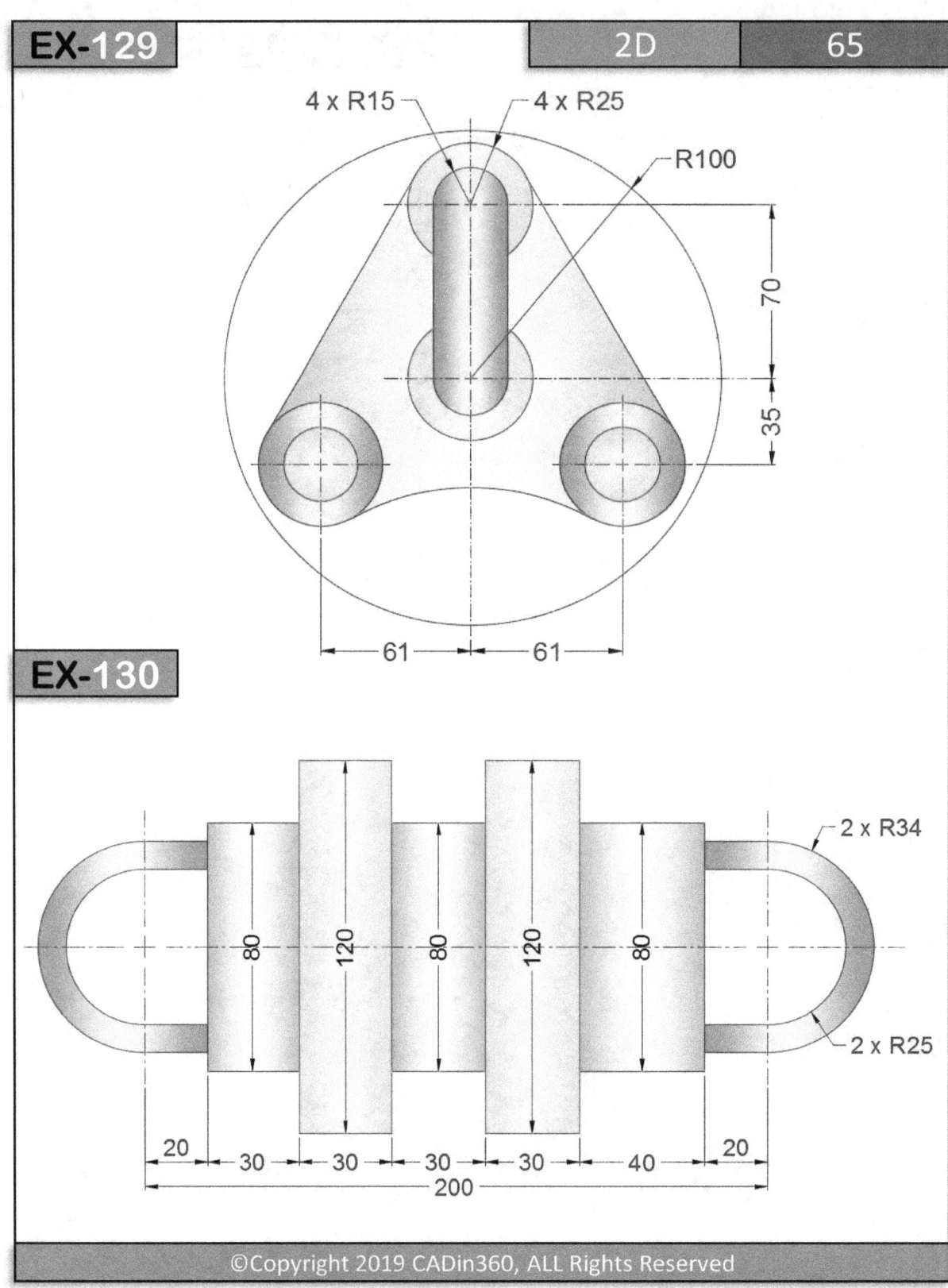

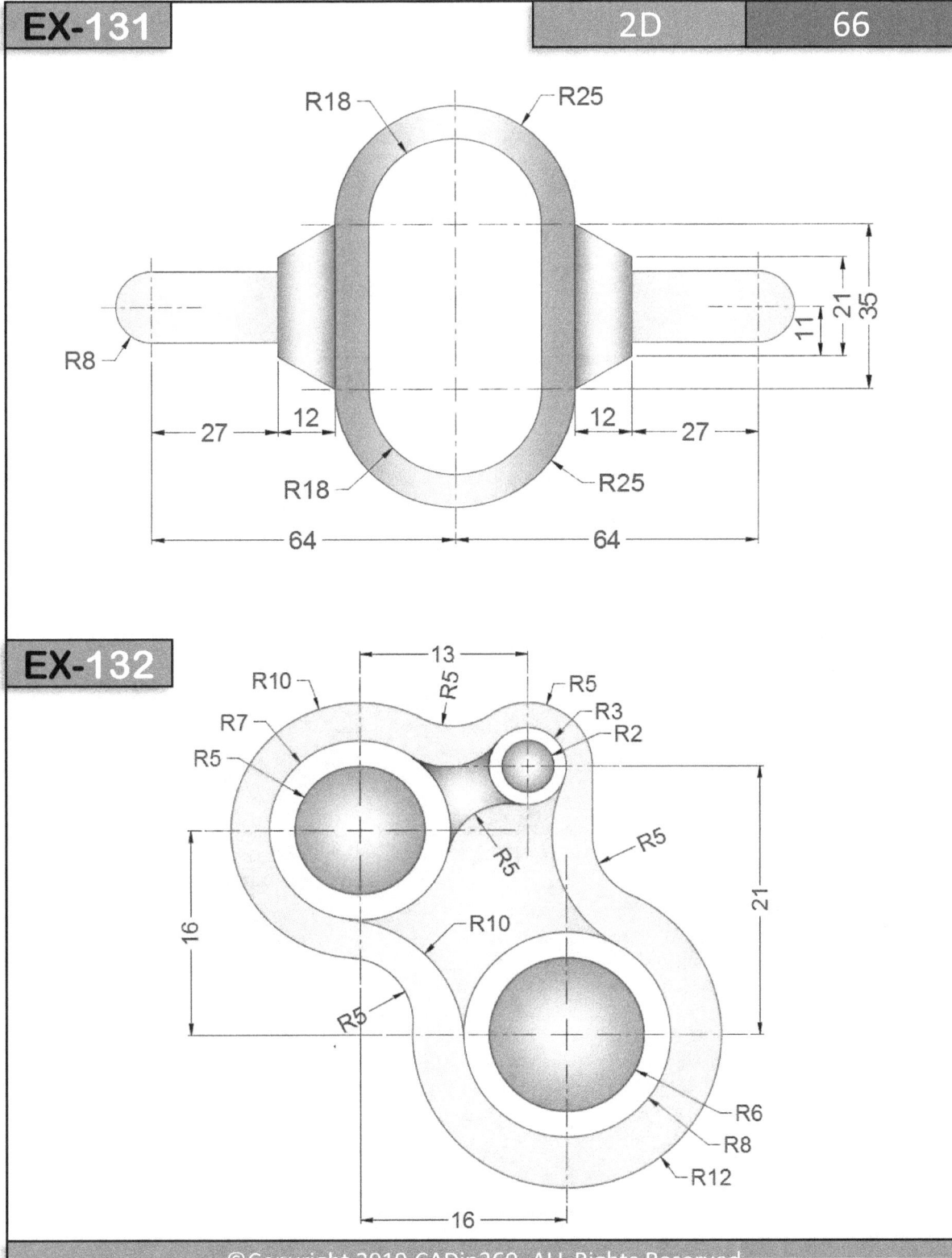

EX-131   2D   66

R18   R25
R8
27   12   12   27
R18   R25
64   64
21   35
11

EX-132
R10   R5   R5
R7   R3
R5   R2
13
R5
R5
16   R10
21
R5
R5
R6
R8
16   R12

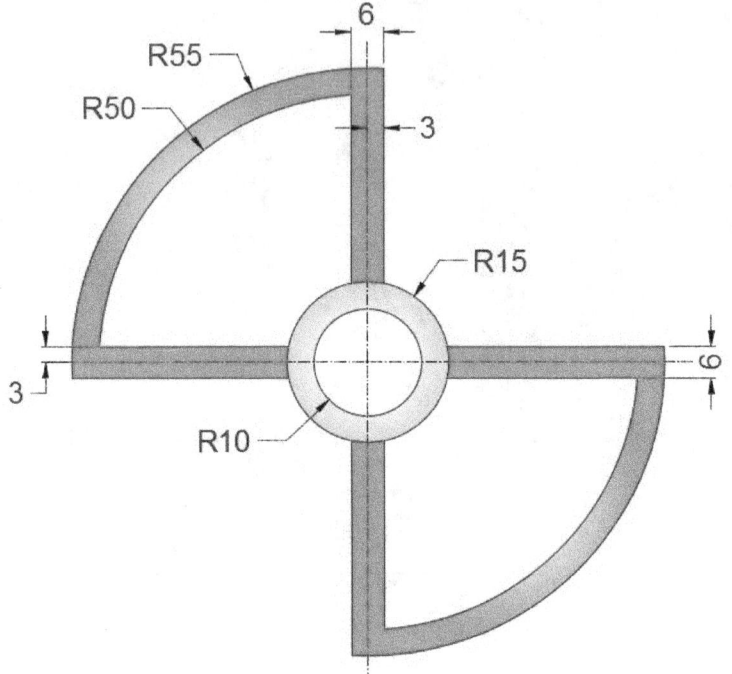

EX-134

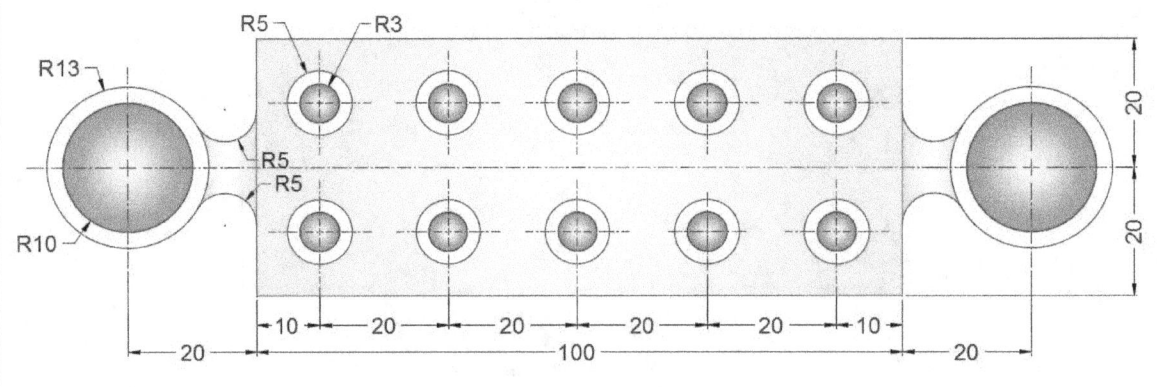

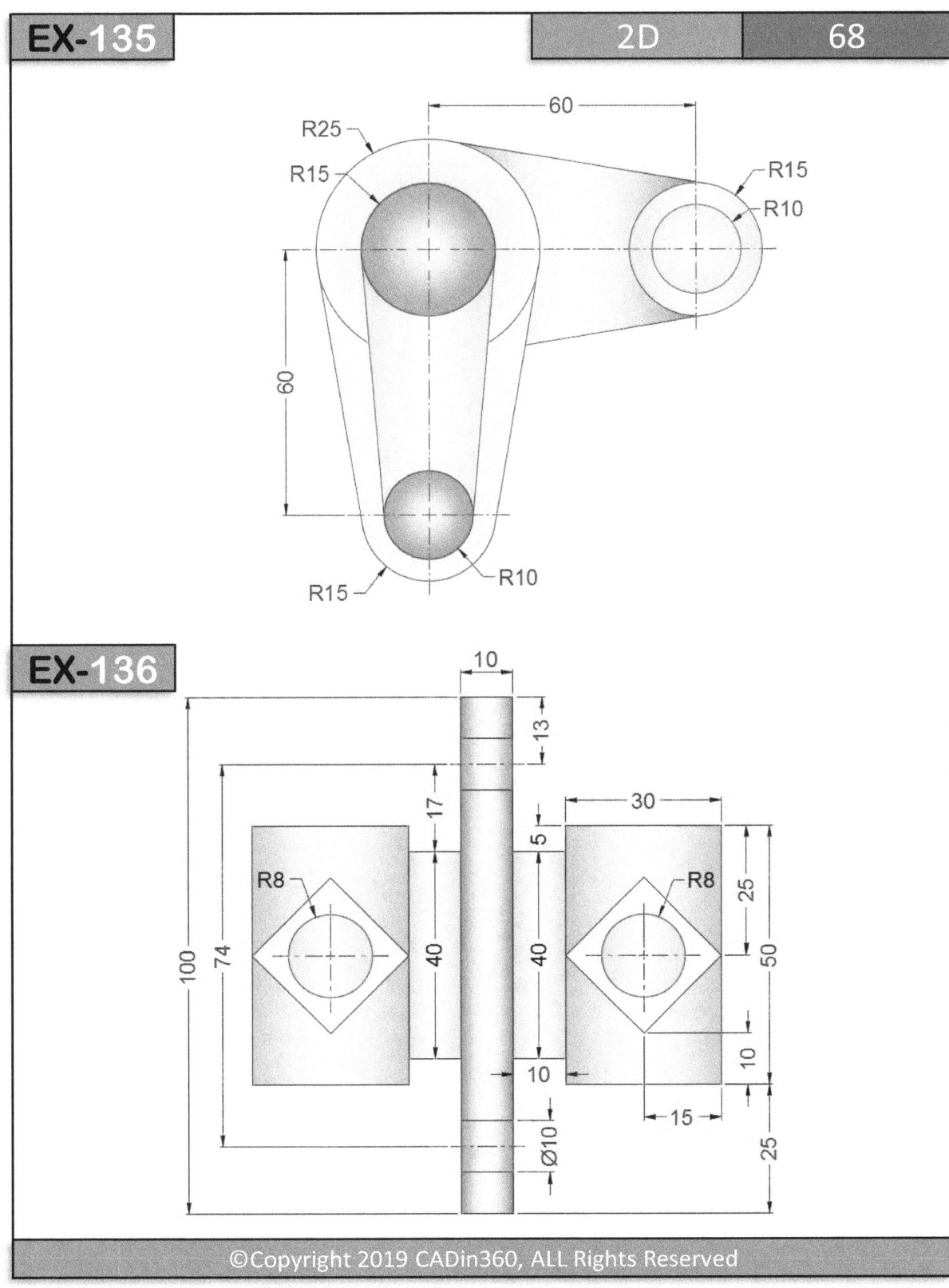

EX-136

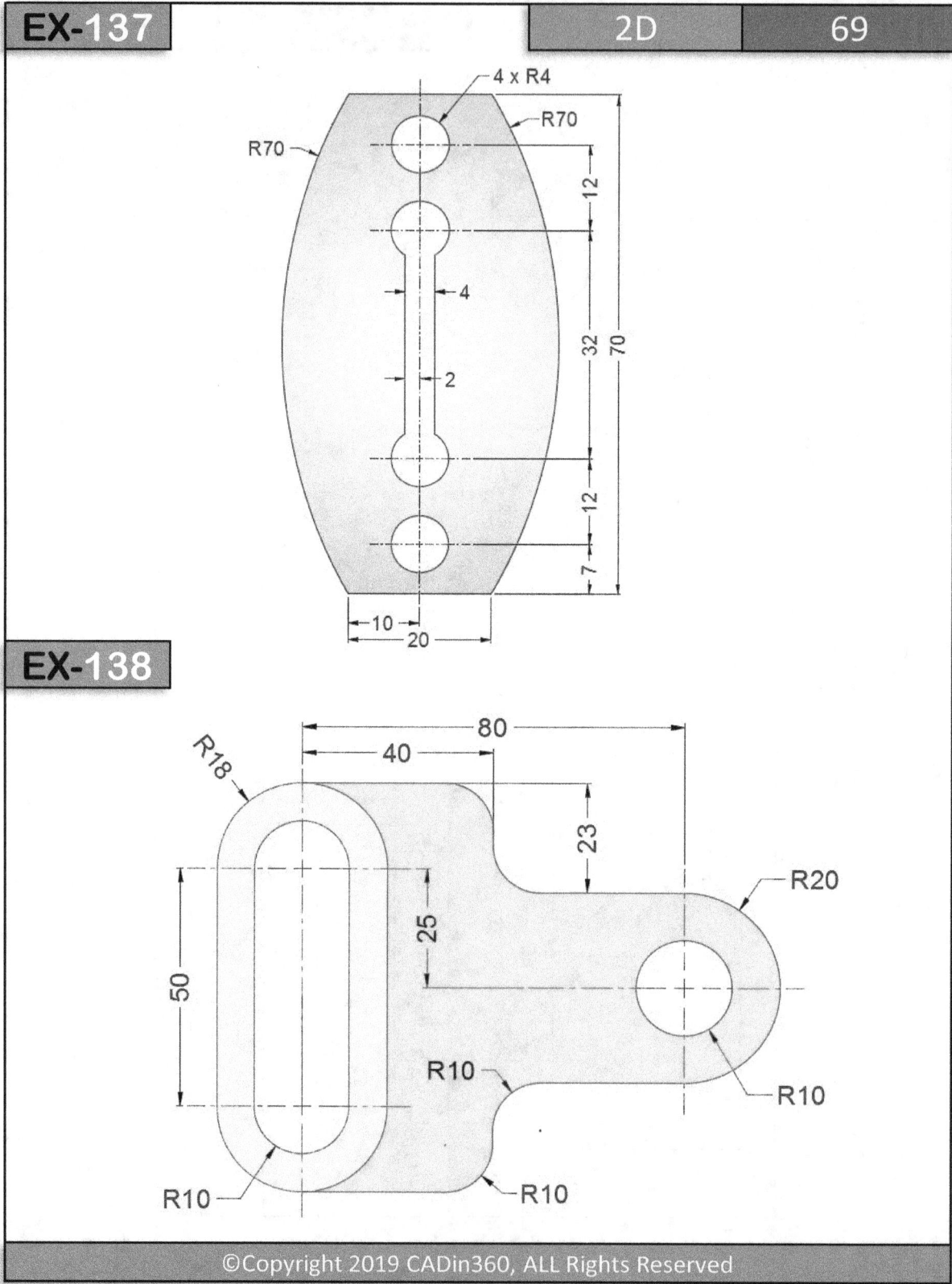

EX-138

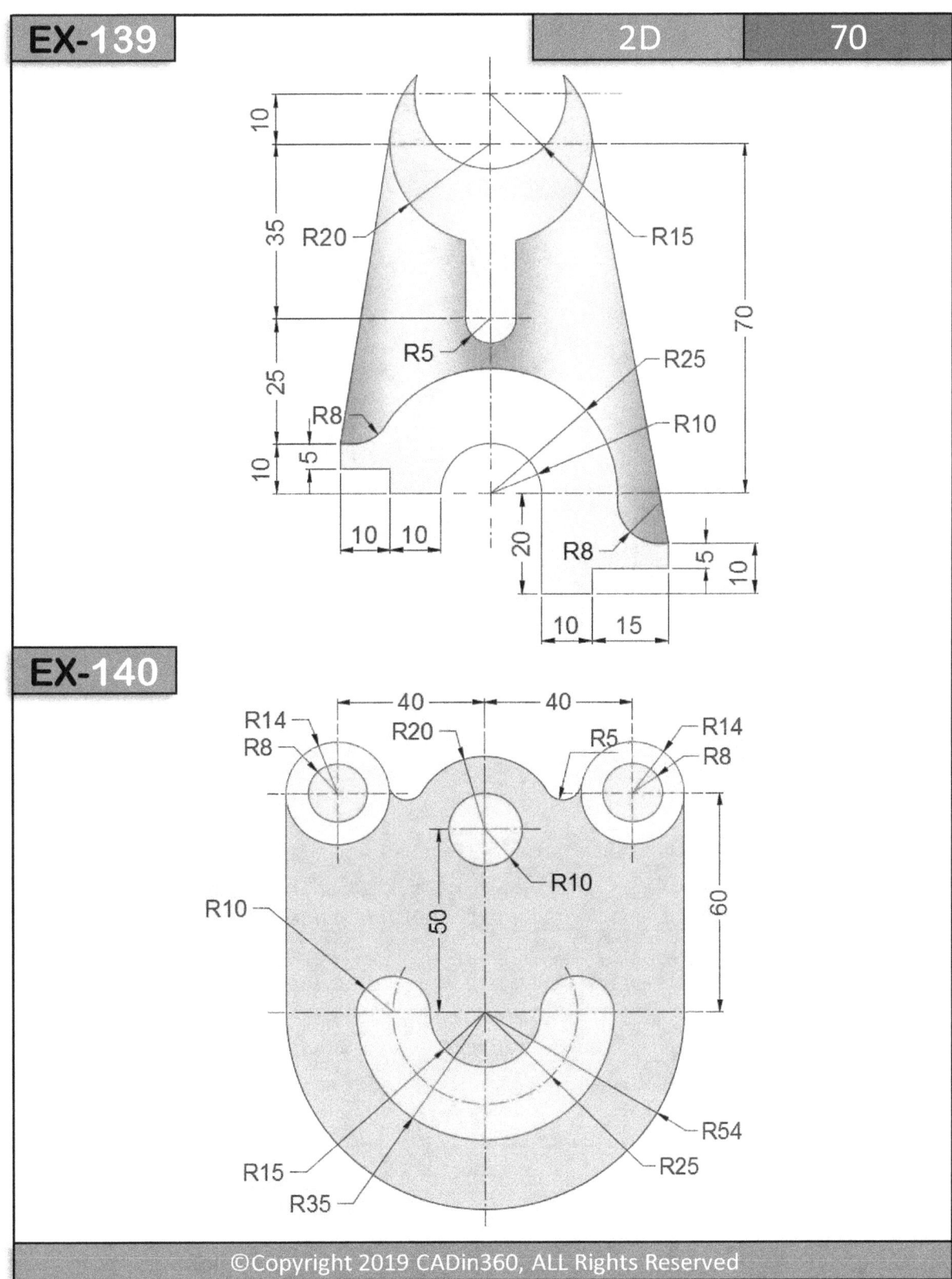

EX-140

EX-141

100
R75
R10
R50
R5
R5
R40
R5
R20
65
65
R10
R20

EX-142

R50
R40
R10
R10
R30
10
20
R20
R10
R10
23
60
70
47

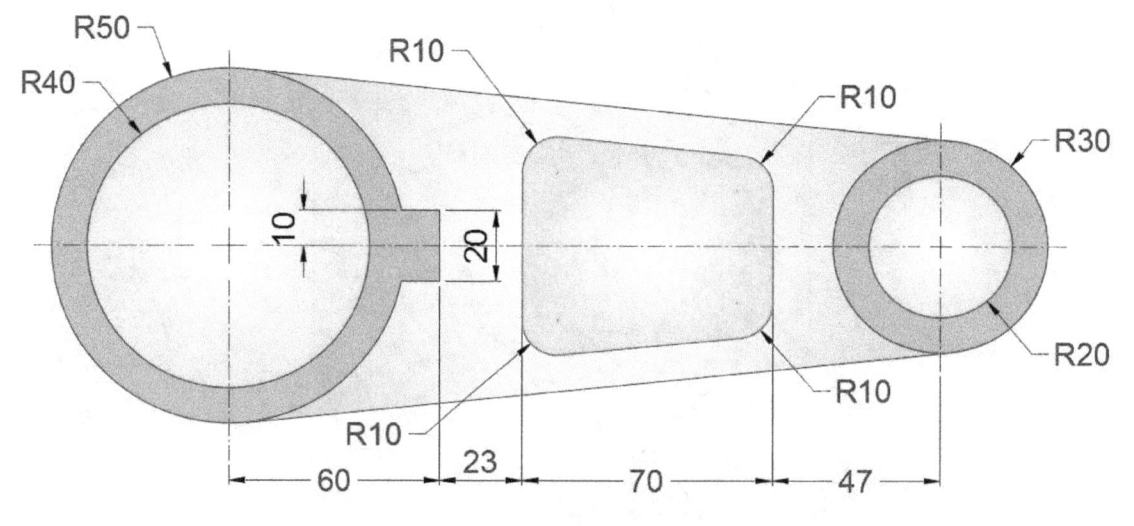

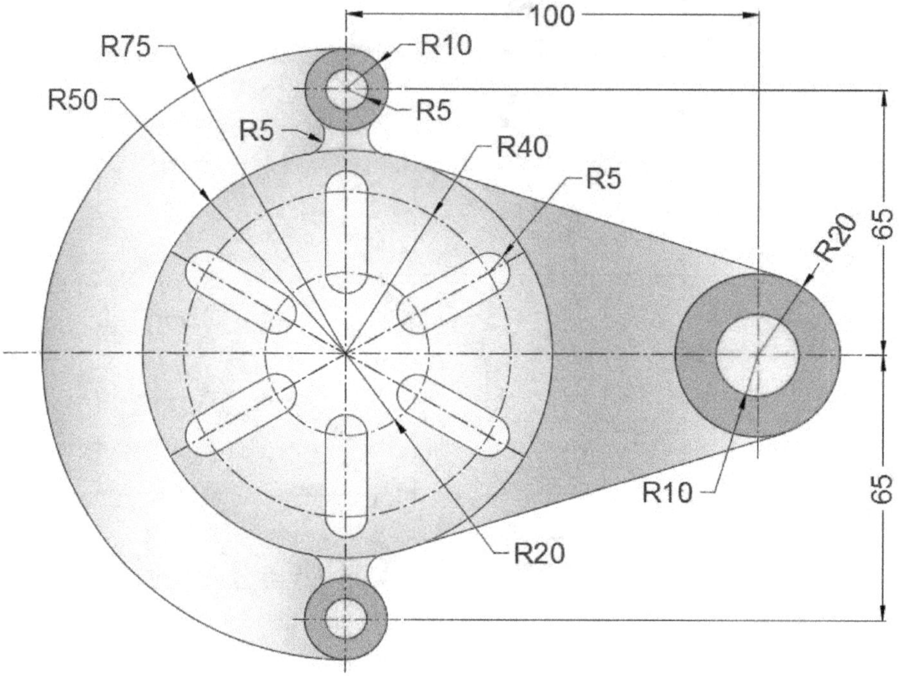

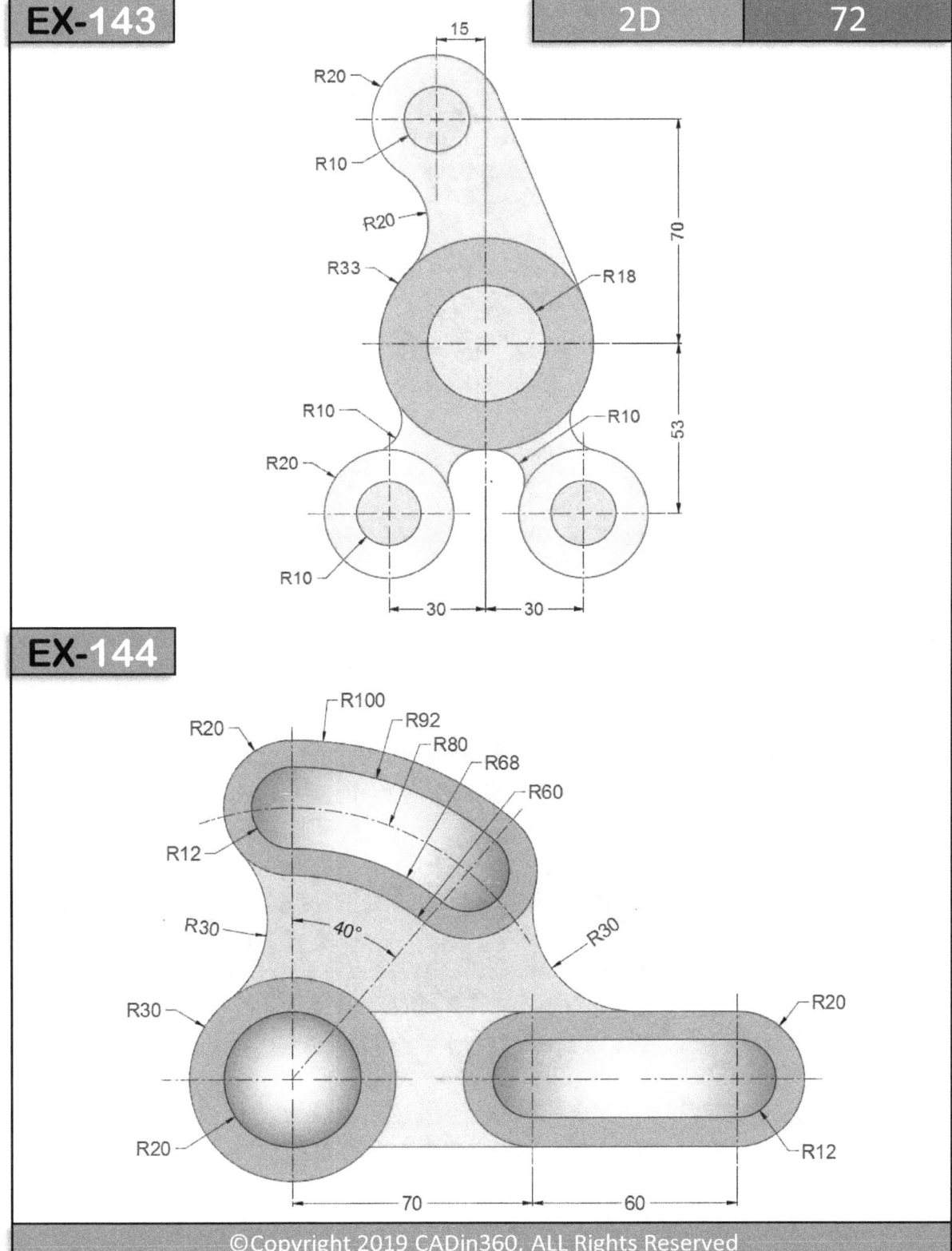

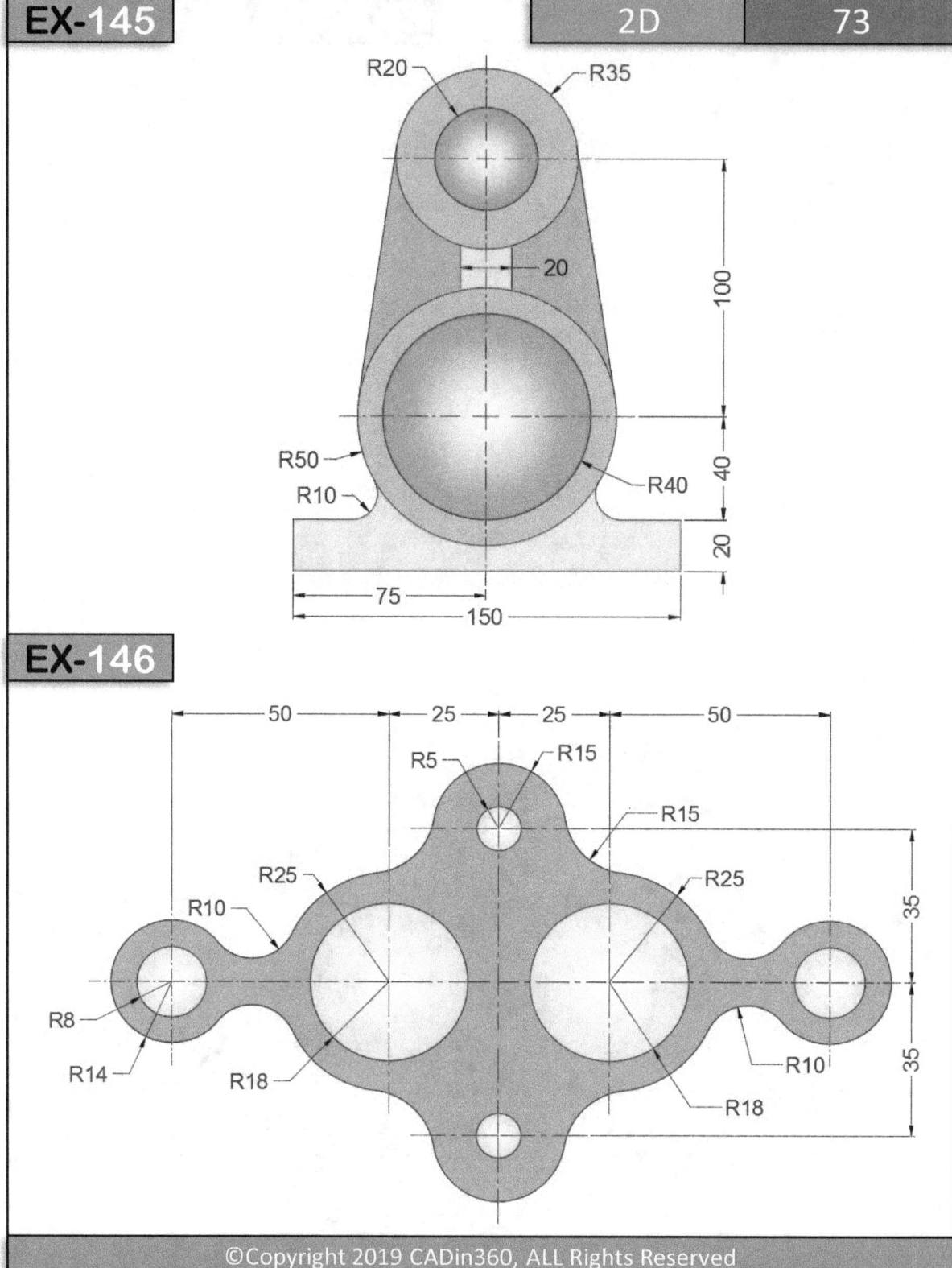

EX-145

R20
R35
20
R50
R10
R40
100
40
20
75
150

EX-146

50   25   25   50
R5   R15
R15
R25   R25
R10
R15
35
R8
R14   R18   R10
35
R18

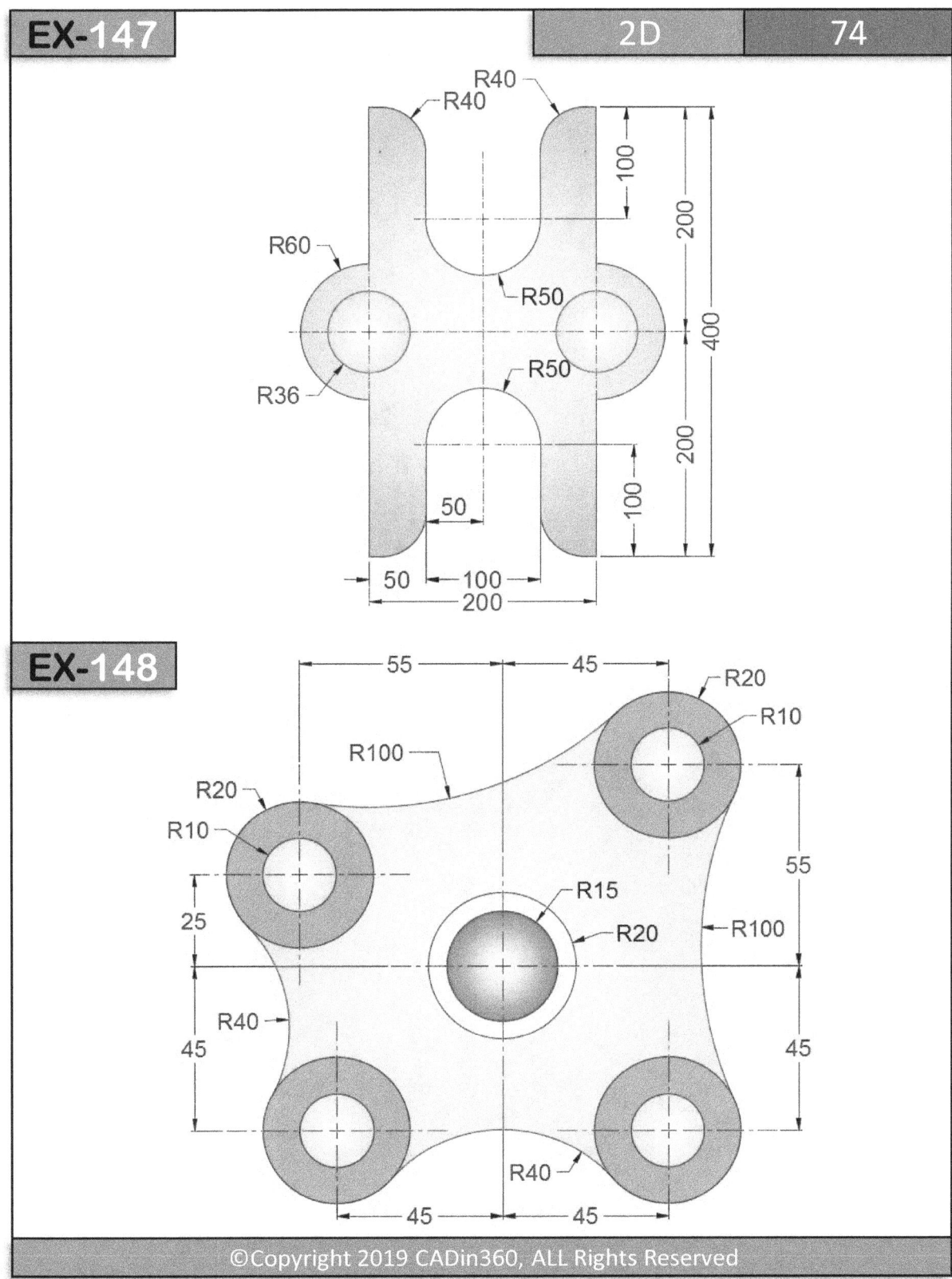

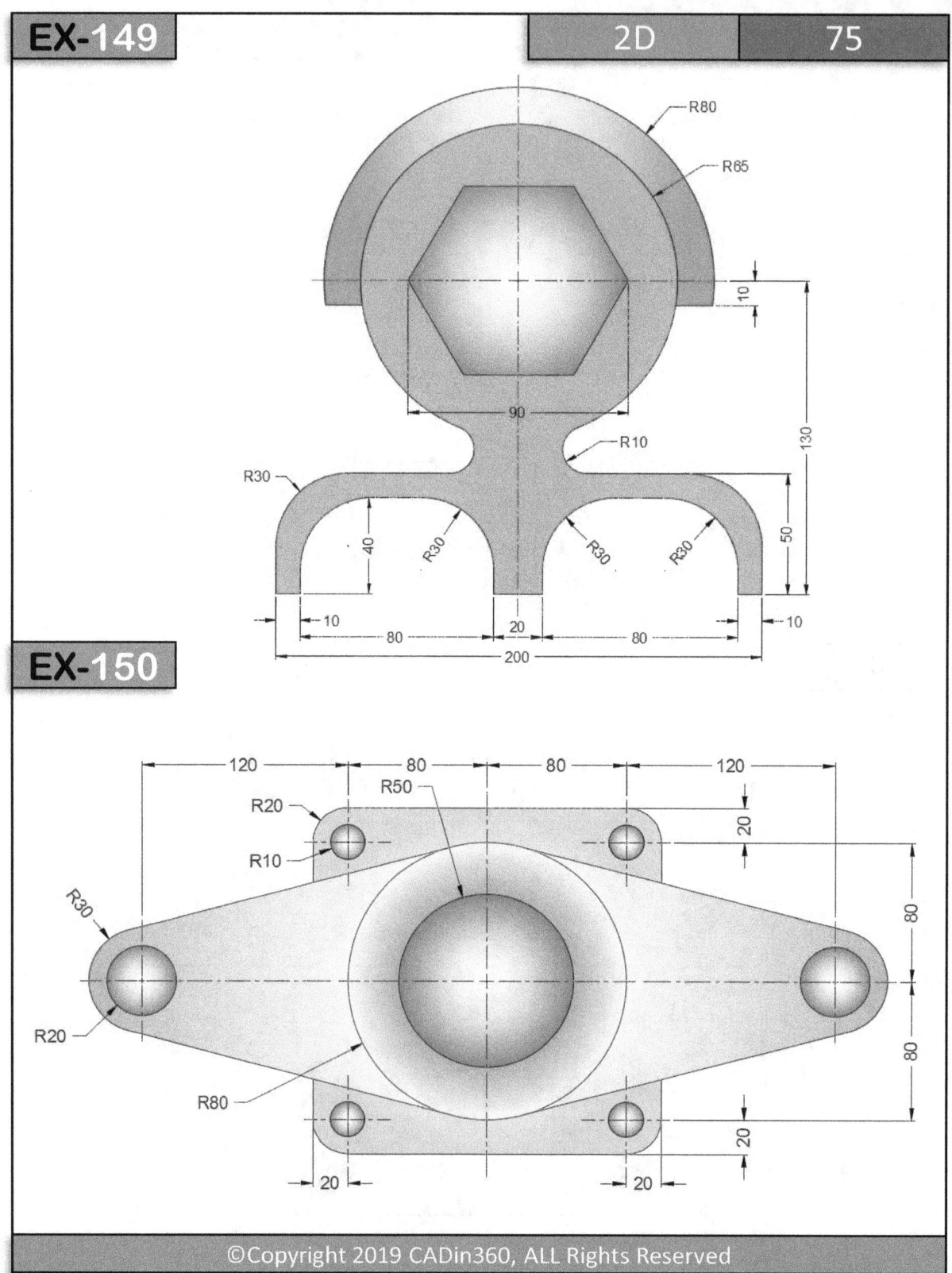

EX-149

2D   75

R80
R65
10
130
50
90
R10
R30
R30
R30
R30
40
10
80
20
80
10
200

EX-150

120
80
80
120
R20
R50
20
R10
80
R30
R20
80
R80
20
20
20

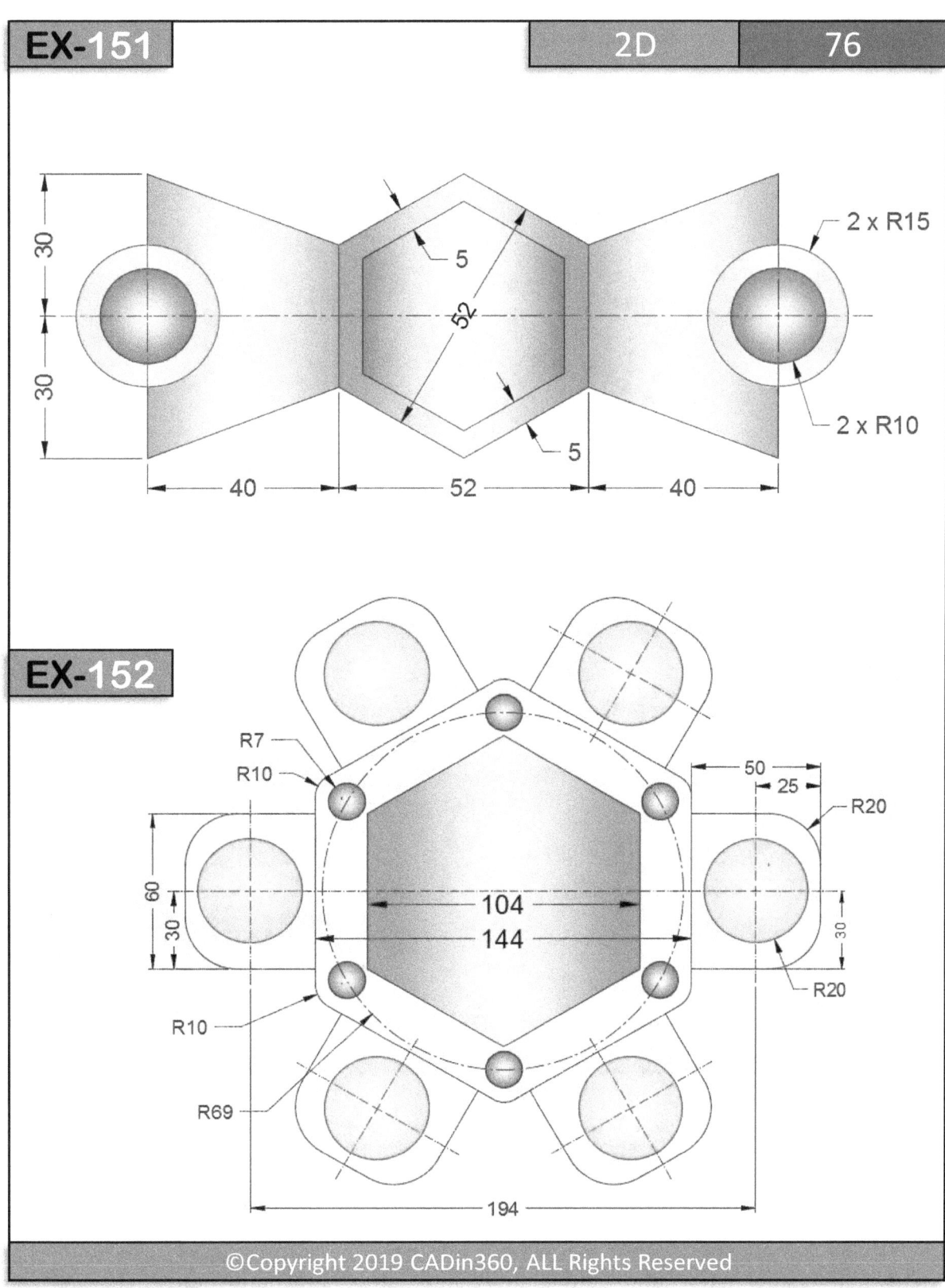

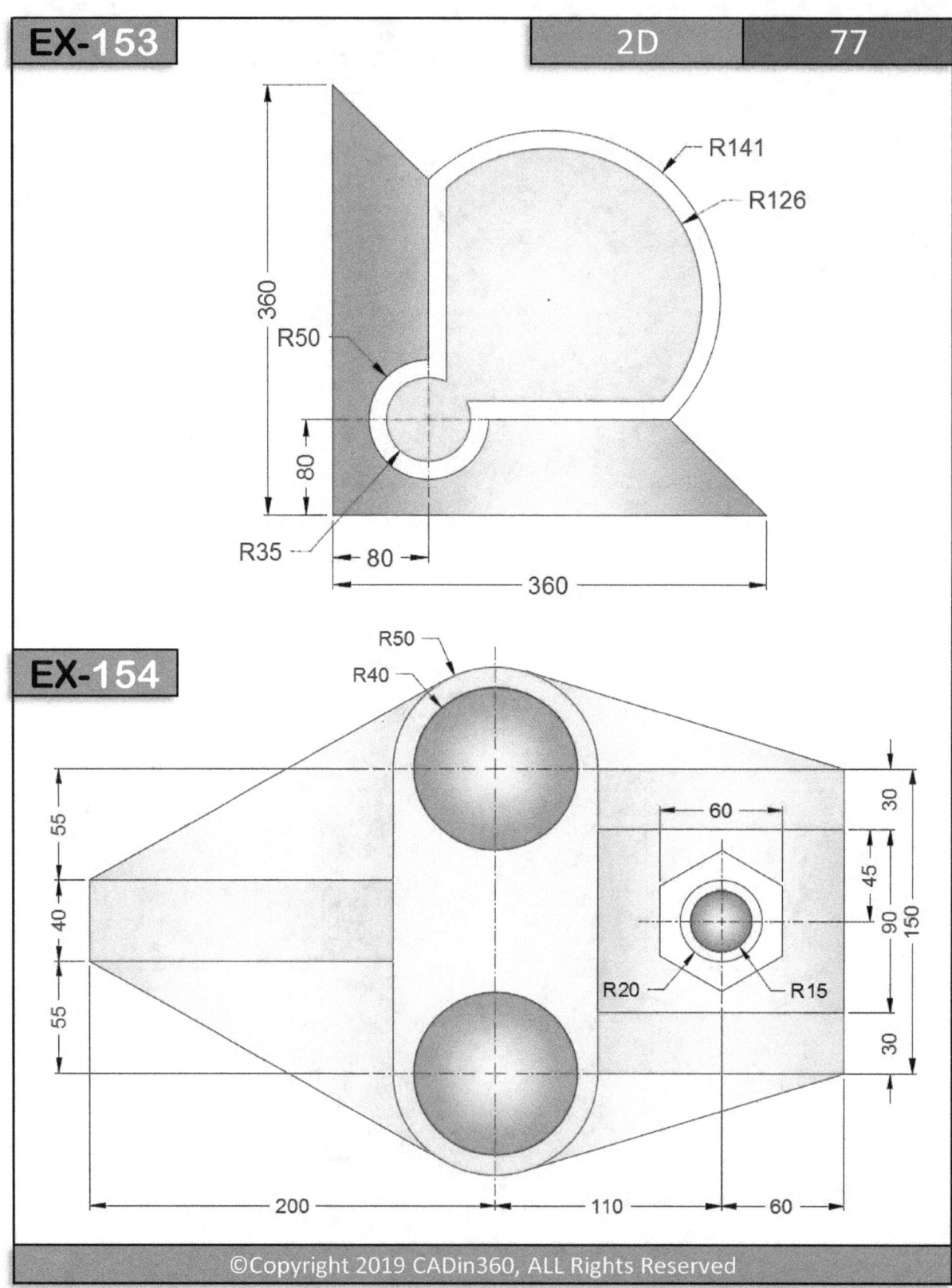

EX-154

R141

R126

R50

360

80

R35

80

360

R50

R40

55

40

55

60

30

45

90

150

30

R20

R15

200

110

60

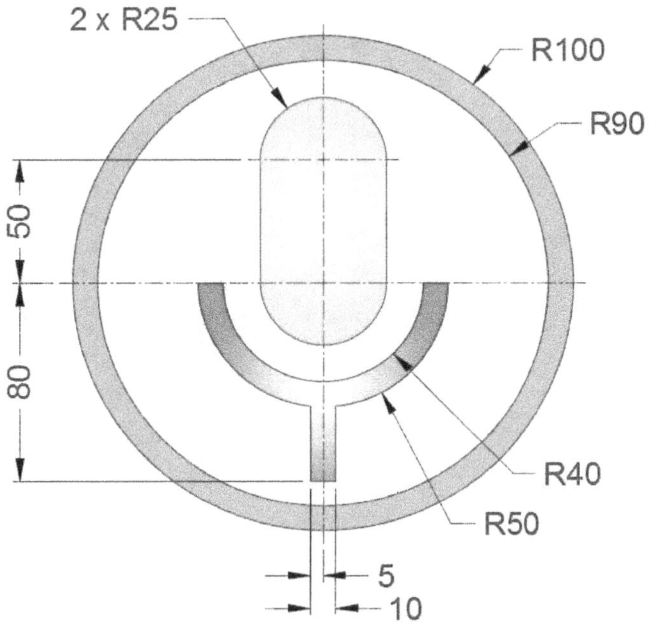

2 x R25
R100
R90
50
80
R40
R50
5
10

EX-156

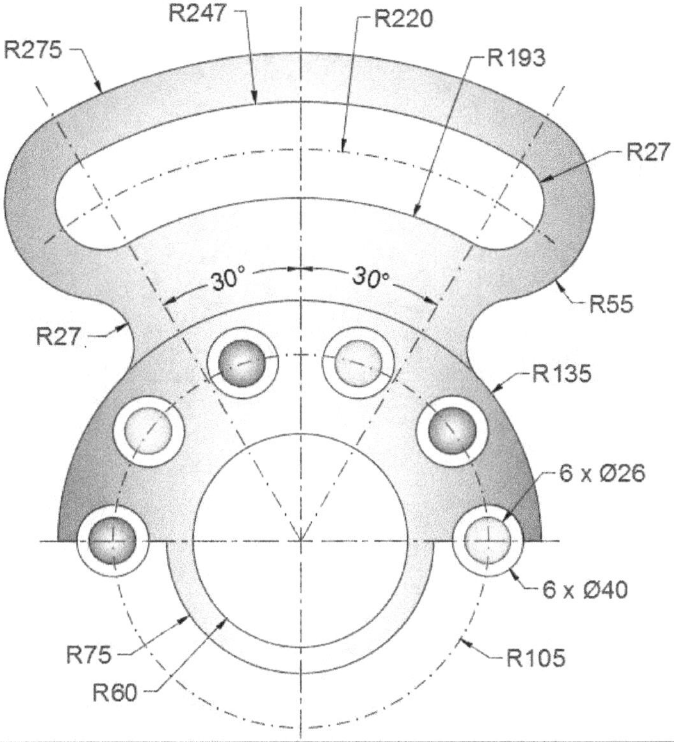

R247
R220
R275
R193
R27
30° 30°
R27
R55
R135
6 x Ø26
6 x Ø40
R75
R60
R105

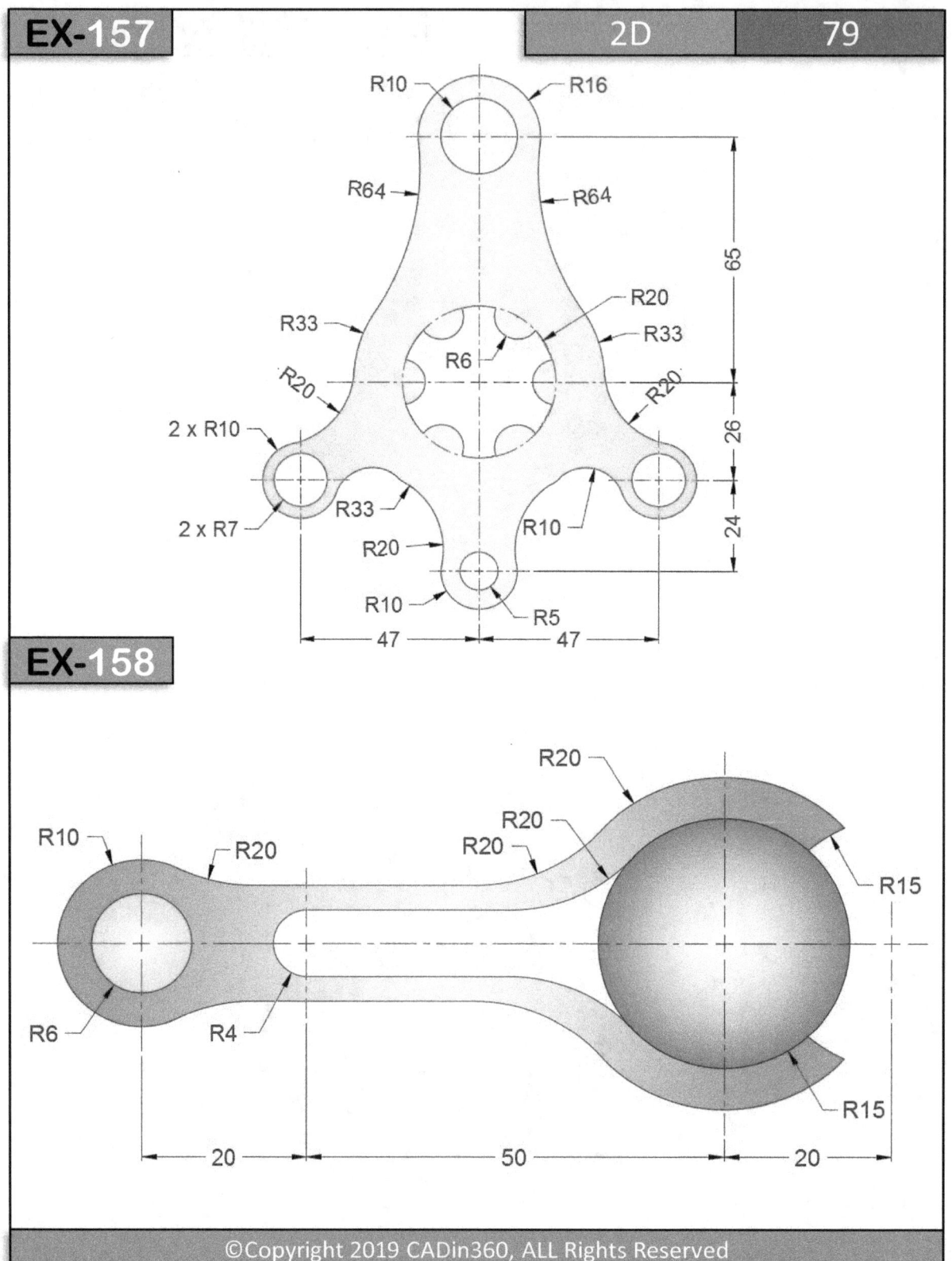

EX-158

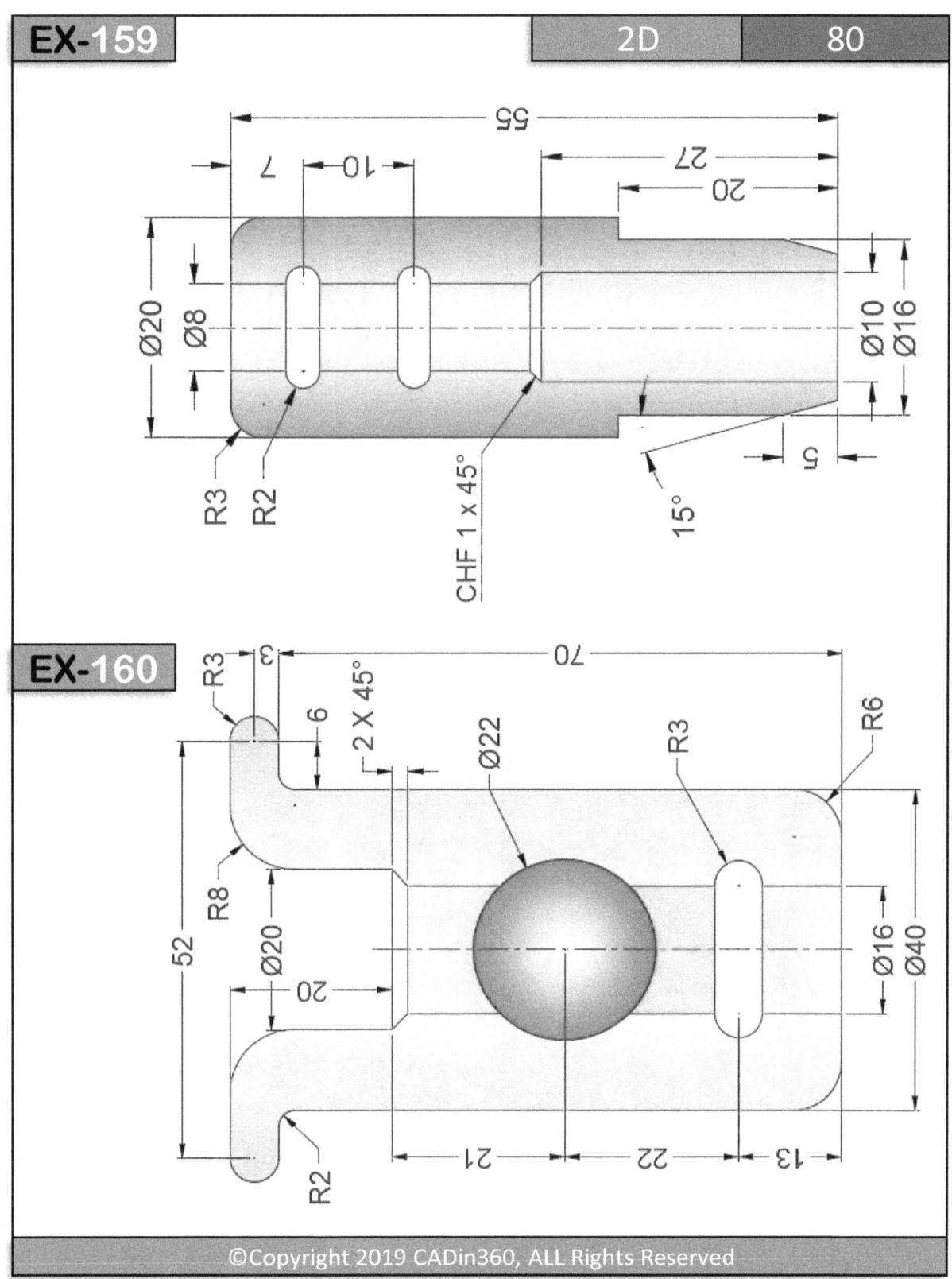

EX-159

Ø20
Ø8
55
7
10
27
20
Ø10
Ø16
5
15°
R3
R2
CHF 1 x 45°

EX-160

R3
3
2 X 45°
6
70
Ø22
R3
R6
R8
52
Ø20
20
Ø16
Ø40
21
22
13
R2

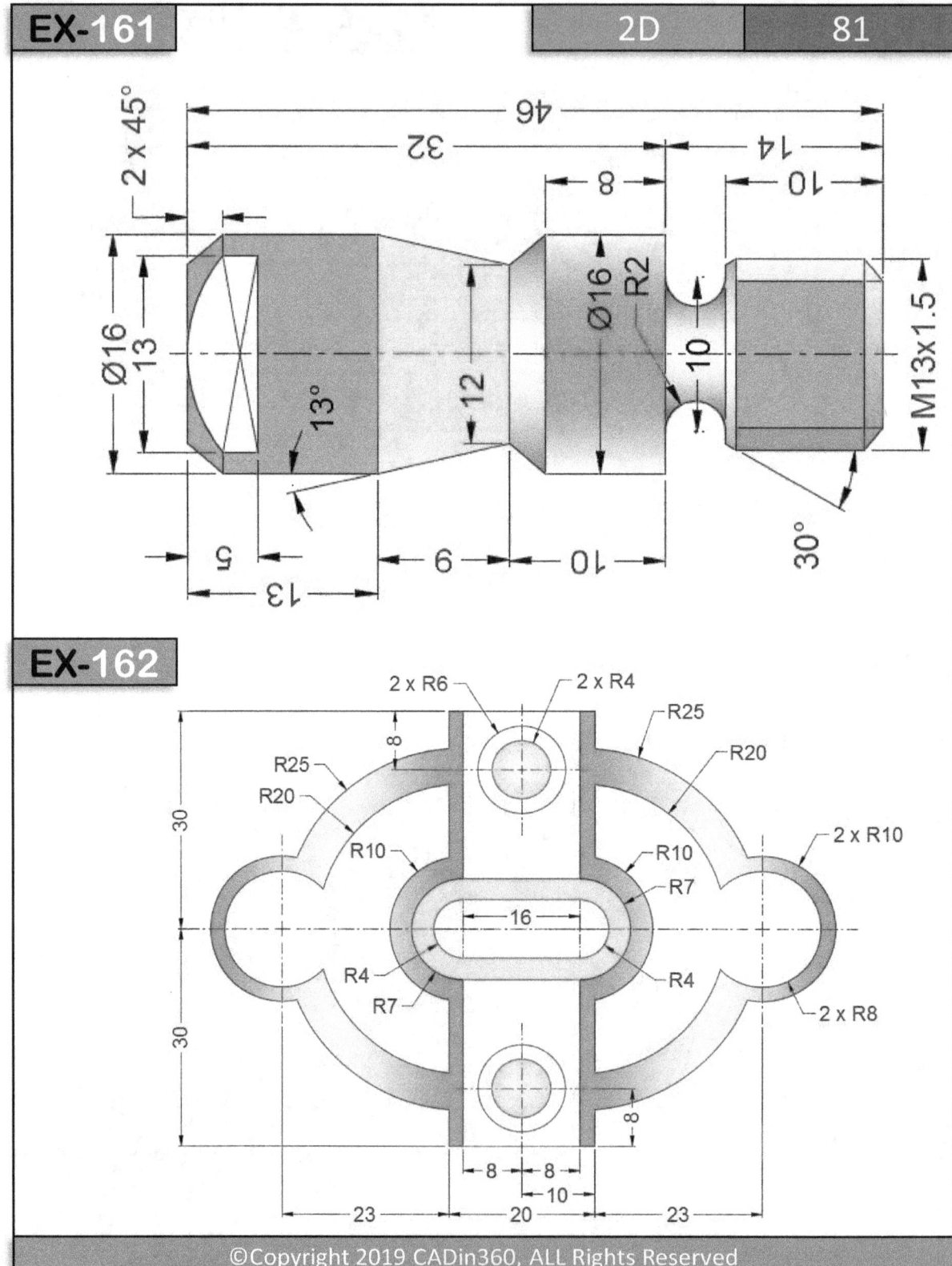

EX-161   2D   81

EX-162

EX-163

Ø20
Ø14
30
90°
10
Ø28
Ø10
R2
13
10
40
135°
Ø12
20
10
15
40
15

EX-164

2 x R10
2 x R5
2 x R4
Ø28
Ø10
10
5
R2
13
10
40
45°
Ø12
30
60

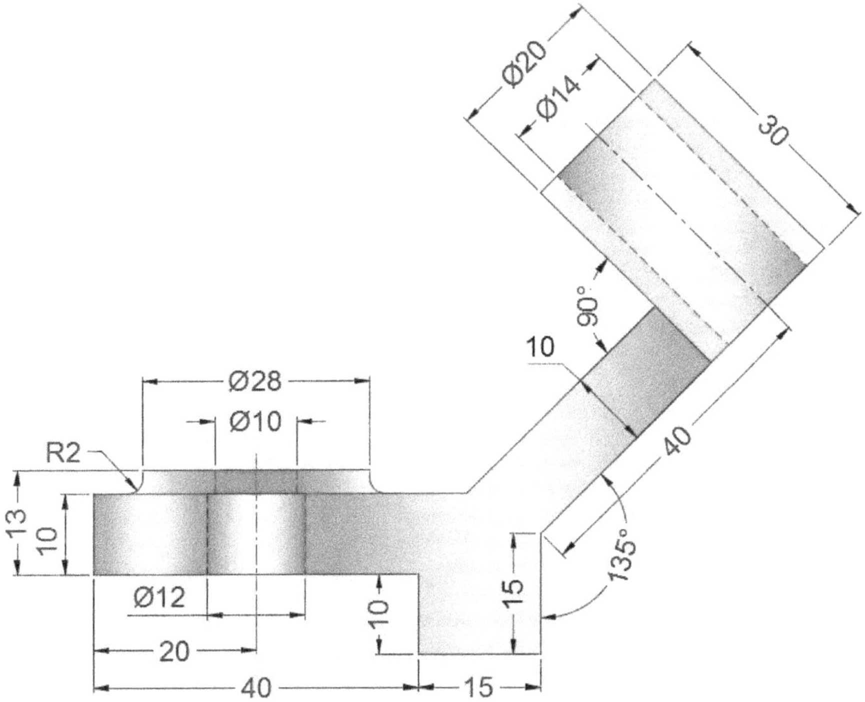

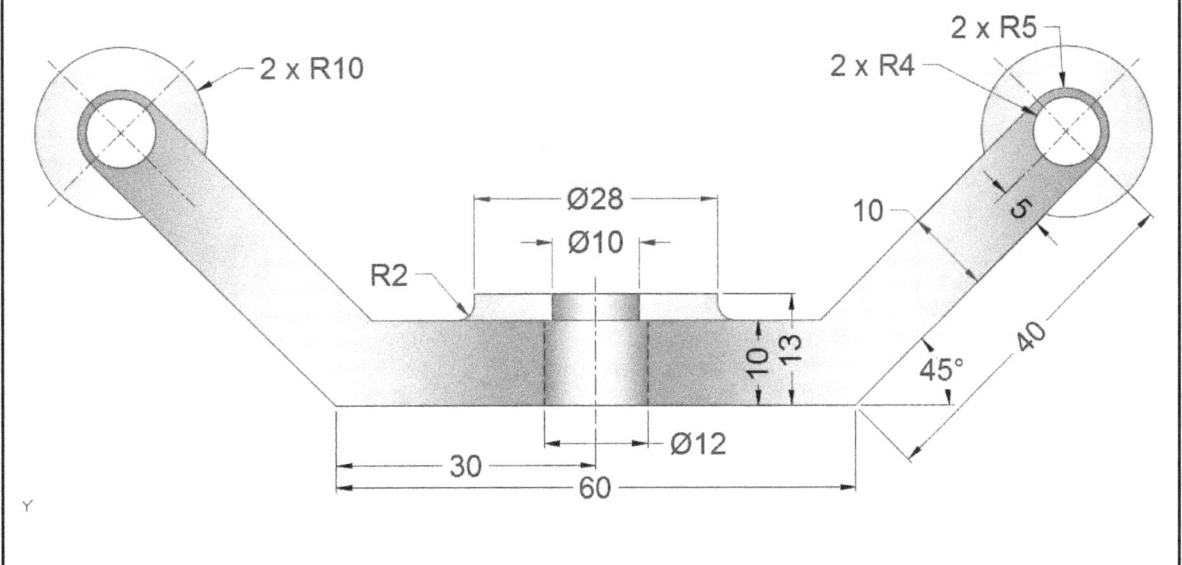

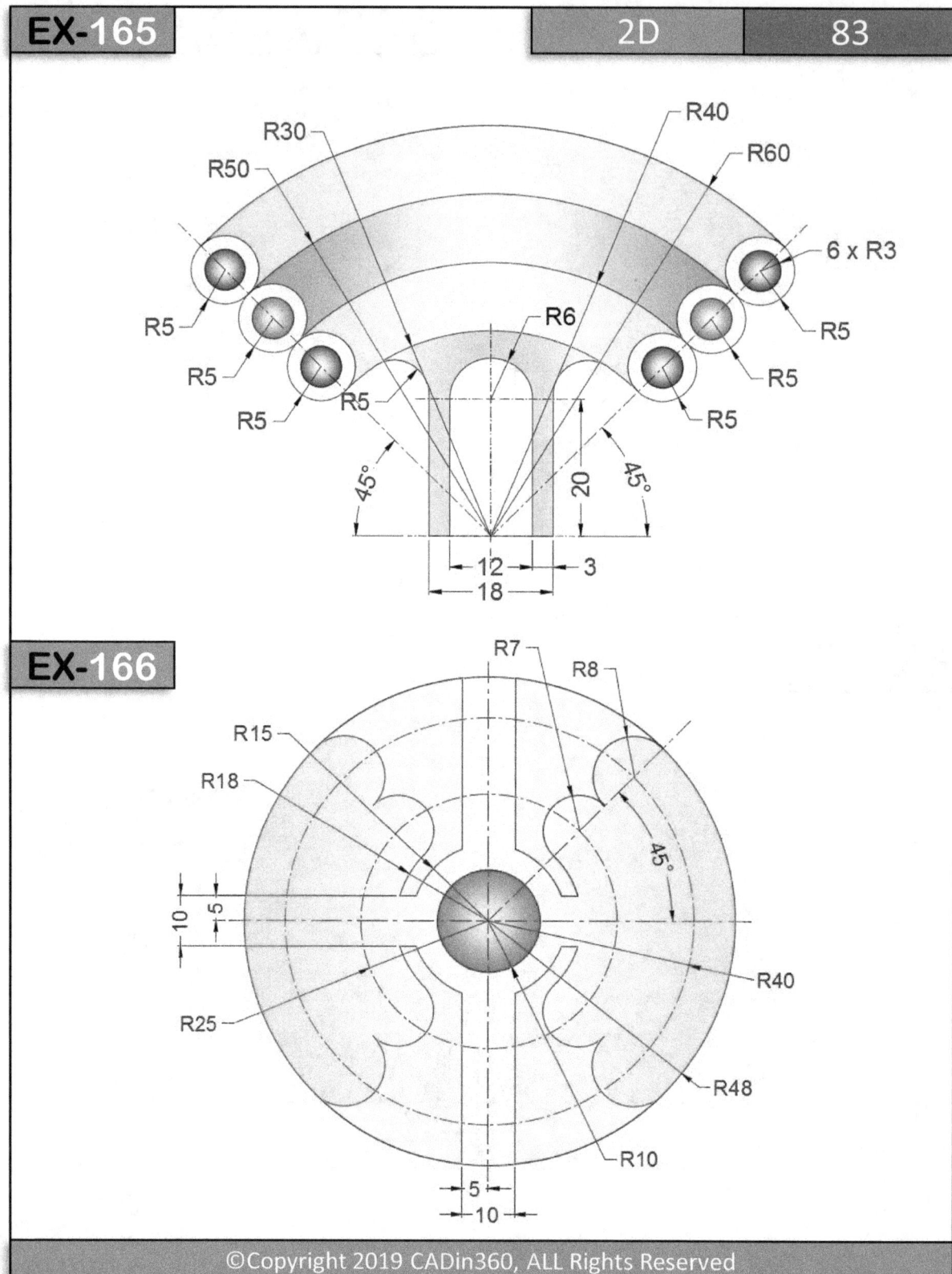

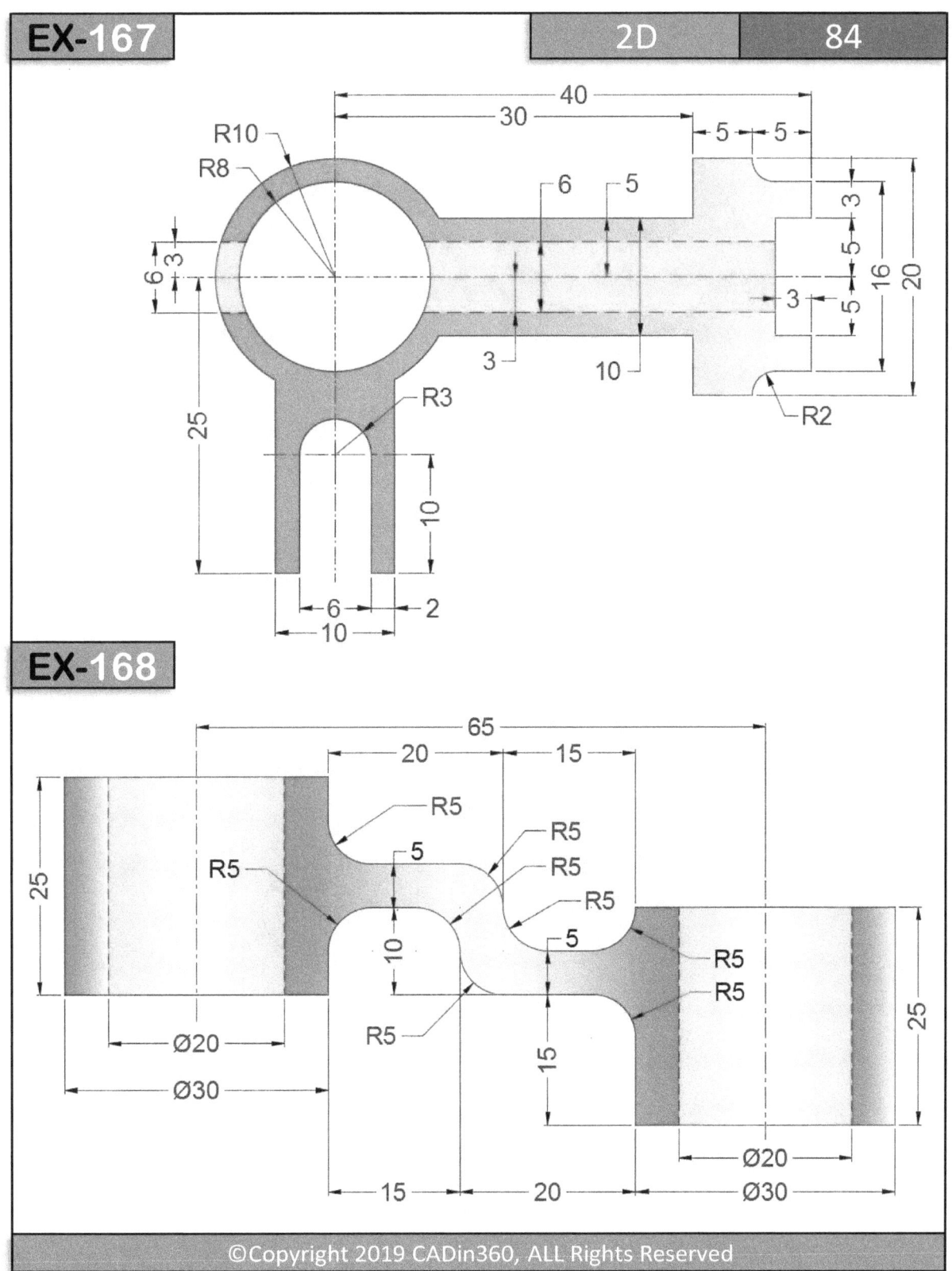

EX-168

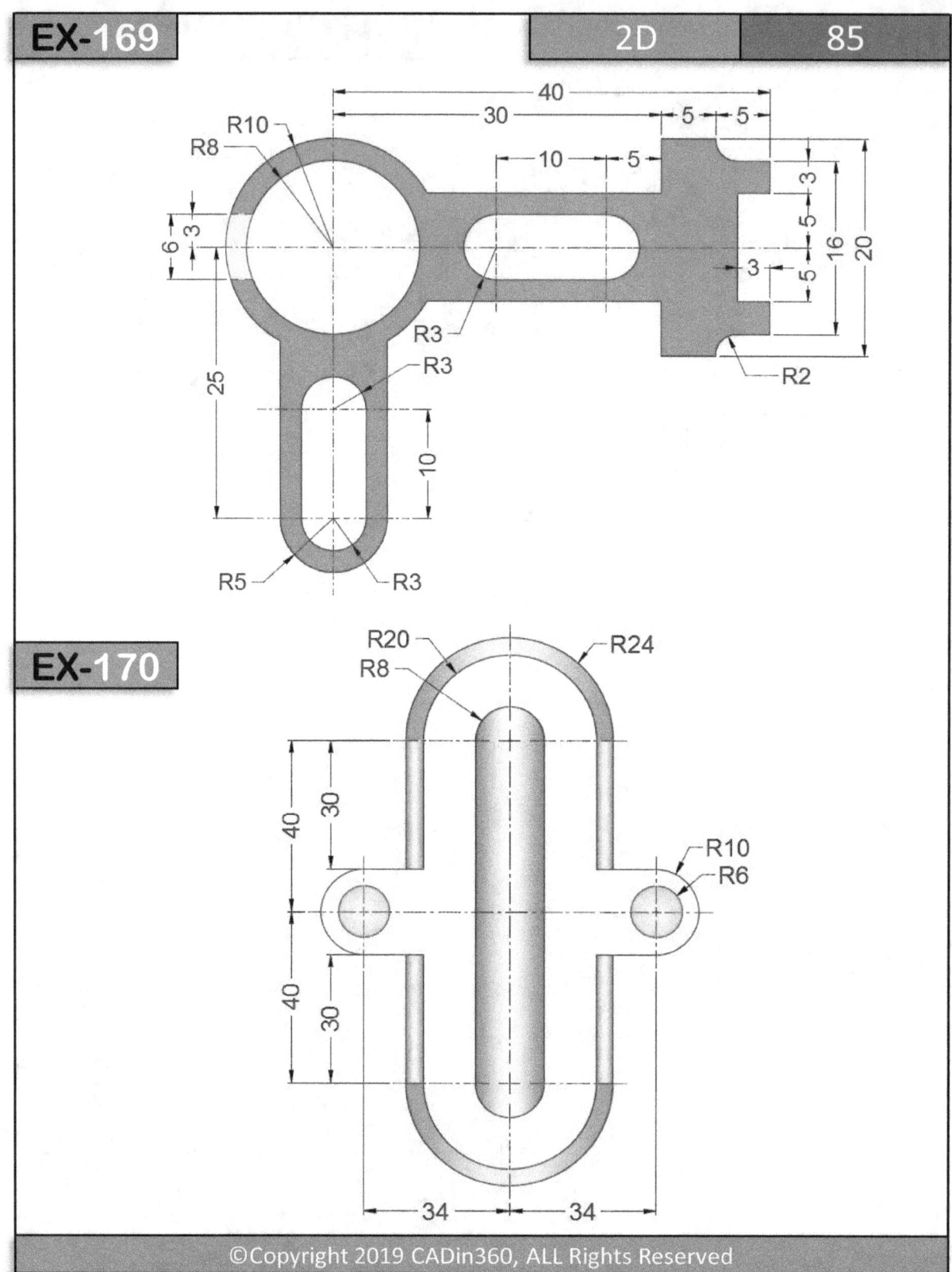

EX-169

2D

85

EX-170

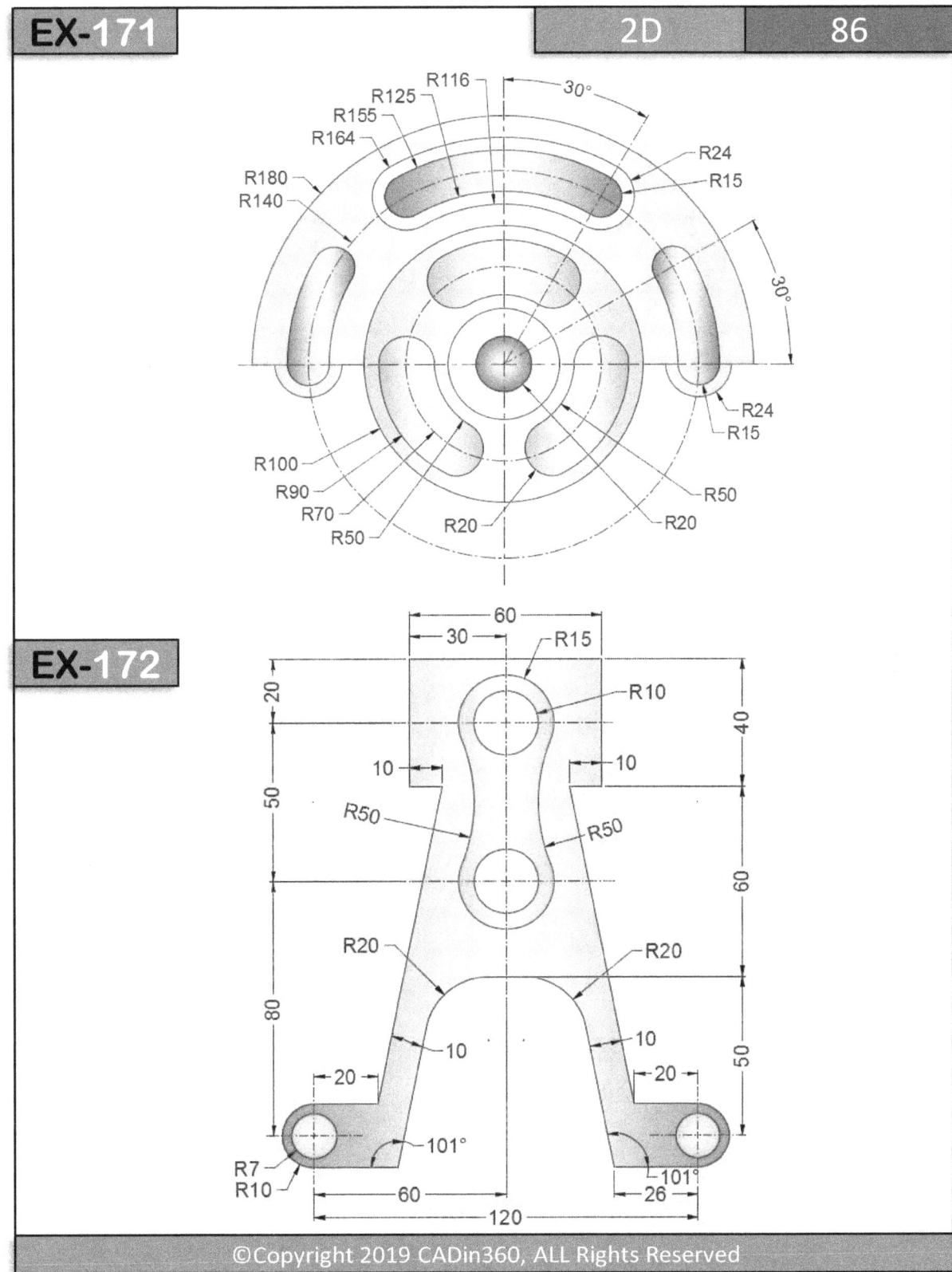

EX-172

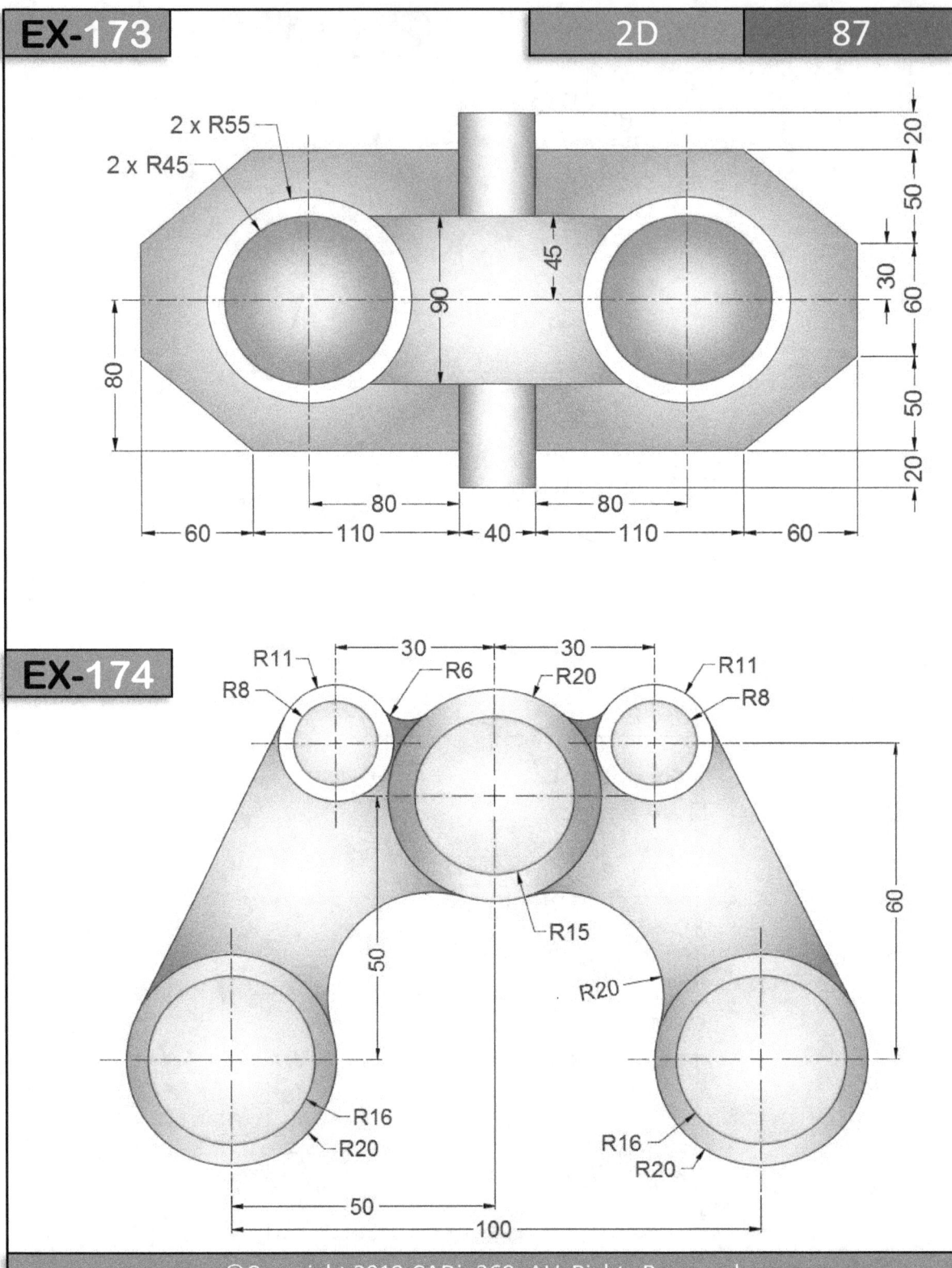

EX-176

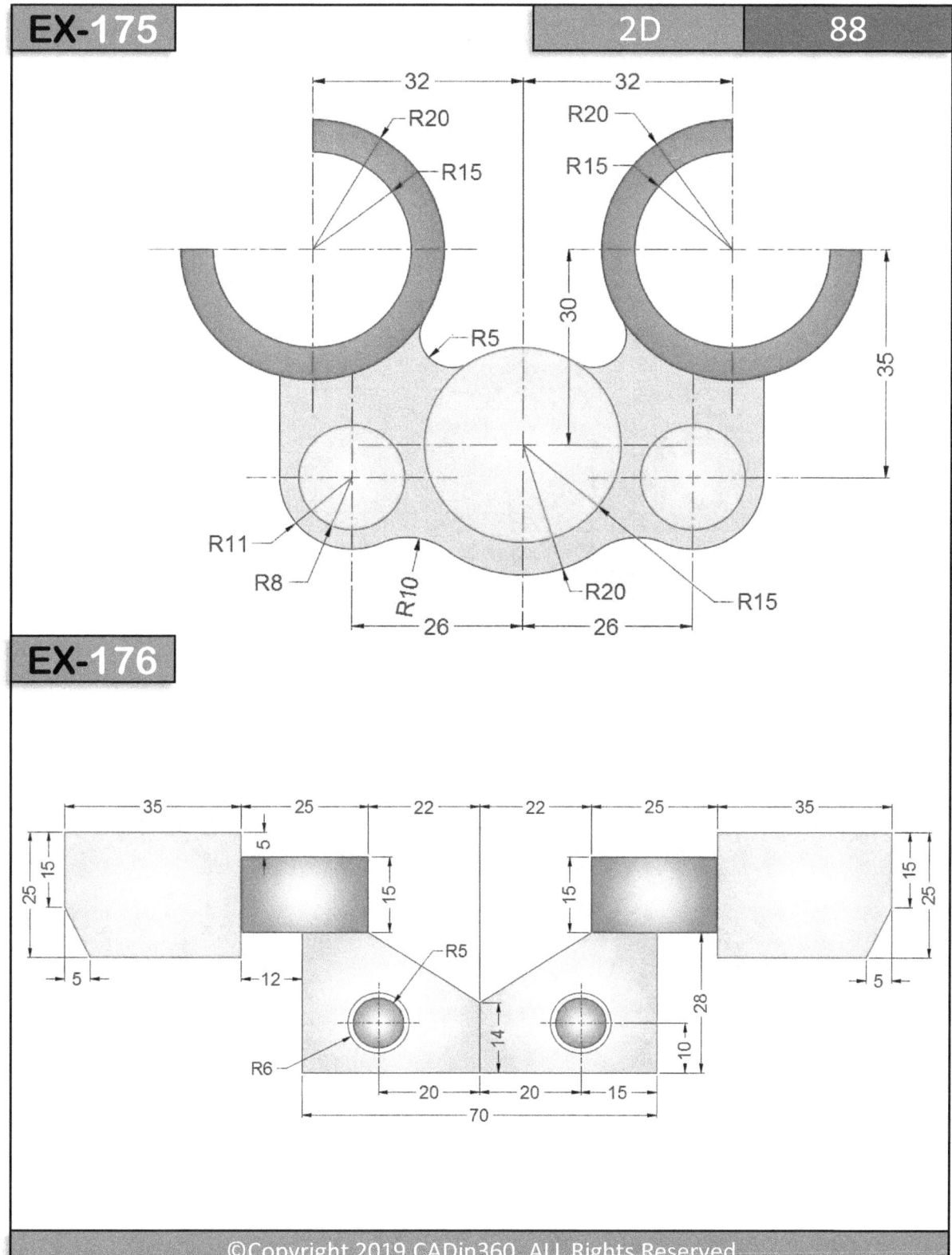

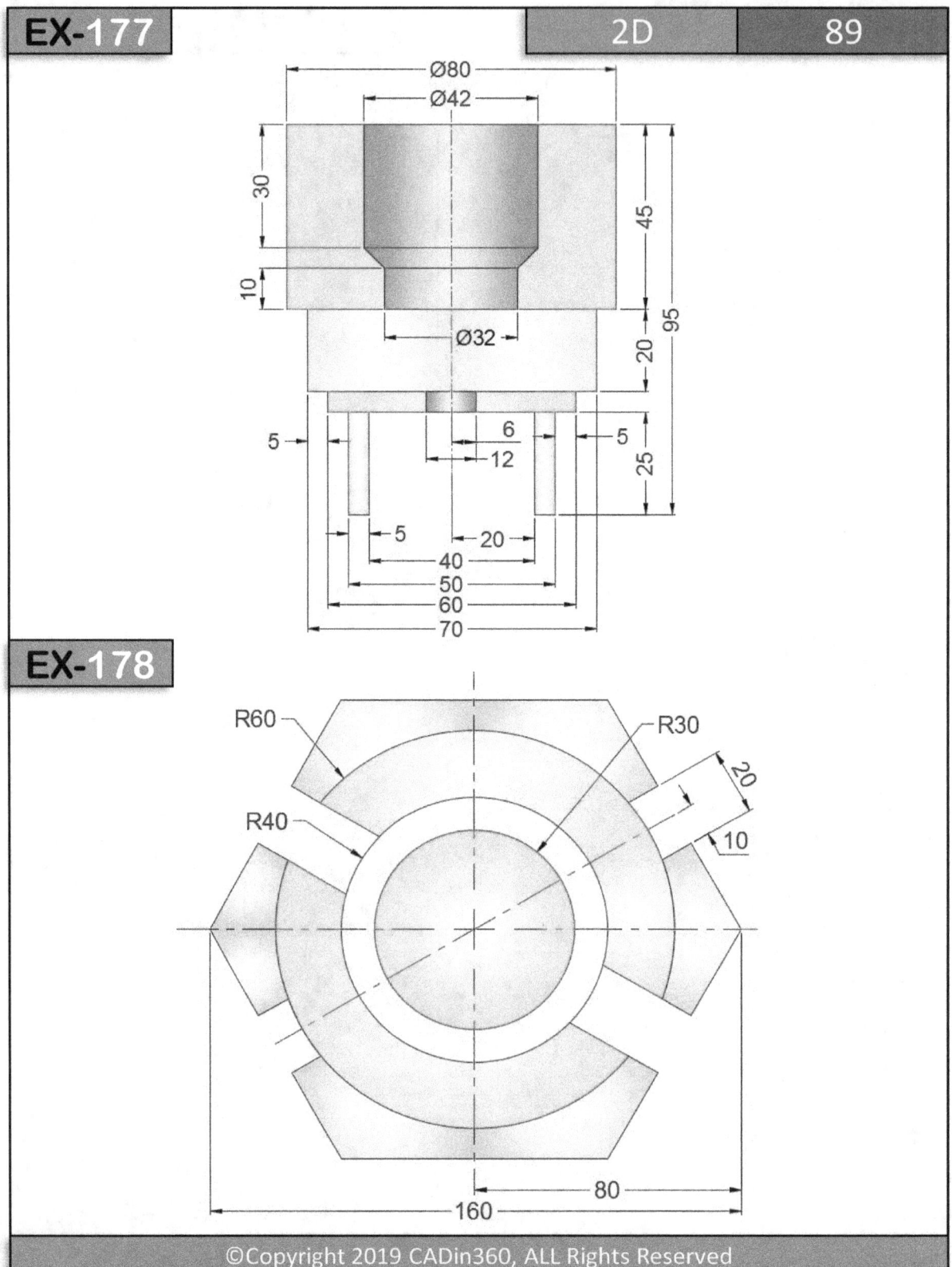

EX-178

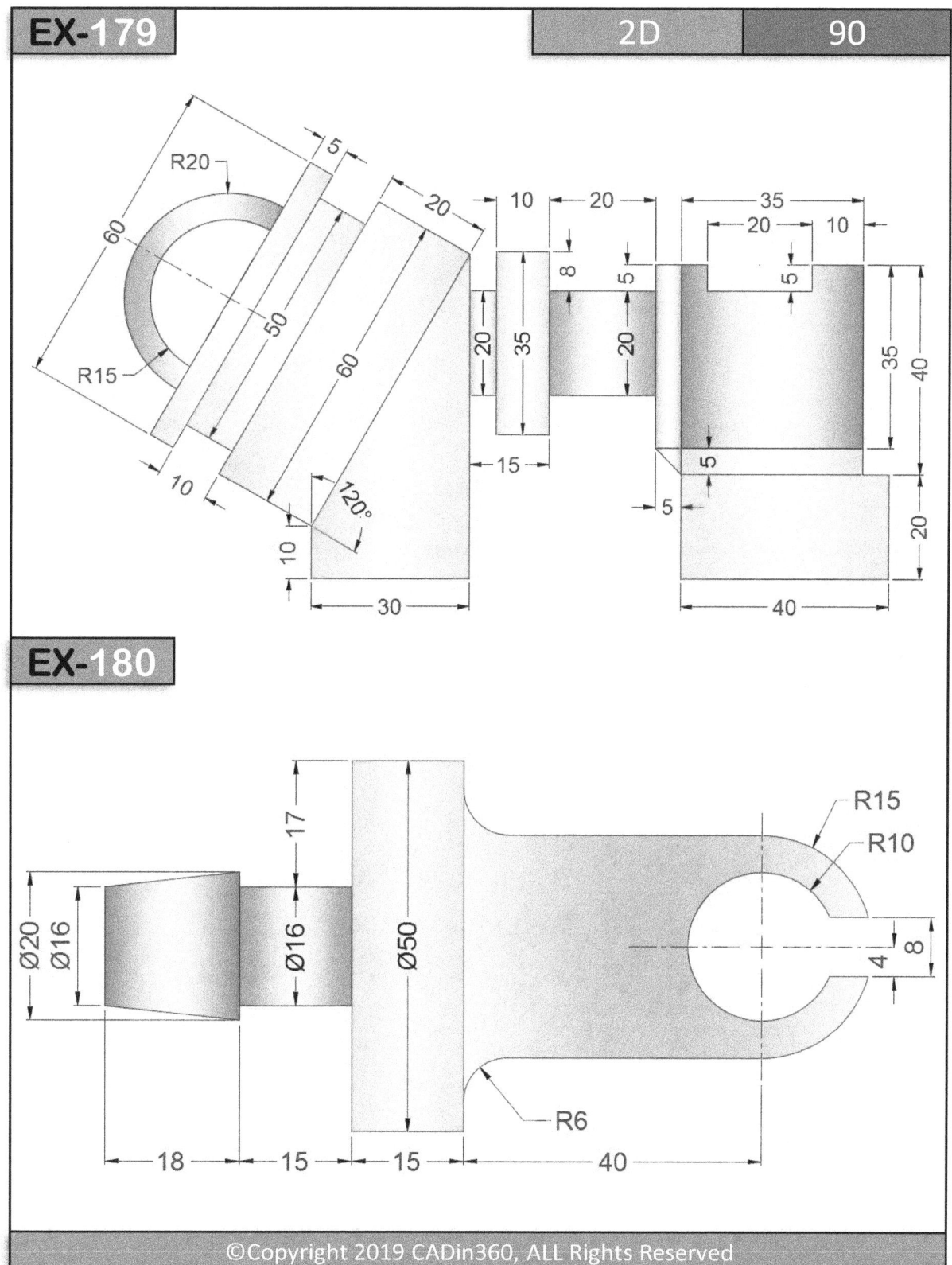

EX-180

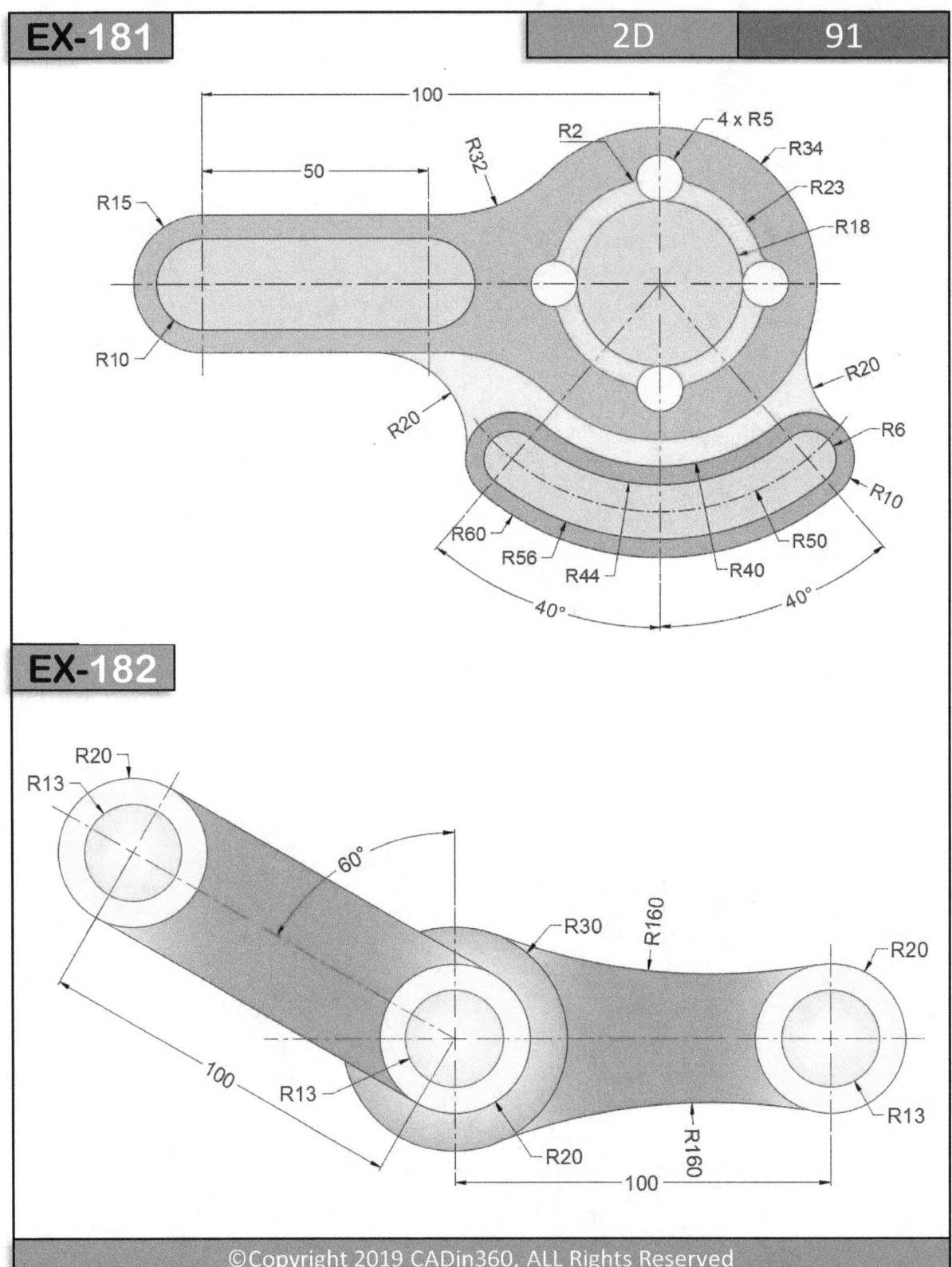

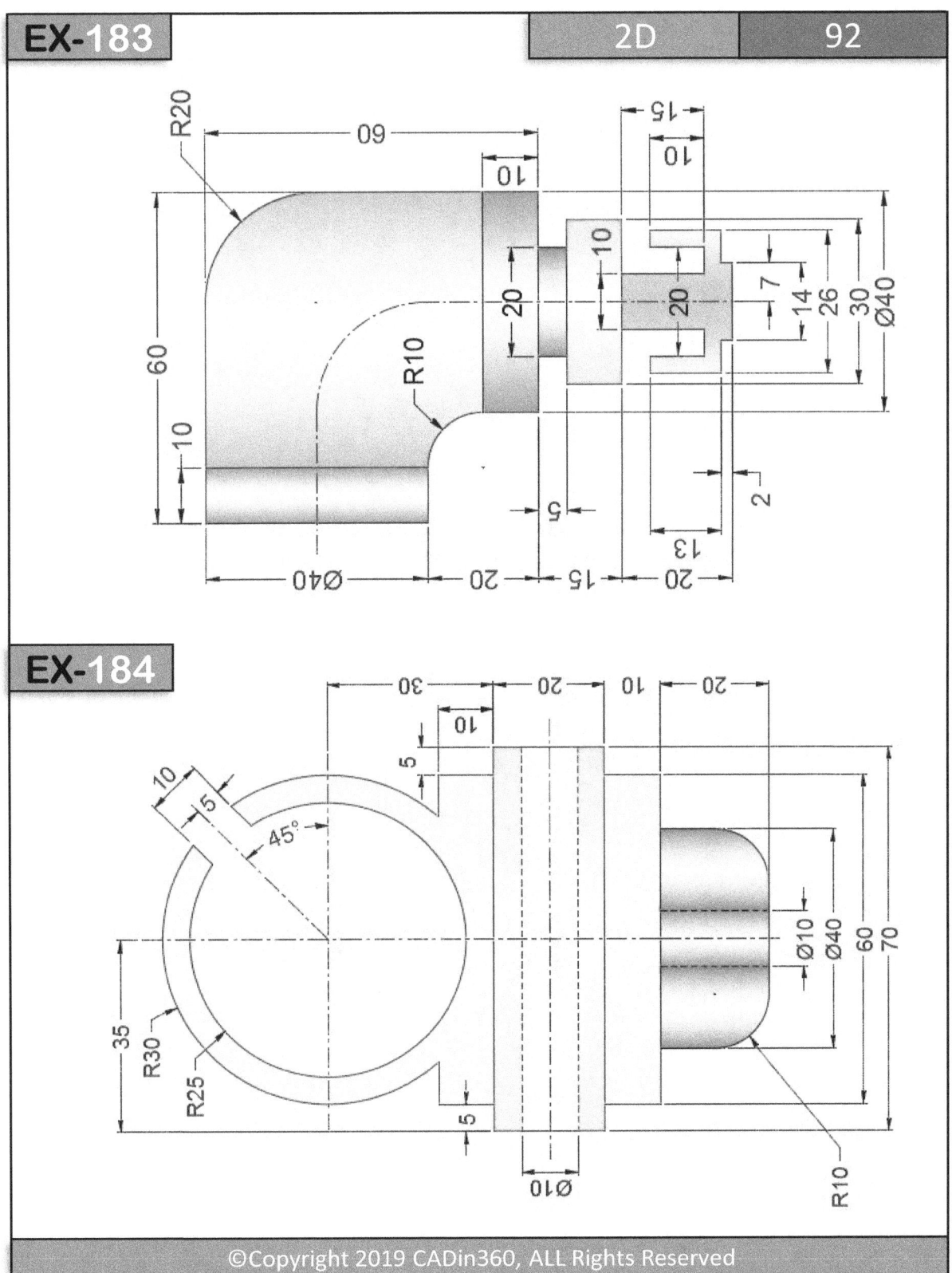

EX-184

EX-185

2 x R20
2 x R8
80
17
23
23
17
2 x R15
6 x R4
Ø46
R30
23

EX-186

R10
R6
30
R10
30
R8
R15
R10
R10
R6
R12
R20
60
25
10
20
R10
30
60
20
10
20
R10

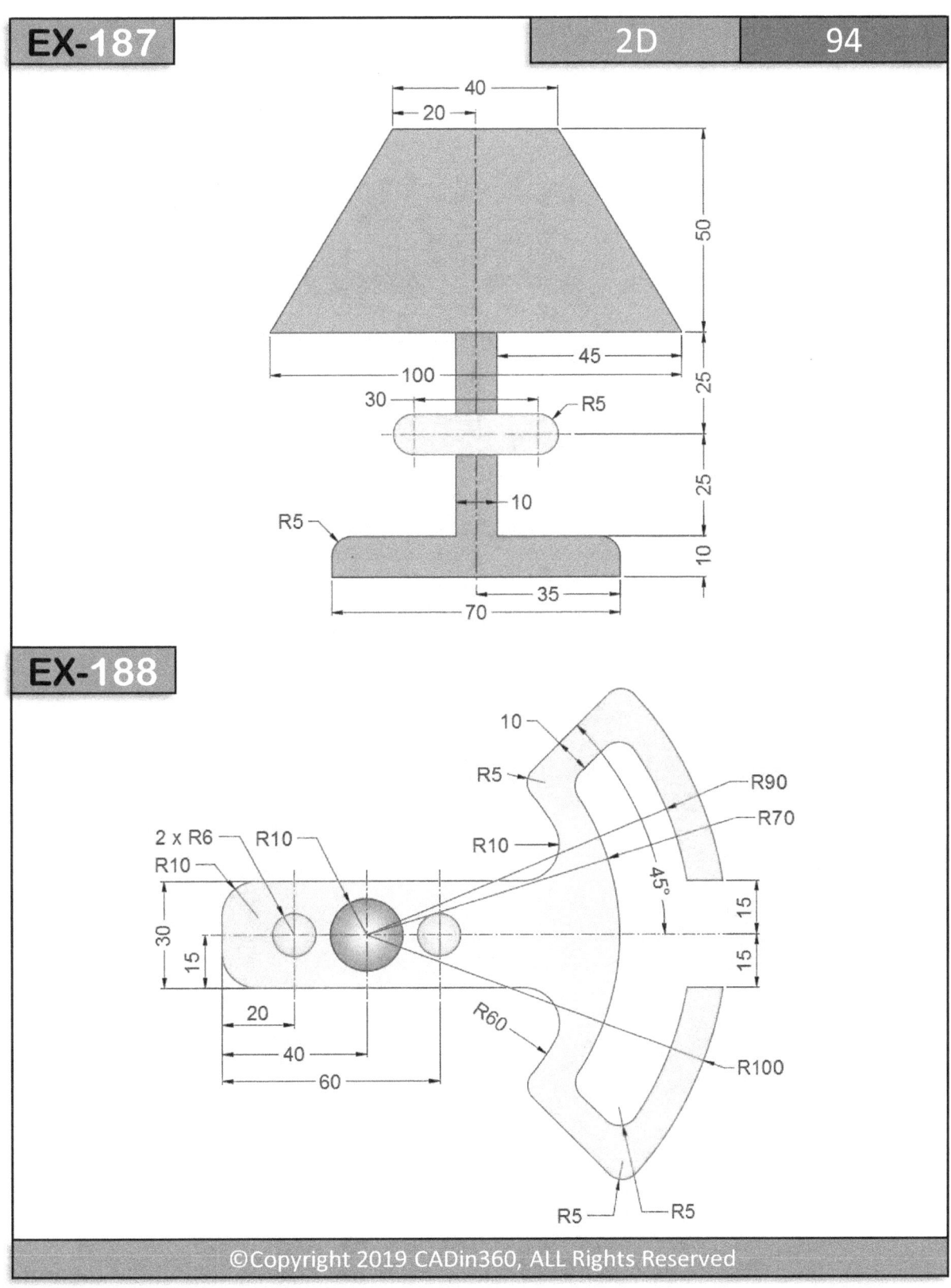

EX-188

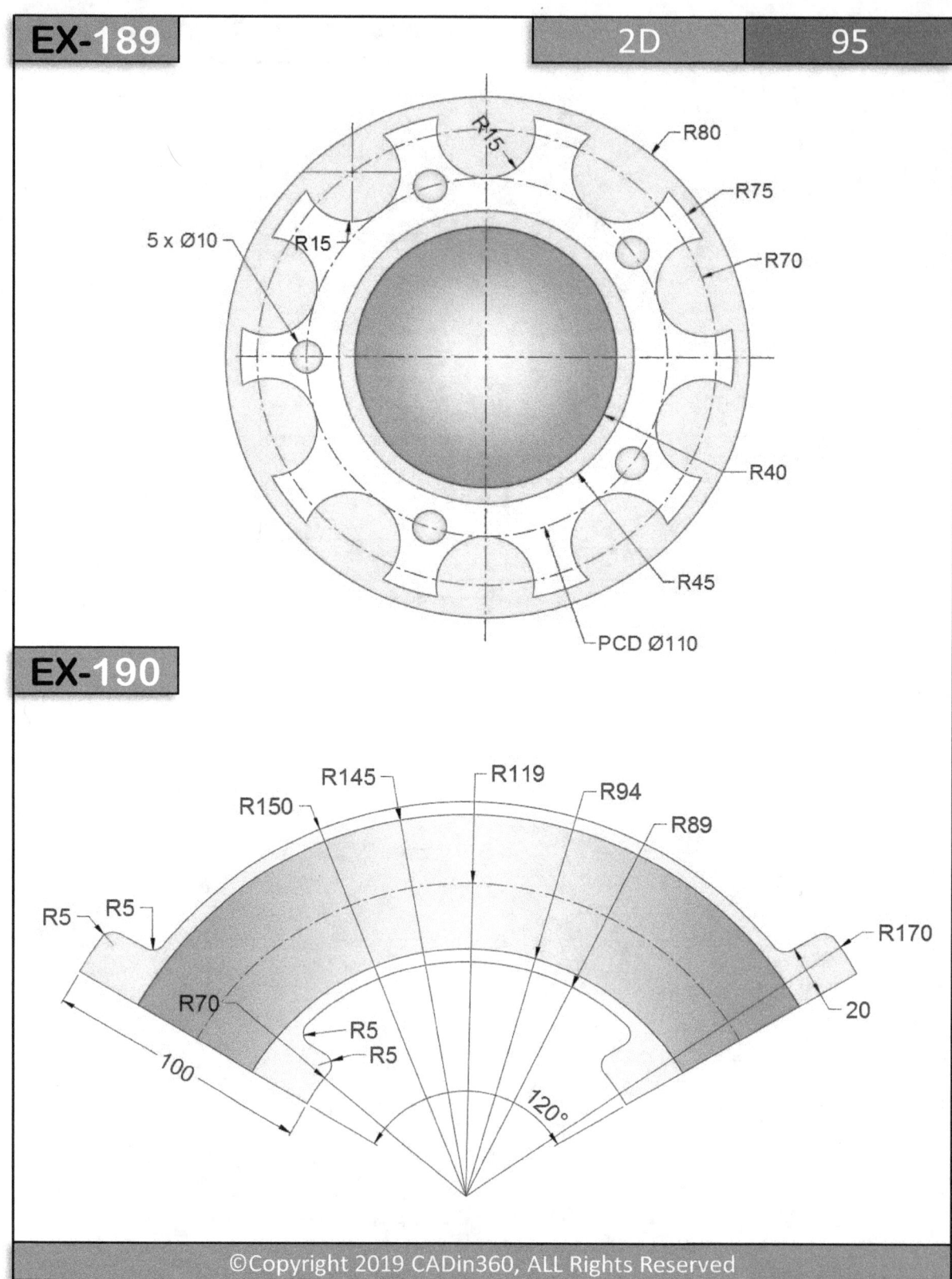

EX-190

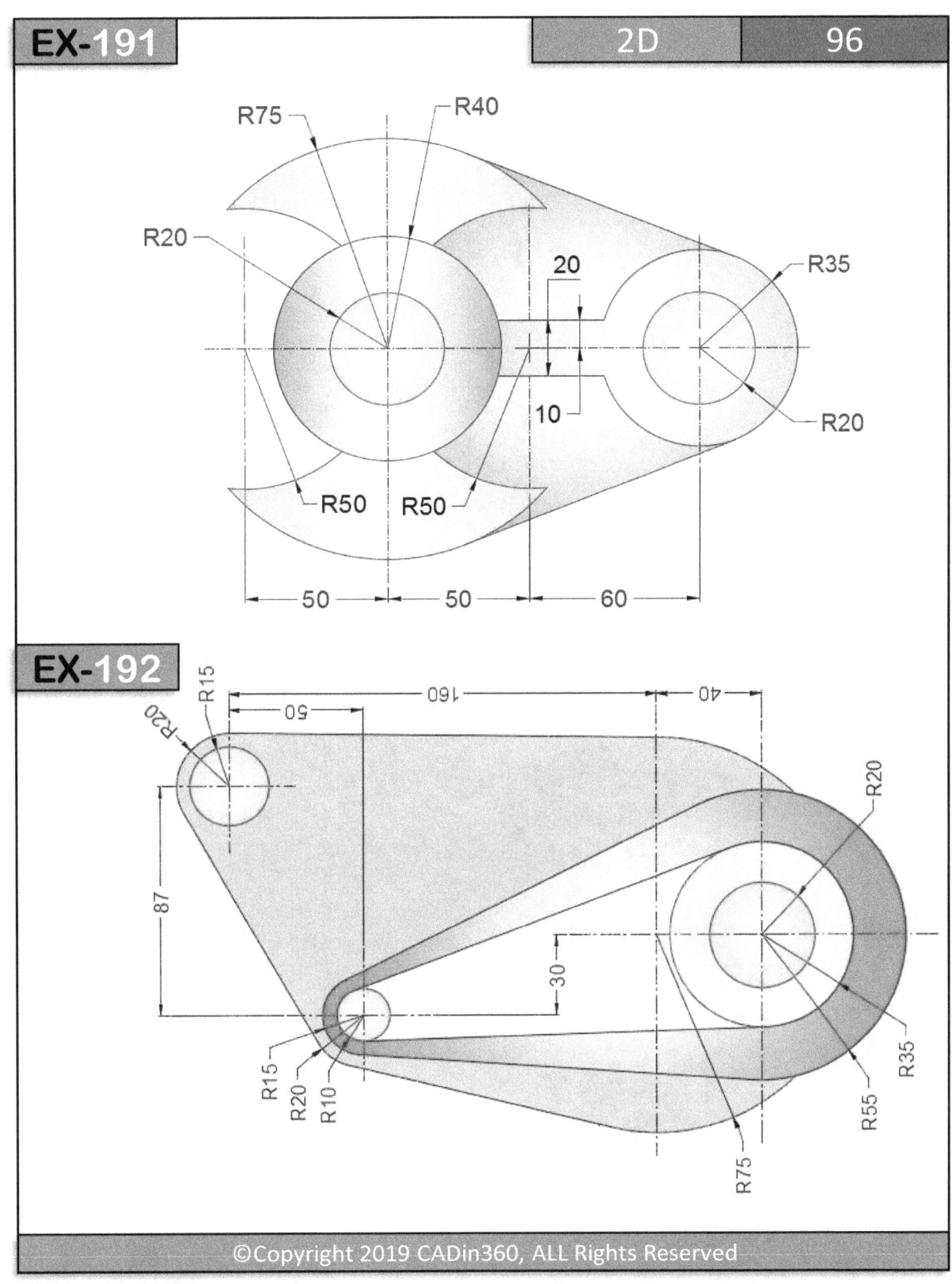

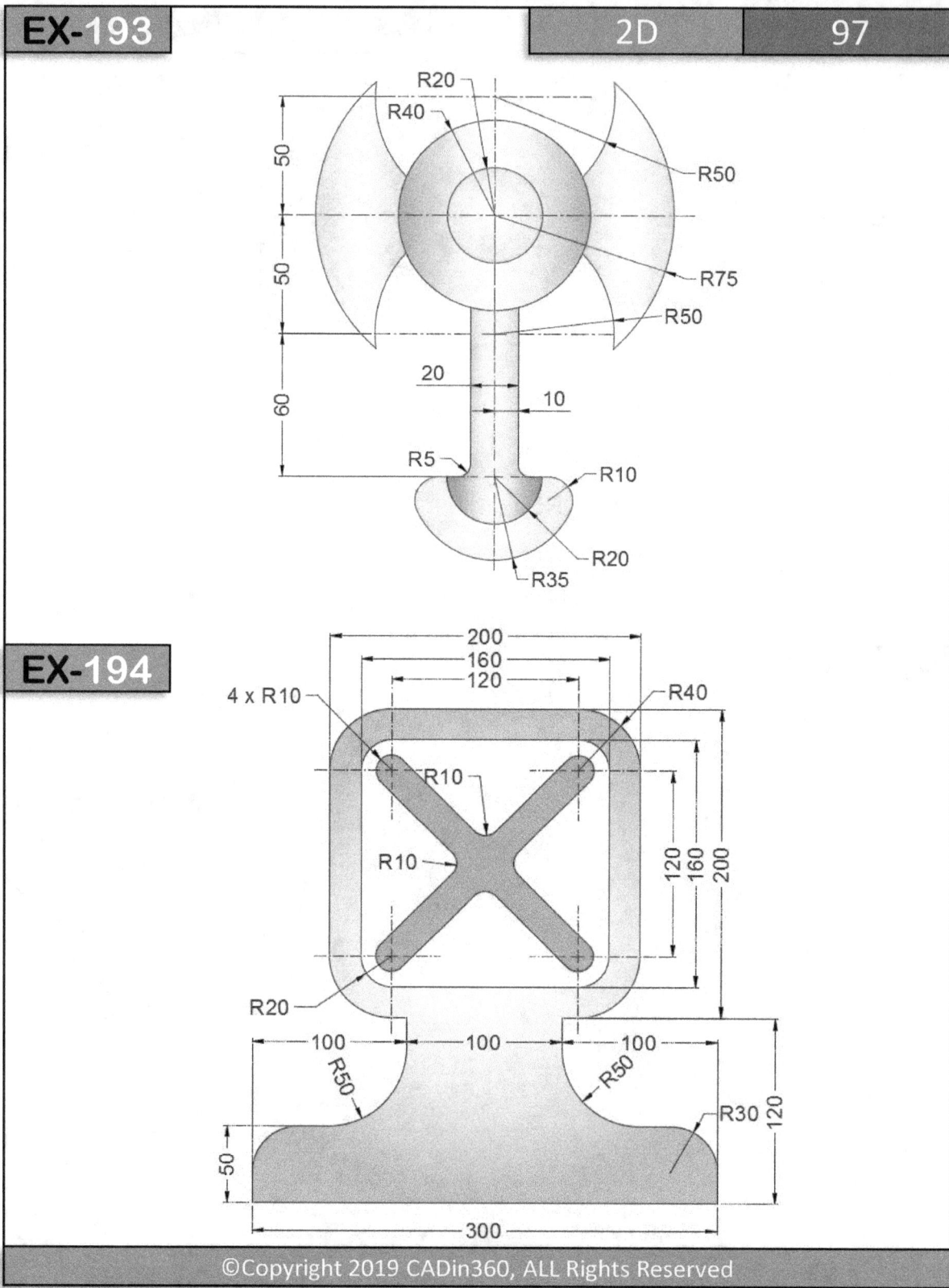

EX-194

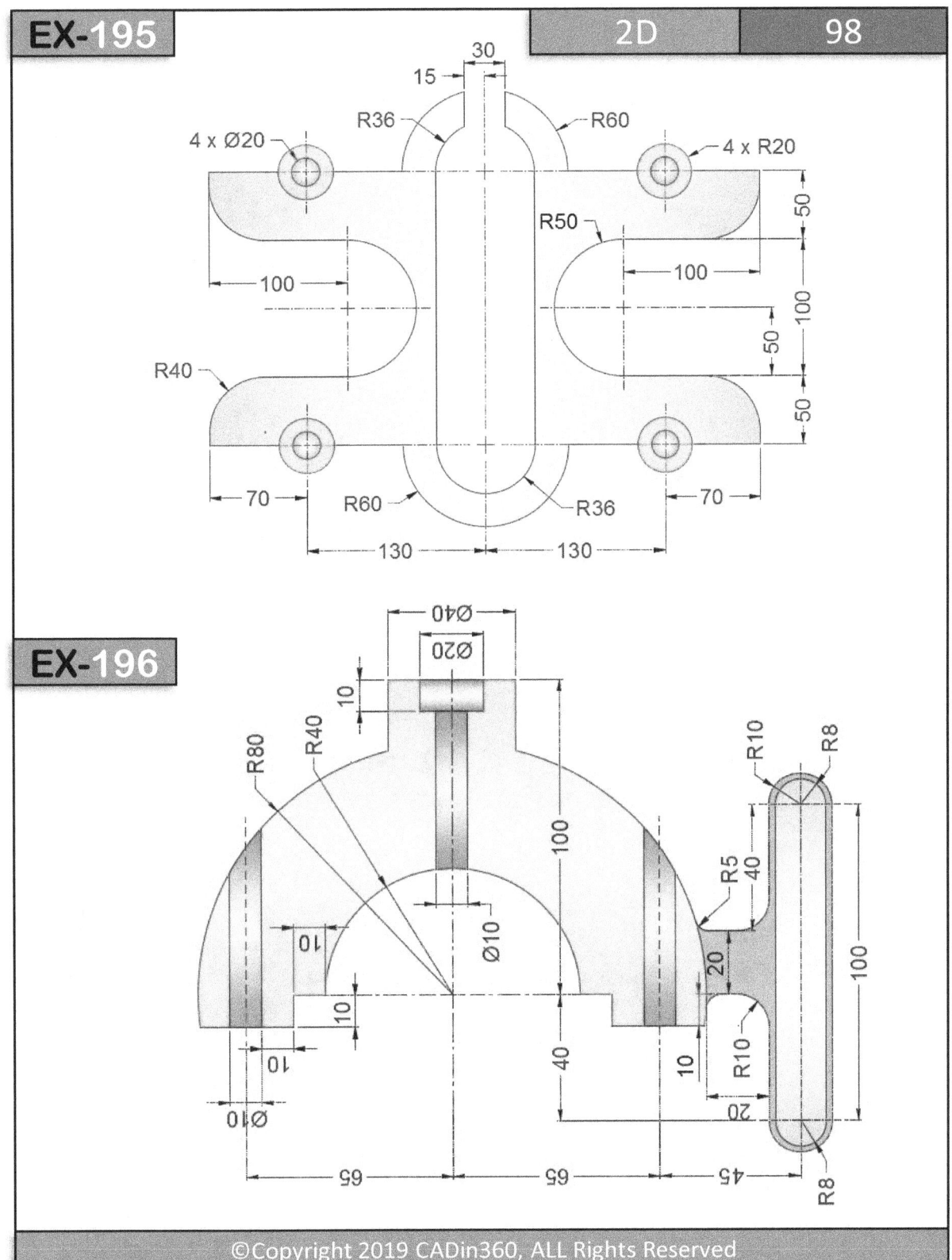

EX-195

2D 98

30
15
R36 R60
4 x Ø20 4 x R20
R50
50
100
100
50
100
R40 50
70 R60 R36 70
130 130

EX-196

Ø40
Ø20
10
R80 R40 R10 R8
R5
40
100
20 R10
Ø10 100
10
10 40
10
Ø10 20
65 65 45 R8

EX-197

50
CHF 3 x 45°
25
12
CHF 5 x 45°
R12
10
20
R5
6
20
15
15
10
15
15
30
20
20
15
100

EX-198

40
35
20
35
40
R10
R5
20
R8
20
R7
30
50
120
R8
10
30
60
R15
R12
R30
20
20
20
20
30
25

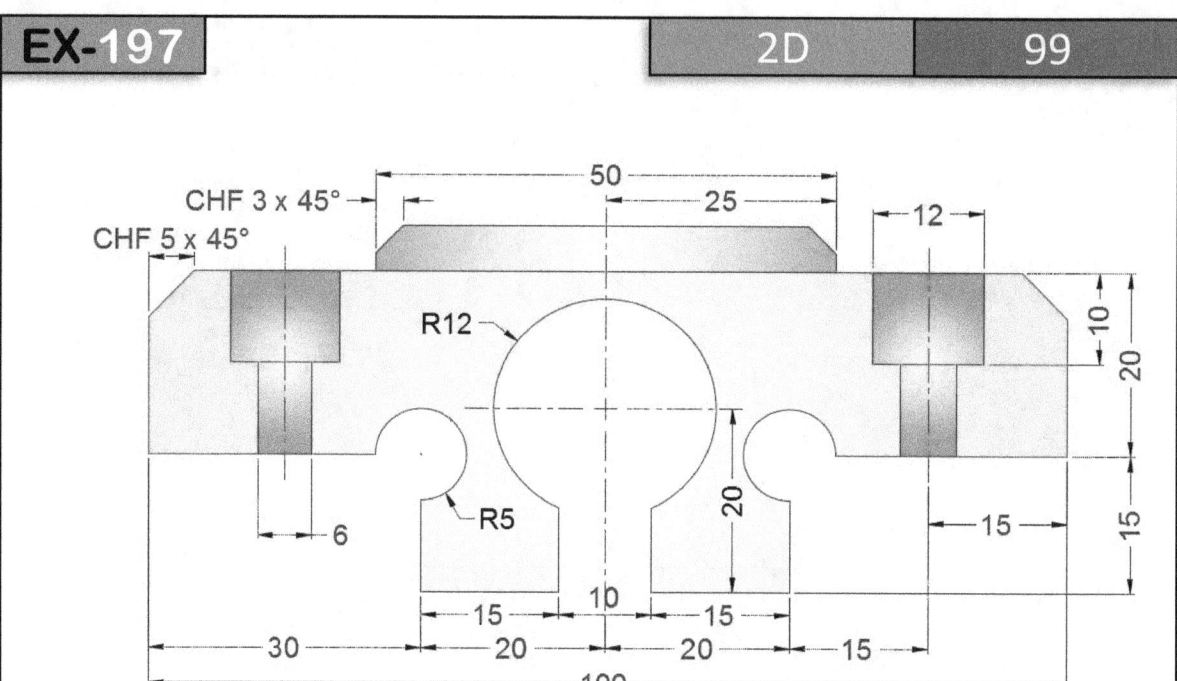

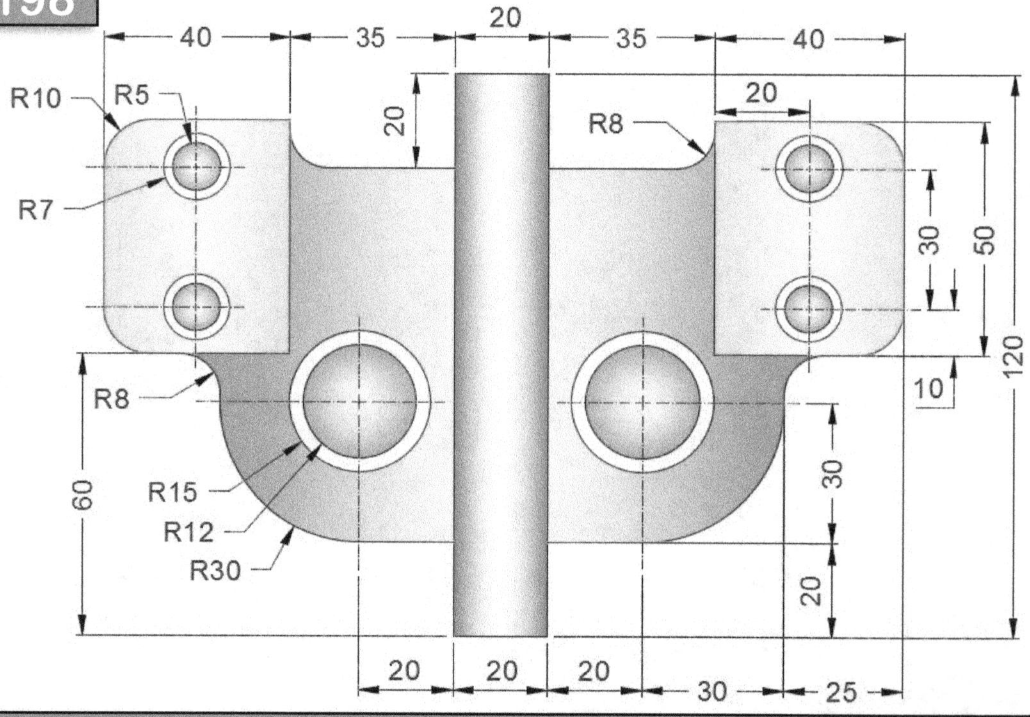

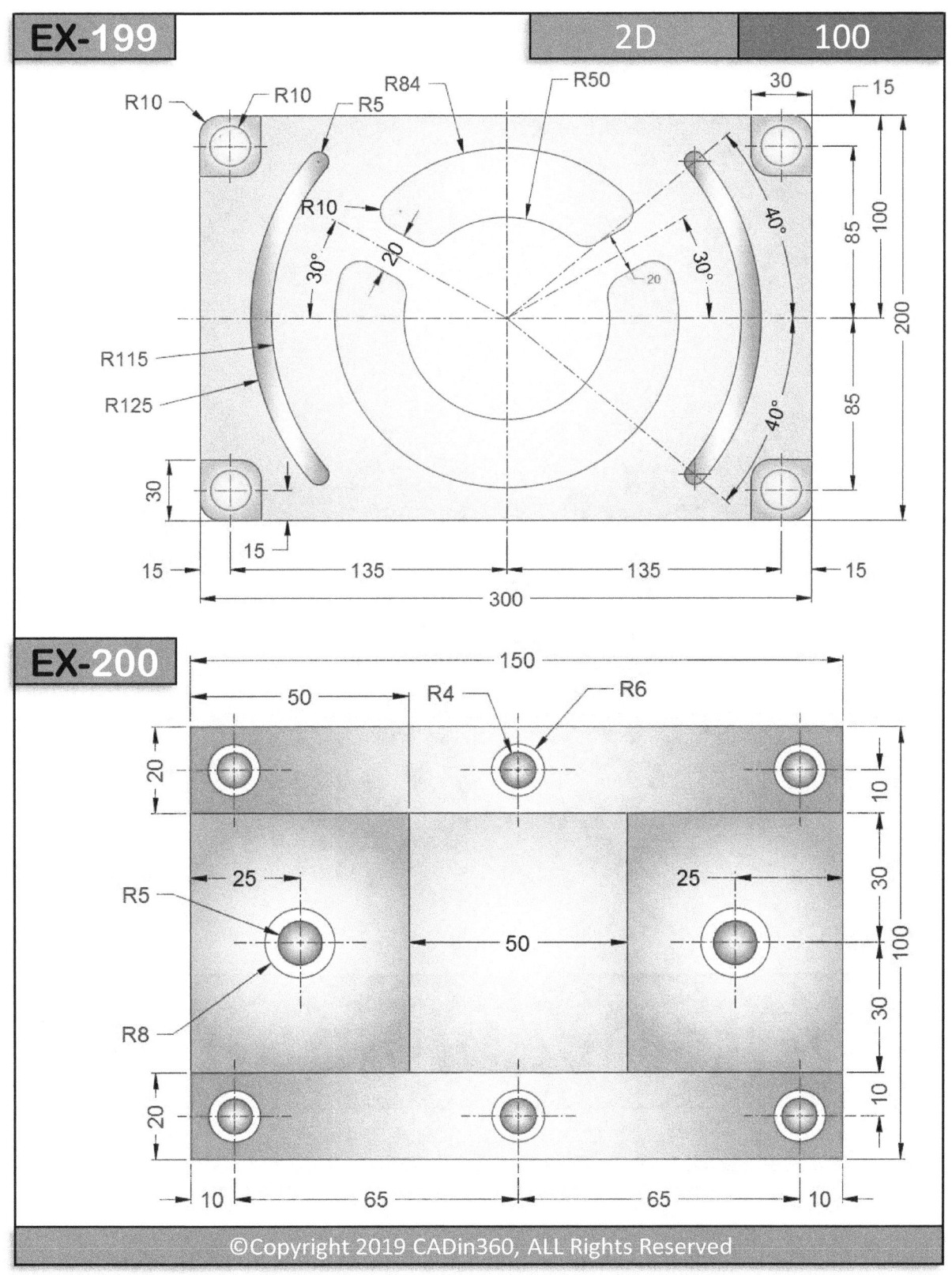

EX-199

2D 100

R10 R10 R5 R84 R50 30 15

R10

40°

30°

30°

20

20 85 100

200

R115

R125

40° 85

30

15

15 135 135 15

300

EX-200

150

50 R4 R6

20

10

R5 25 25

30

50 100

R8

30

20 10

10 65 65 10

PART MODELING

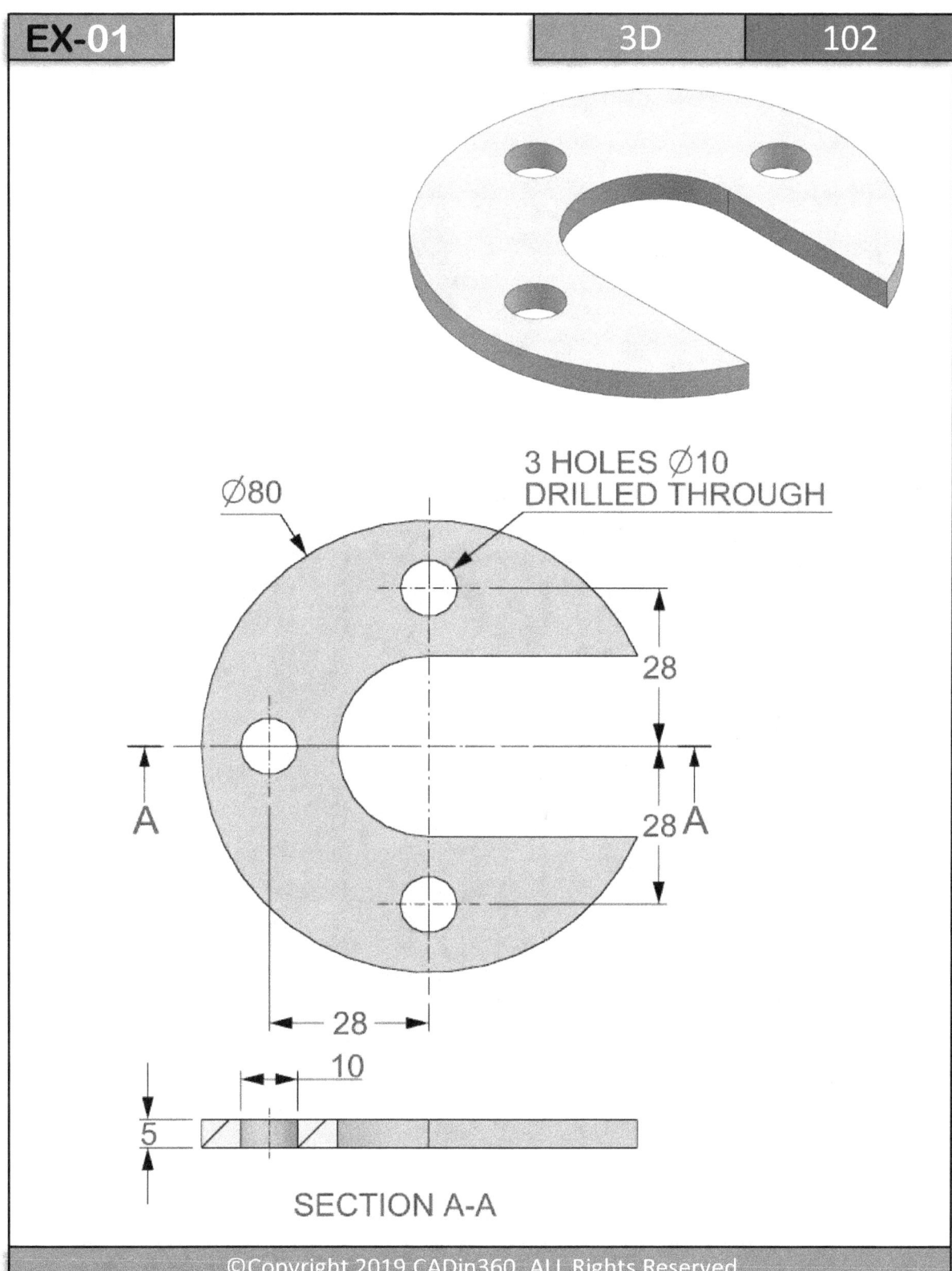

Ø80

3 HOLES Ø10
DRILLED THROUGH

28

28 A

A

28

10

5

SECTION A-A

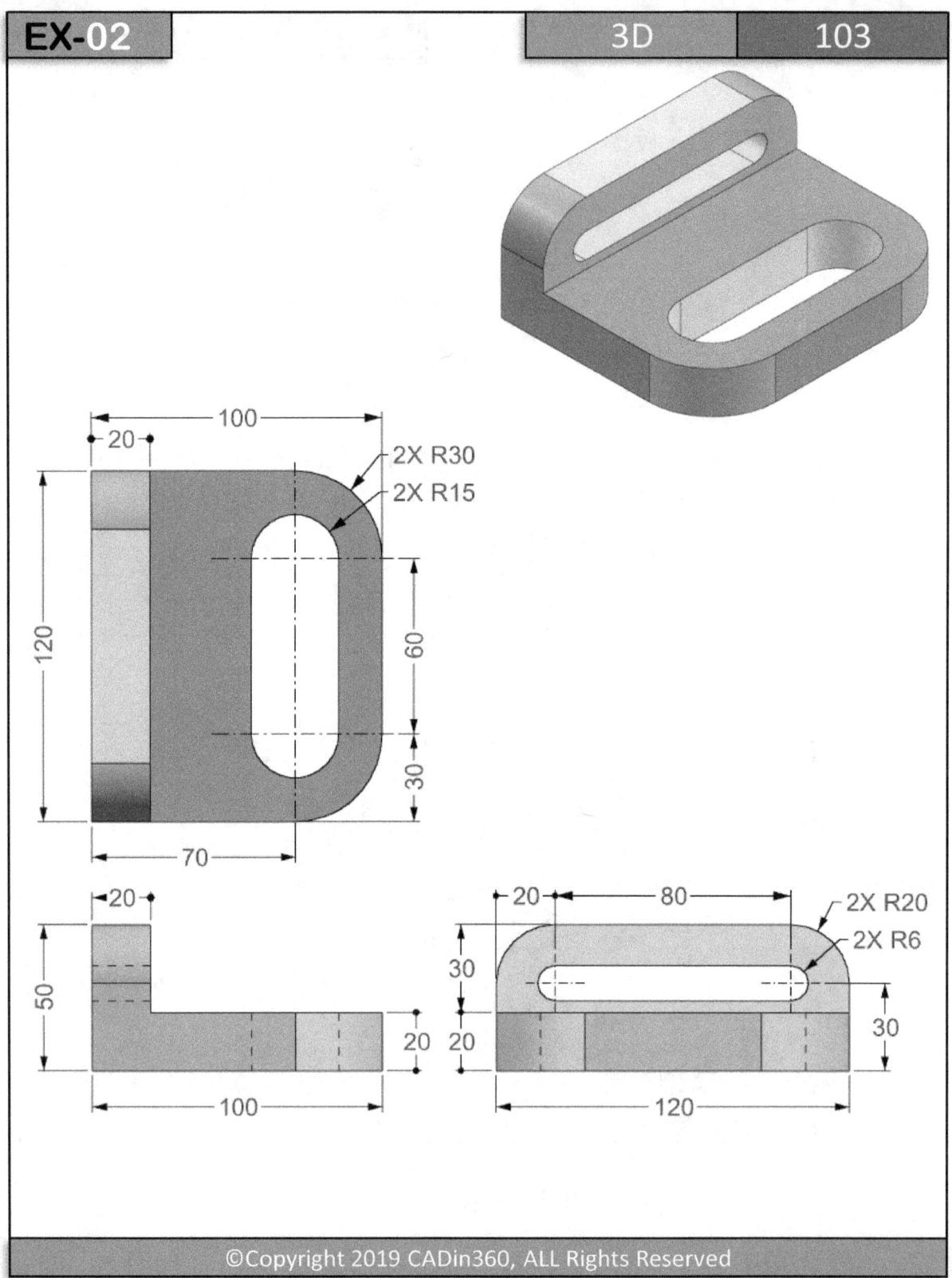

2X R30
2X R15

100
20
120
60
30
70

20
50
20
100

20
80
2X R20
2X R6
30
20
30
120

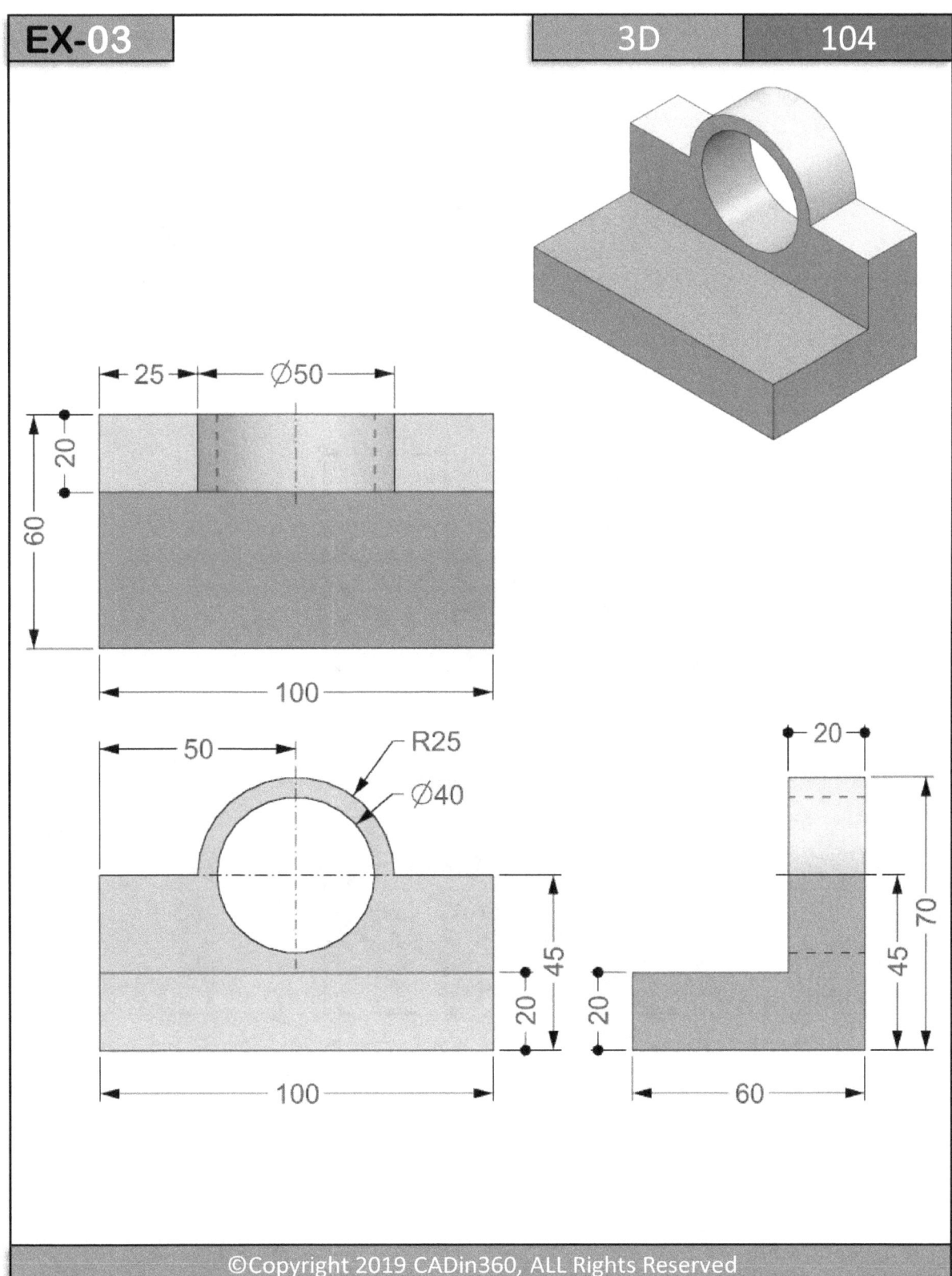

25 ⌀50

20

60

100

50 R25

⌀40

45

20

100

20

70

45

20

60

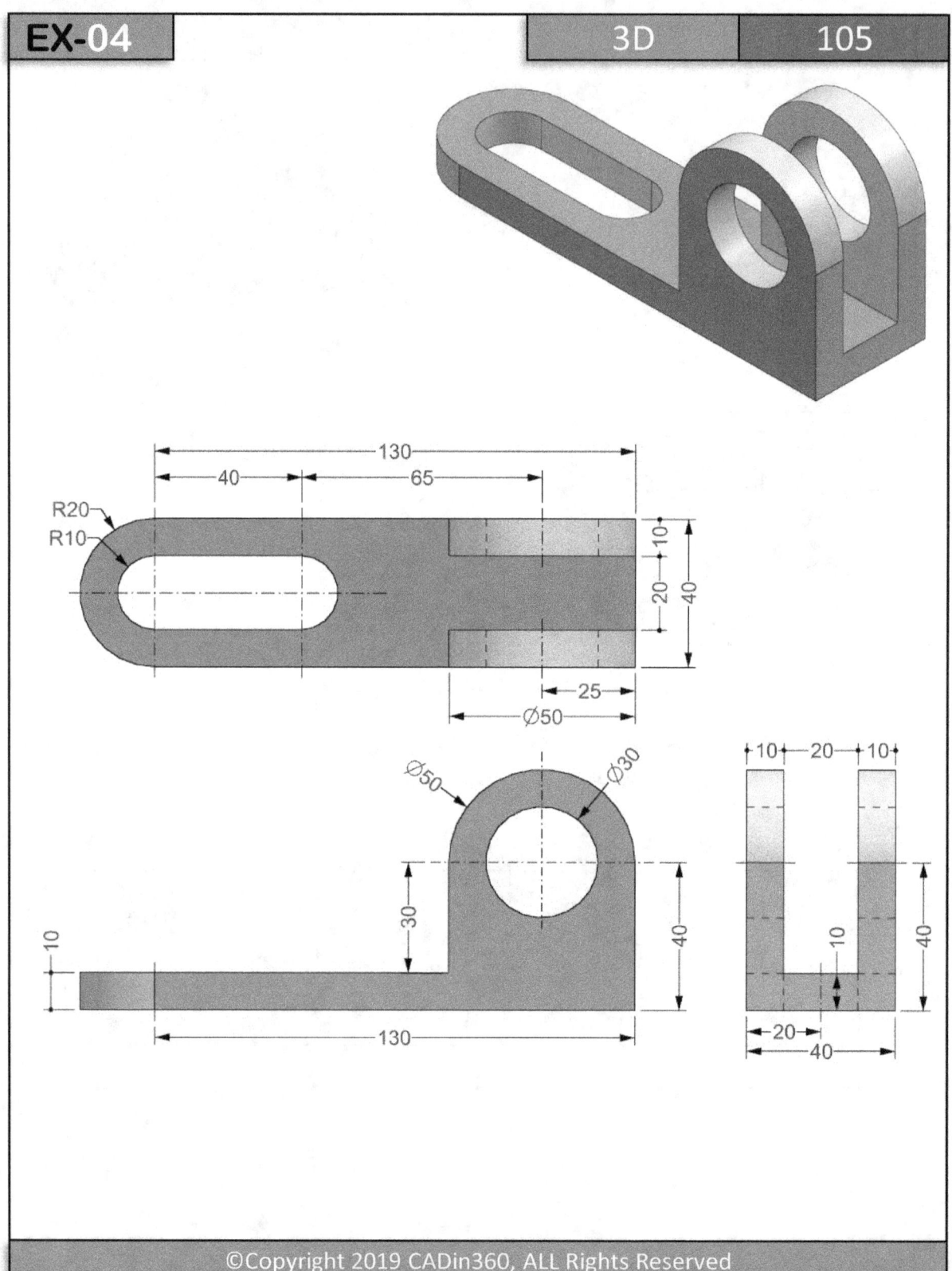

130
40
65

R20
R10

10
20
40

25
Ø50

Ø50
Ø30

30
40

10

130

10
20
10

10
40

20
40

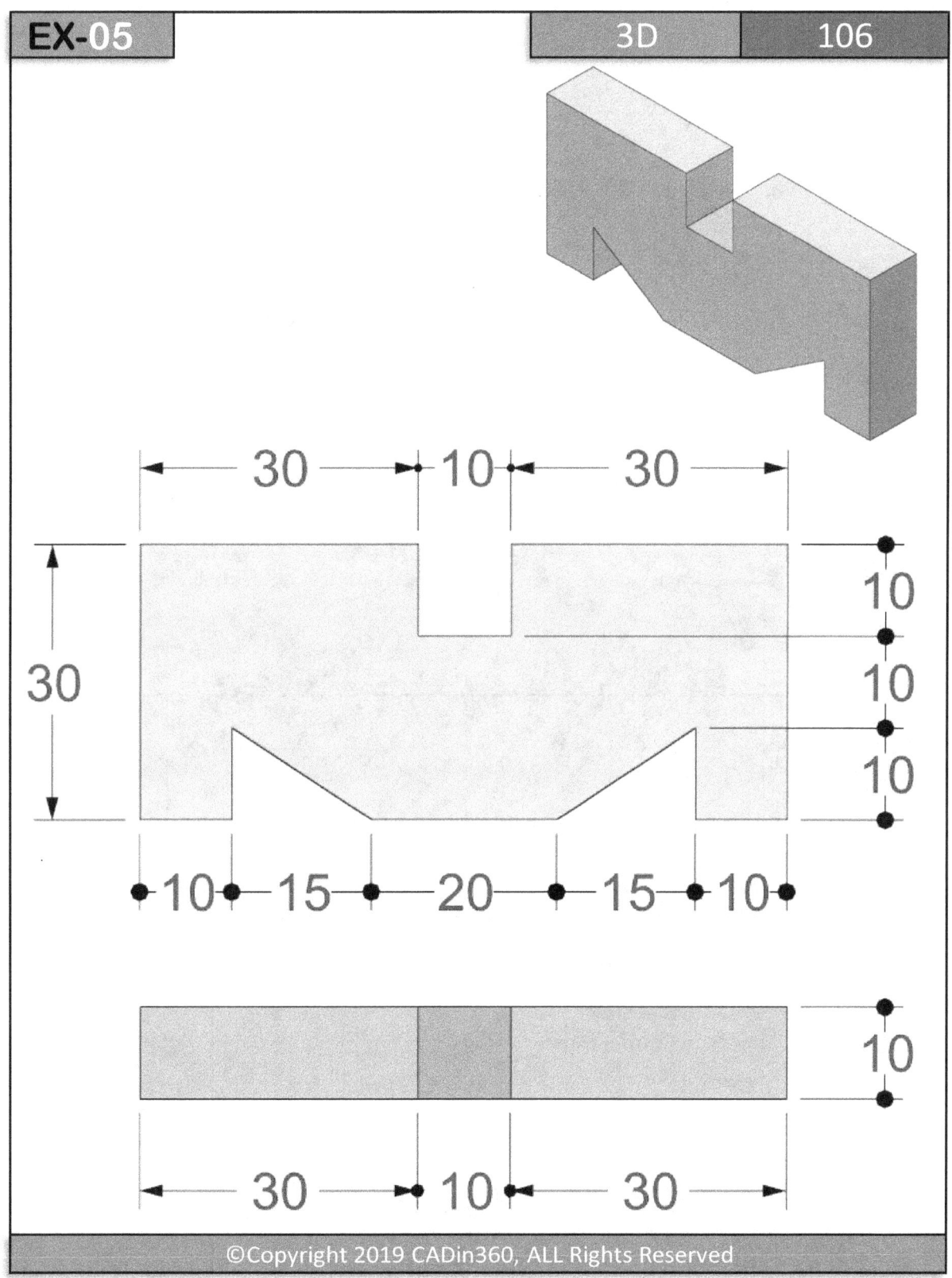

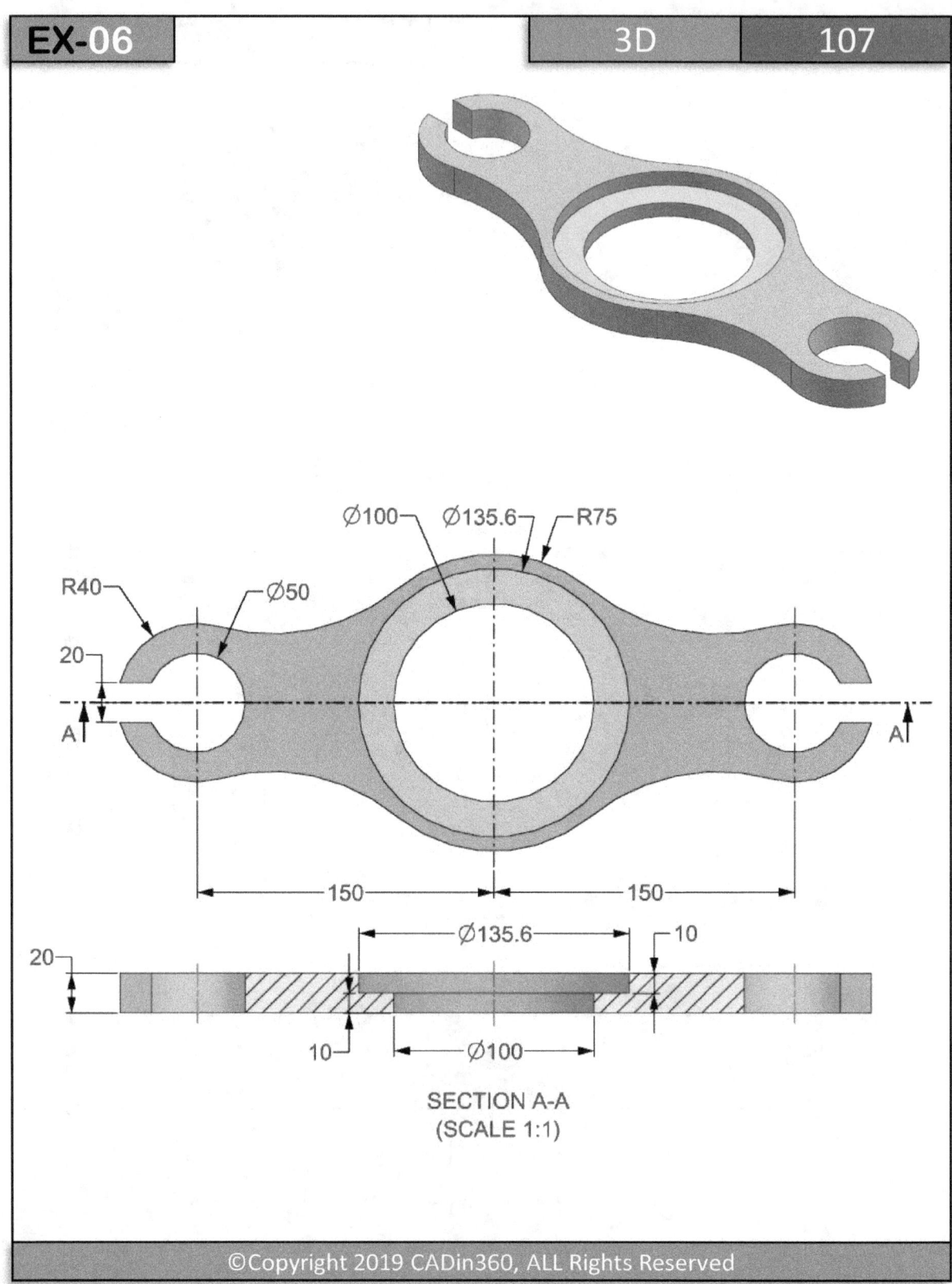

Ø100   Ø135.6   R75

R40   Ø50

20

A

150   150

Ø135.6   10

20

10   Ø100

SECTION A-A
(SCALE 1:1)

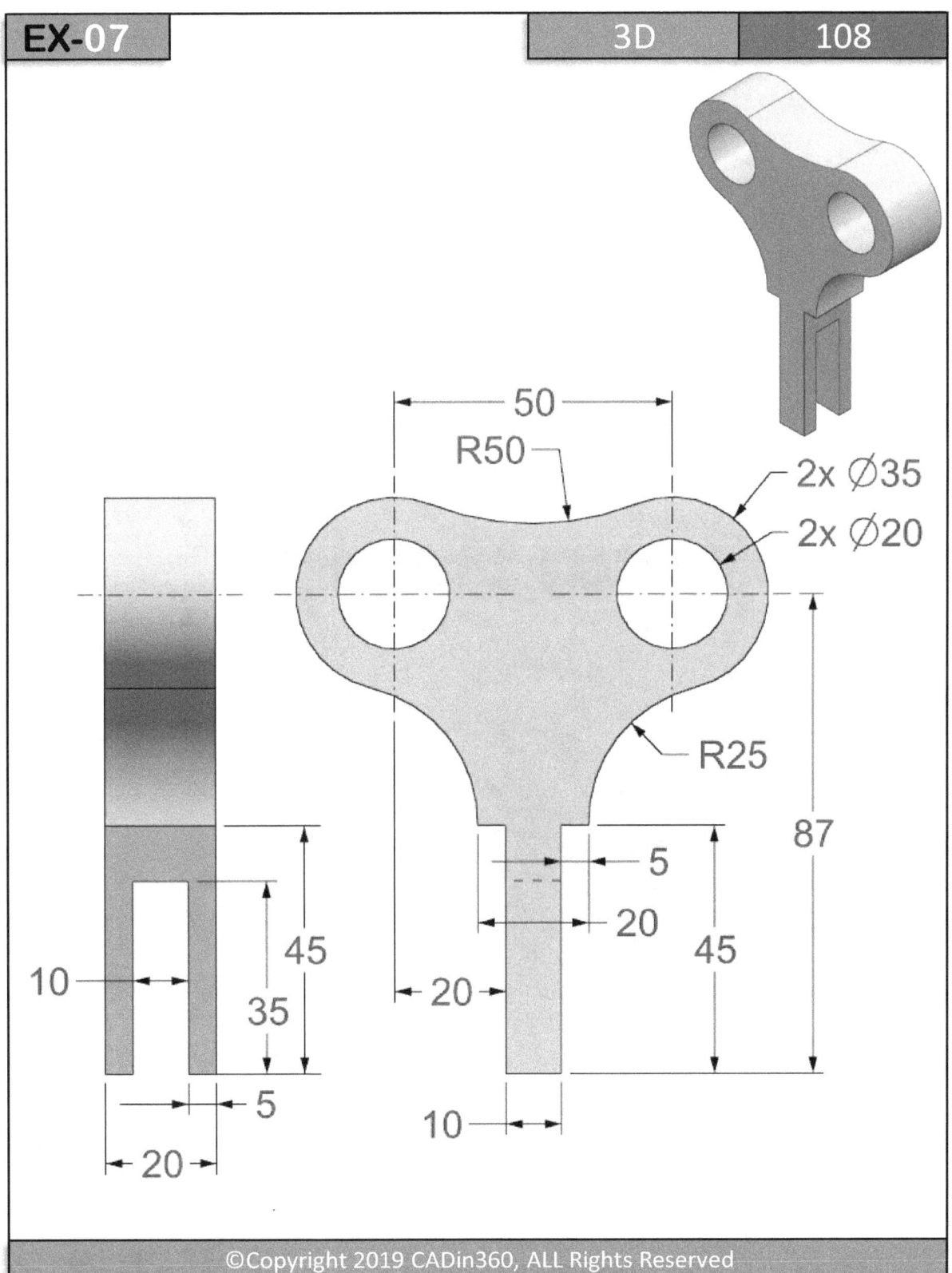

50

R50

2x ⌀35

2x ⌀20

R25

87

5

20

45

45

10

20

20

10

35

5

20

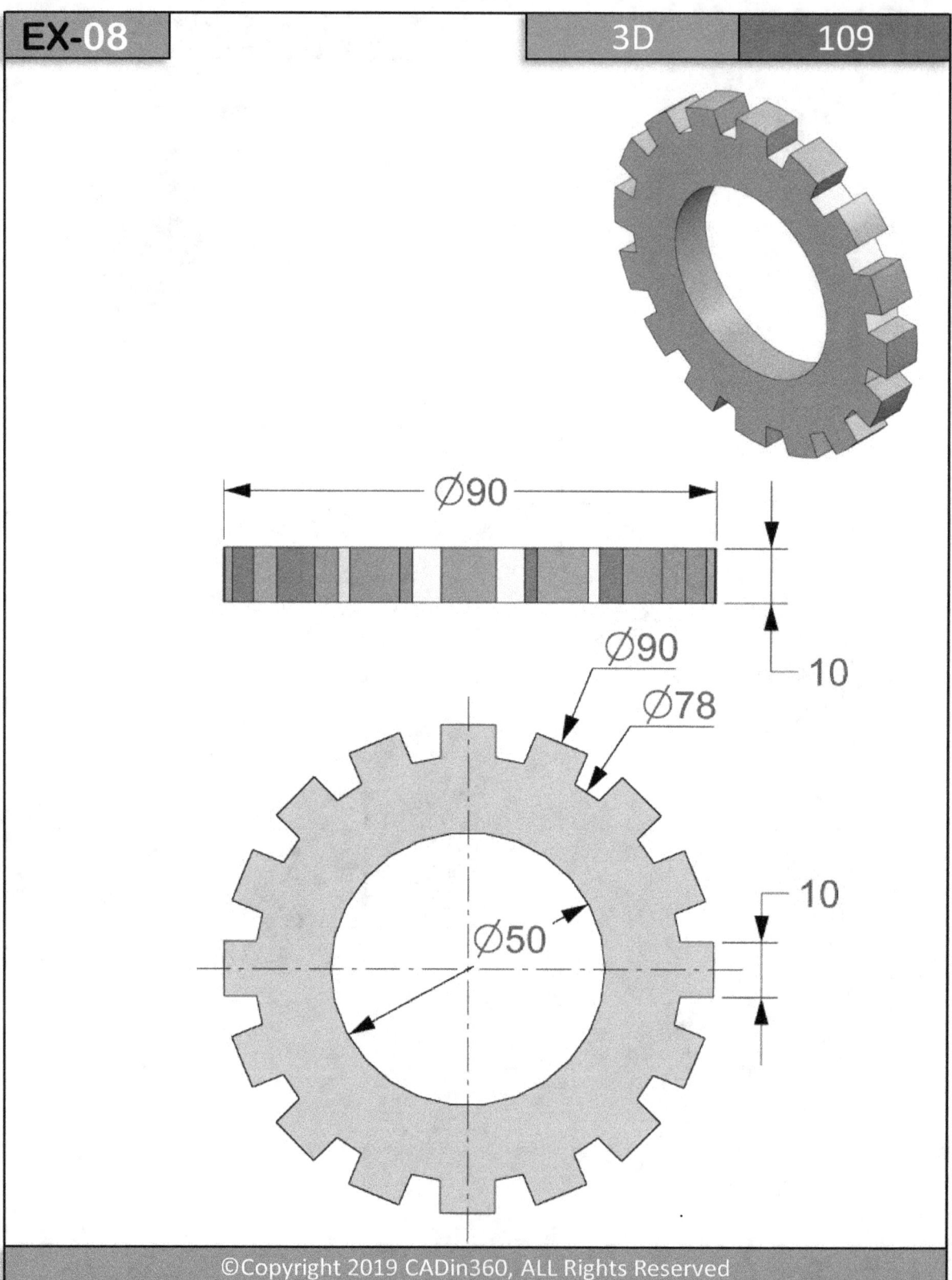

Ø90

10

Ø90

Ø78

Ø50

10

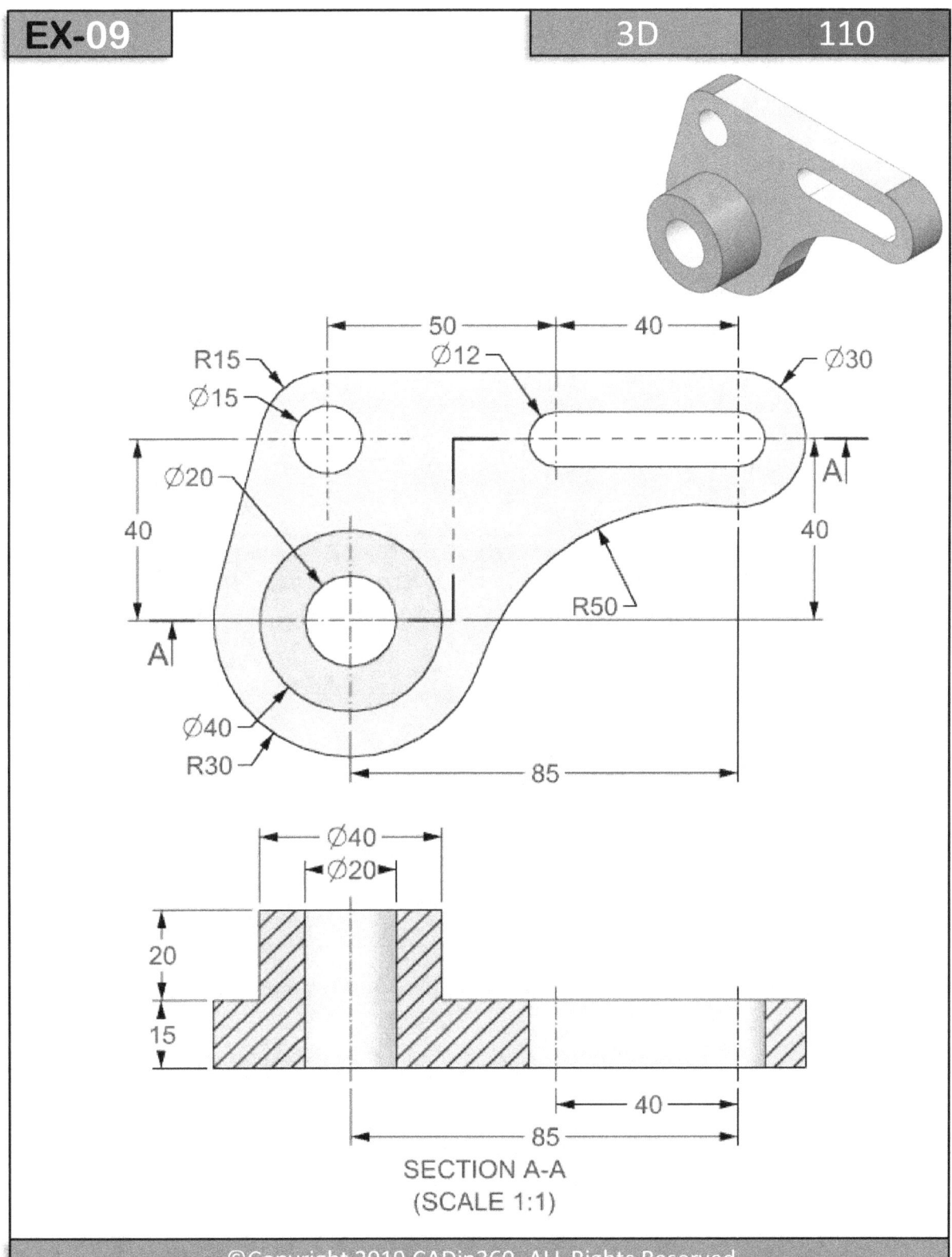

R15
Ø15
Ø20
40
A
Ø40
R30
50
Ø12
40
Ø30
A
40
R50
85

SECTION A-A
(SCALE 1:1)

Ø40
Ø20
20
15
40
85

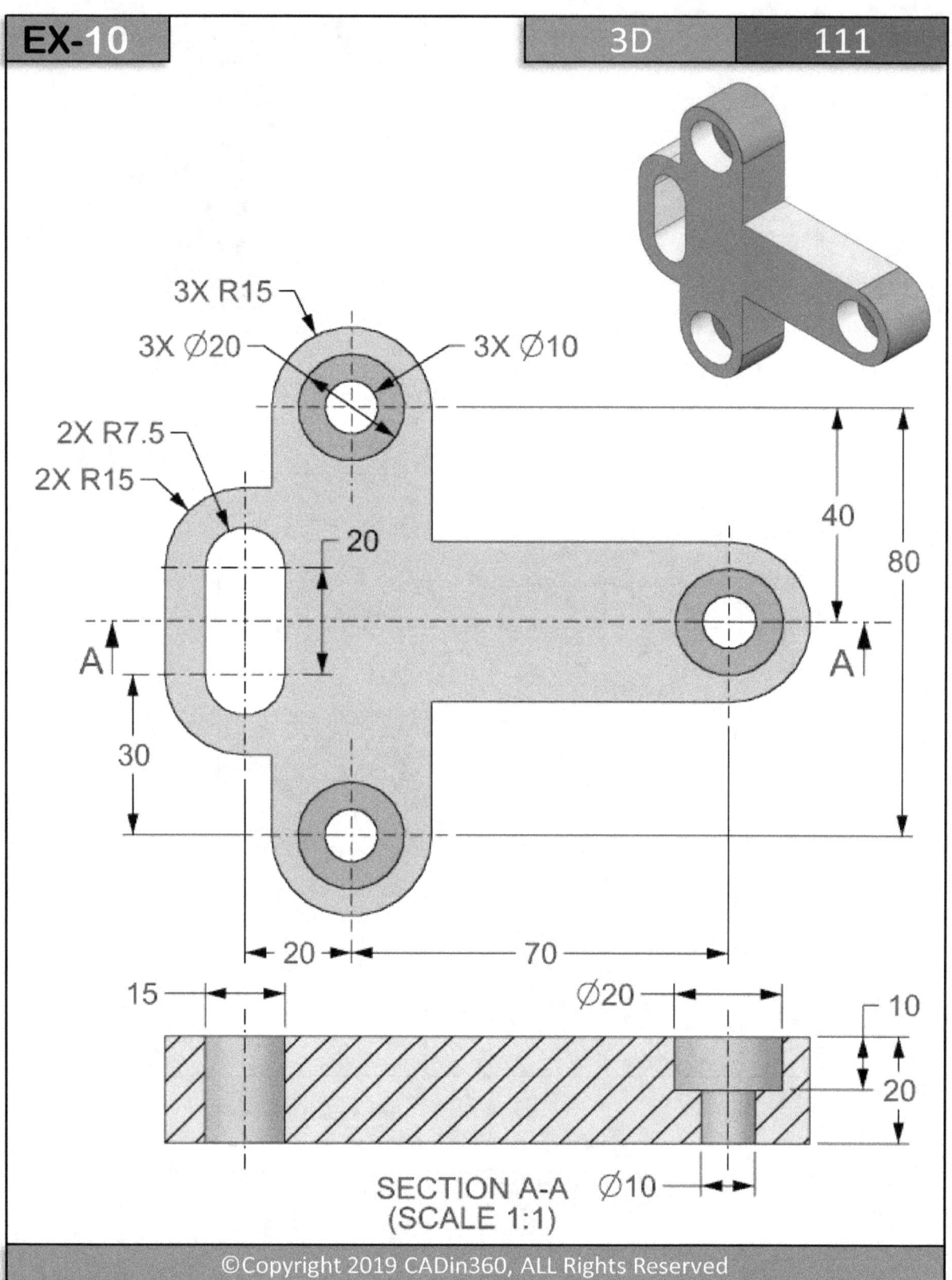

3X R15

3X Ø20

2X R7.5

2X R15

3X Ø10

20

40

80

30

20

70

15

Ø20

10

20

Ø10

SECTION A-A
(SCALE 1:1)

A

A

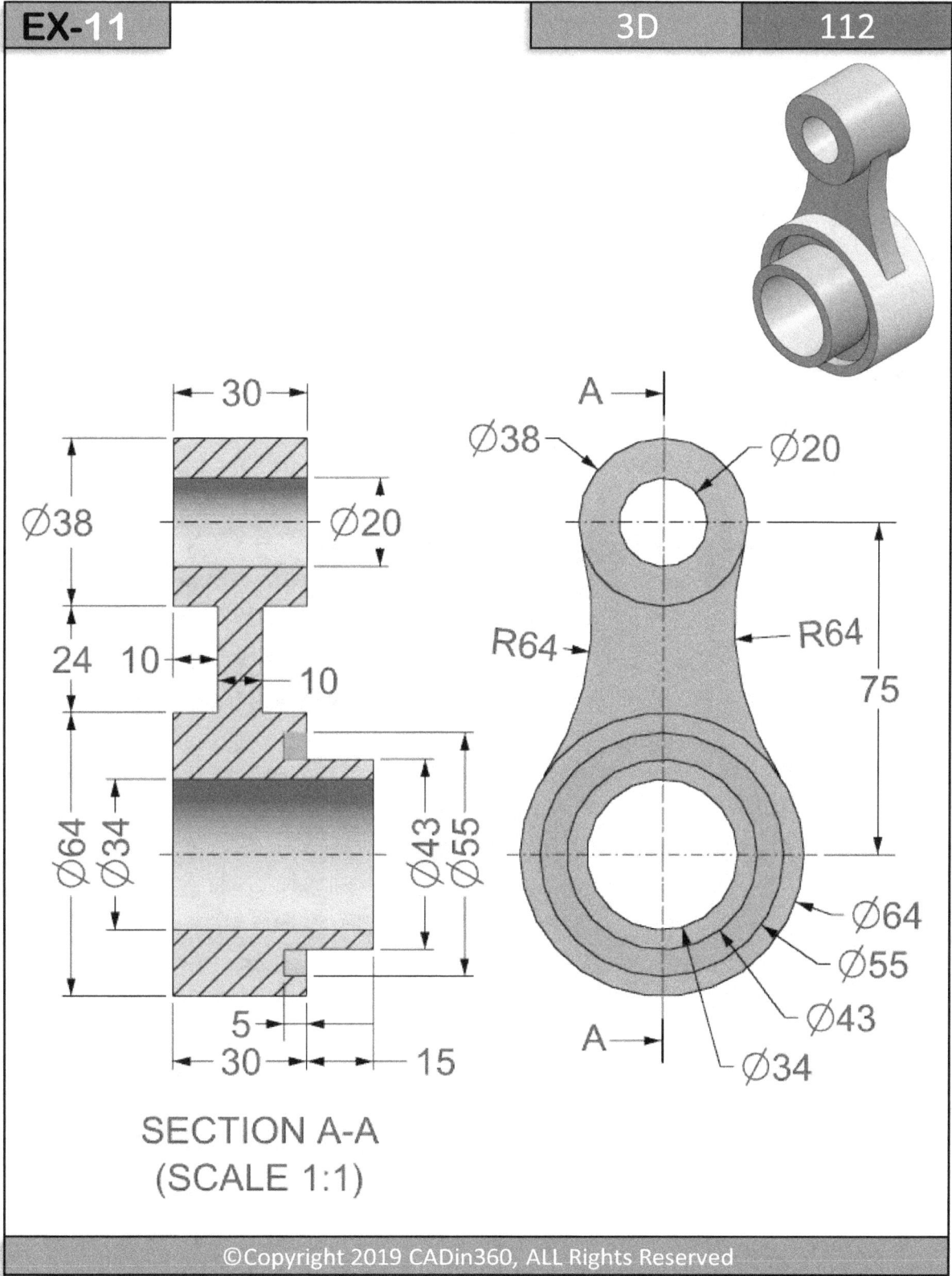

∅38　　　∅20
30
∅38　　　∅20
24　　10
10
∅64　∅34　　∅43　∅55
5
30　　15

SECTION A-A
(SCALE 1:1)

A
∅38　　∅20
R64　　R64
75
∅64
∅55
∅43
A
∅34

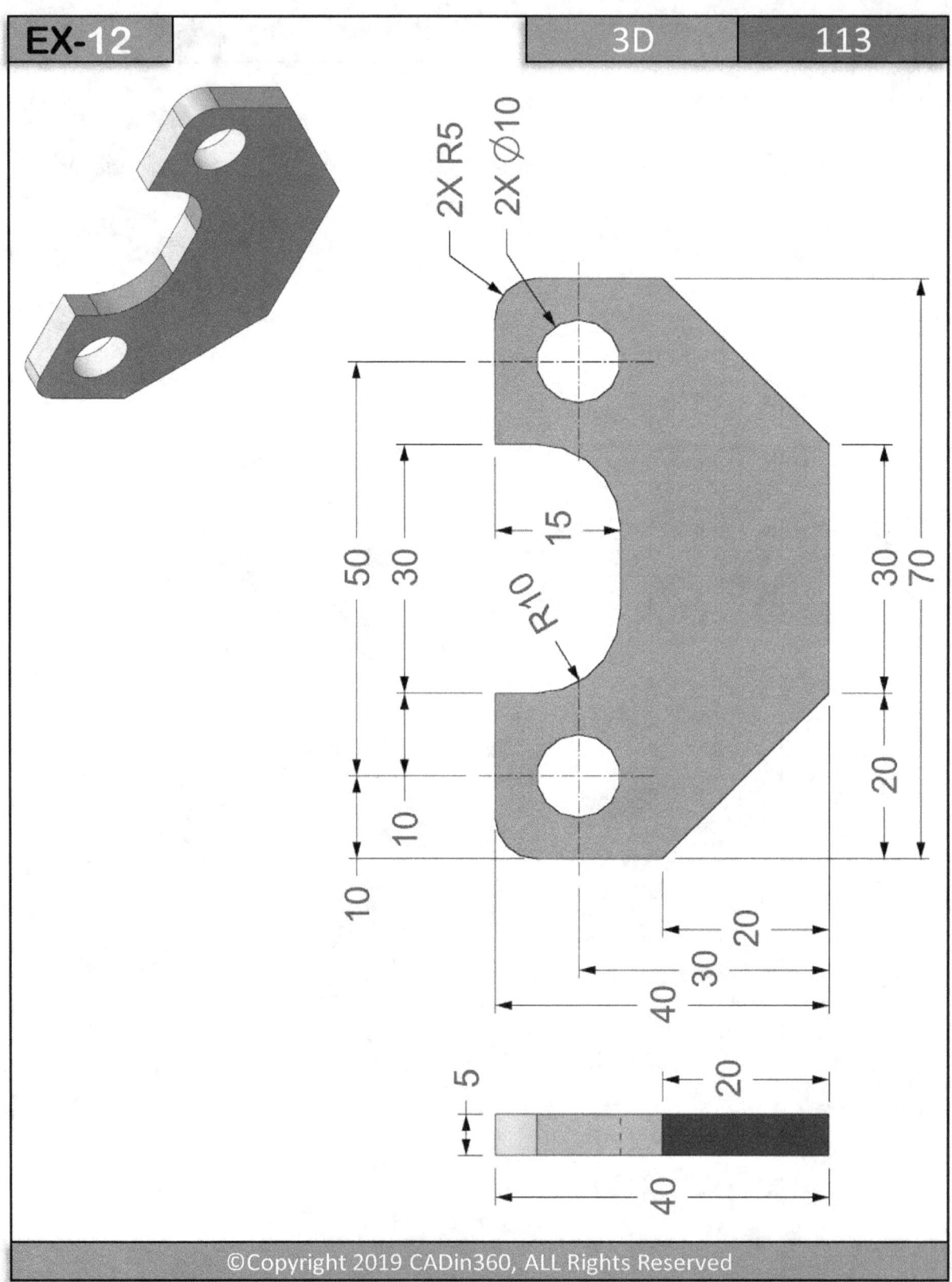

2X R5
2X ⌀10

50
30
15
R10
10
10

30
70
20

40
30
20

5
20
40

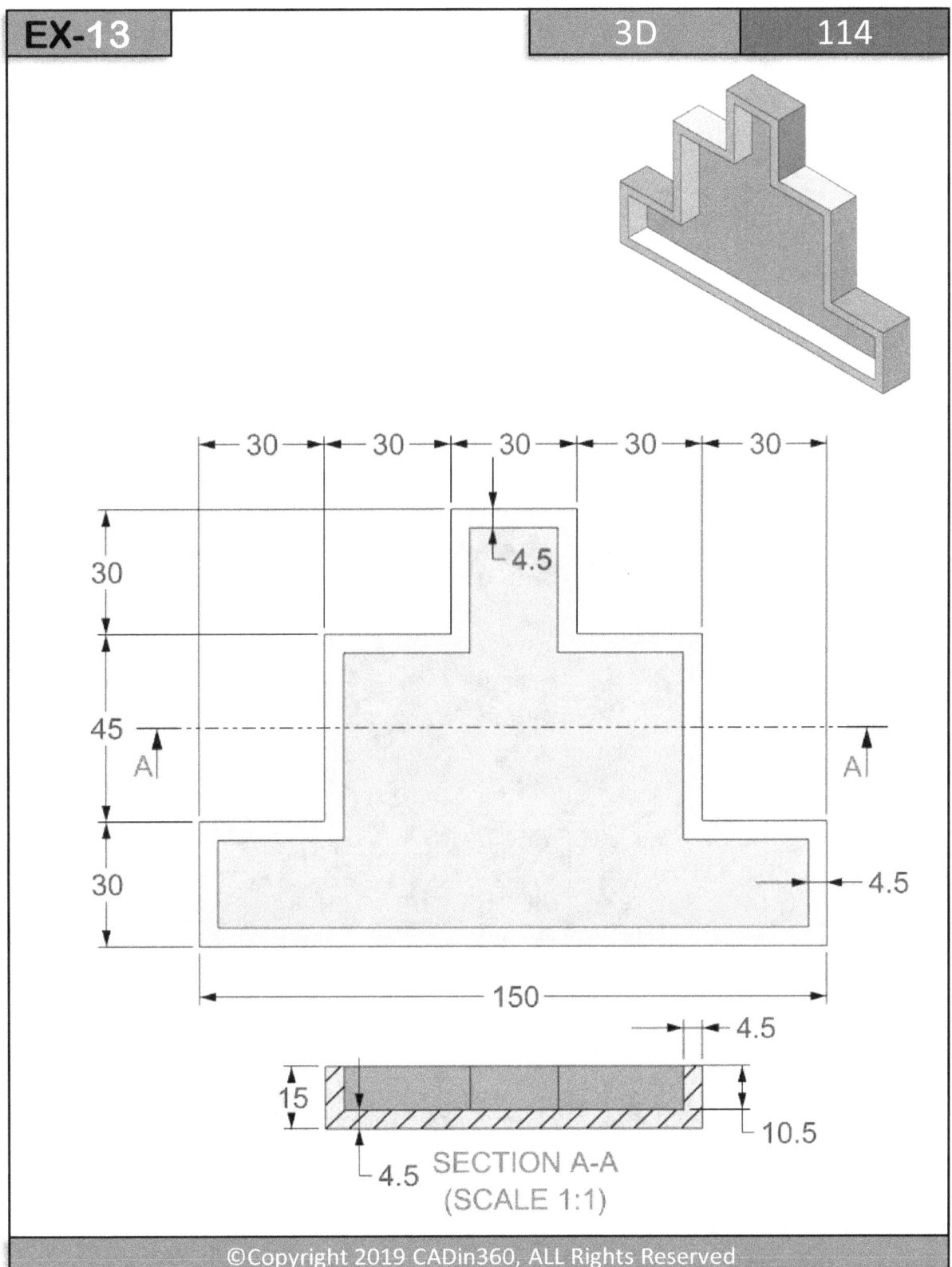

30 | 30 | 30 | 30 | 30

30

45

A|

30

150

A|

4.5

4.5

4.5

15

10.5

4.5

SECTION A-A
(SCALE 1:1)

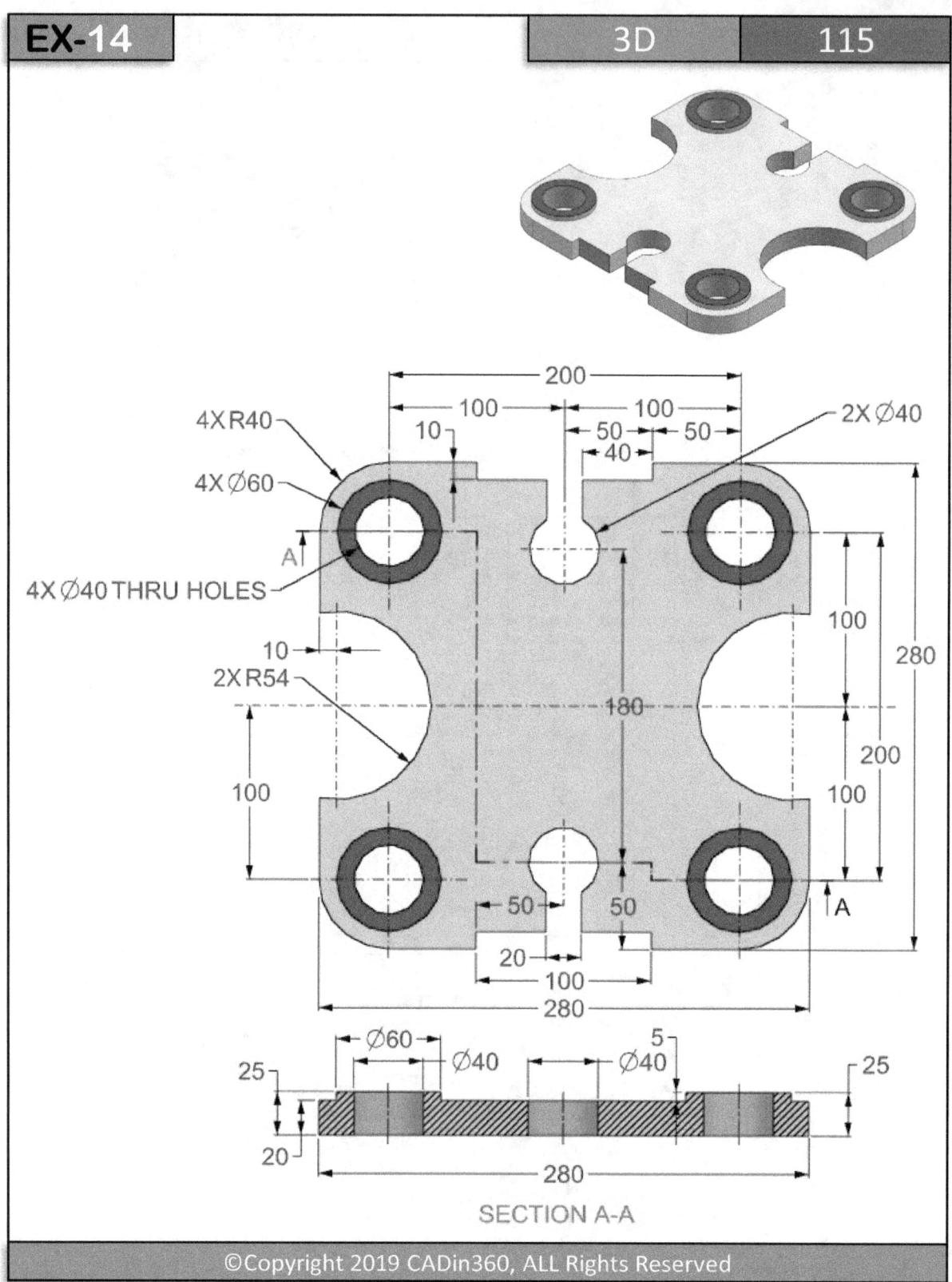

SECTION A-A

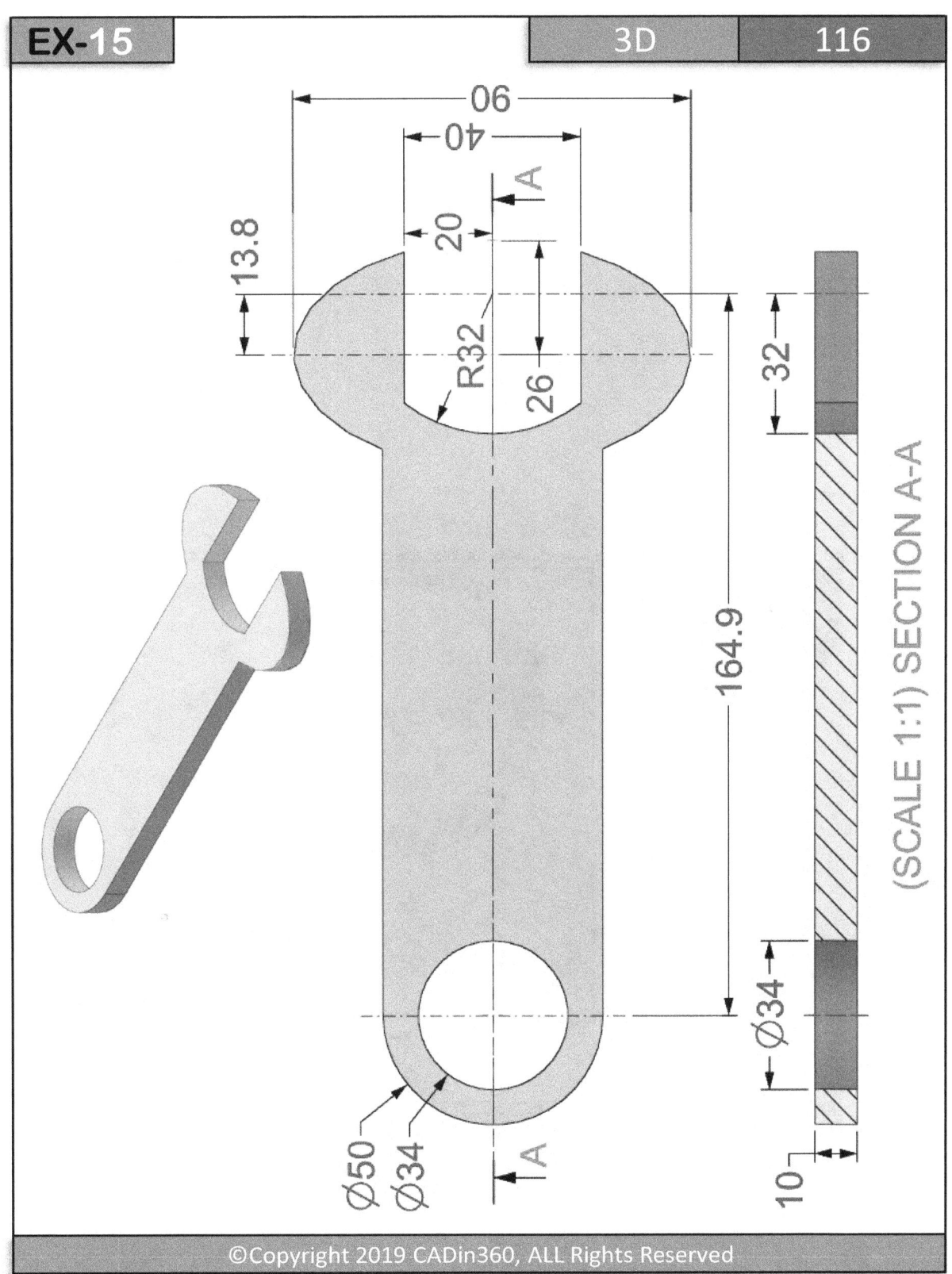

90
40
20
13.8
A
R32
26
164.9
Ø50
Ø34
A
32
Ø34
10
(SCALE 1:1) SECTION A-A

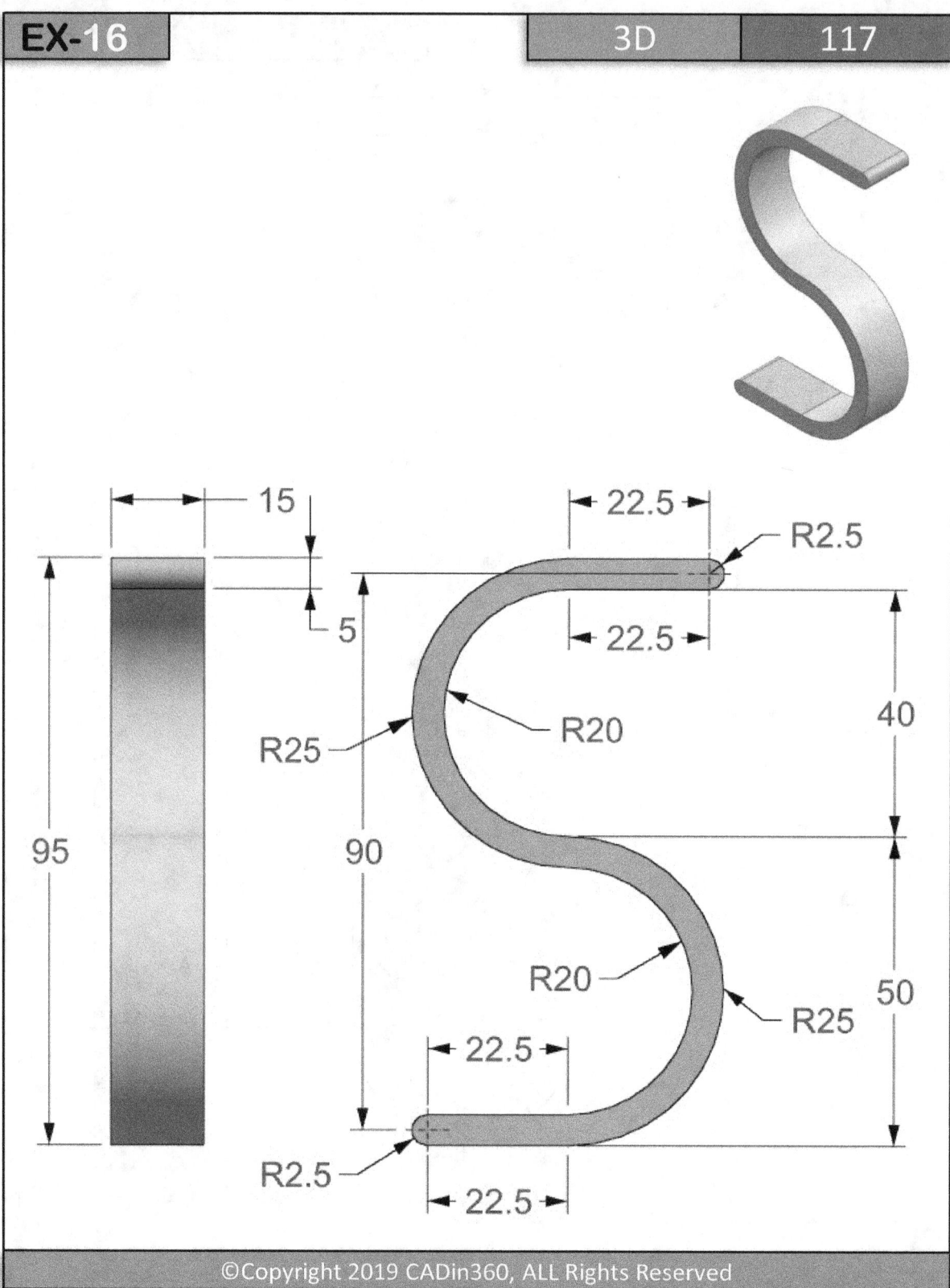

15

5

22.5

R2.5

22.5

R25

R20

40

95

90

R20

R25

50

22.5

R2.5

22.5

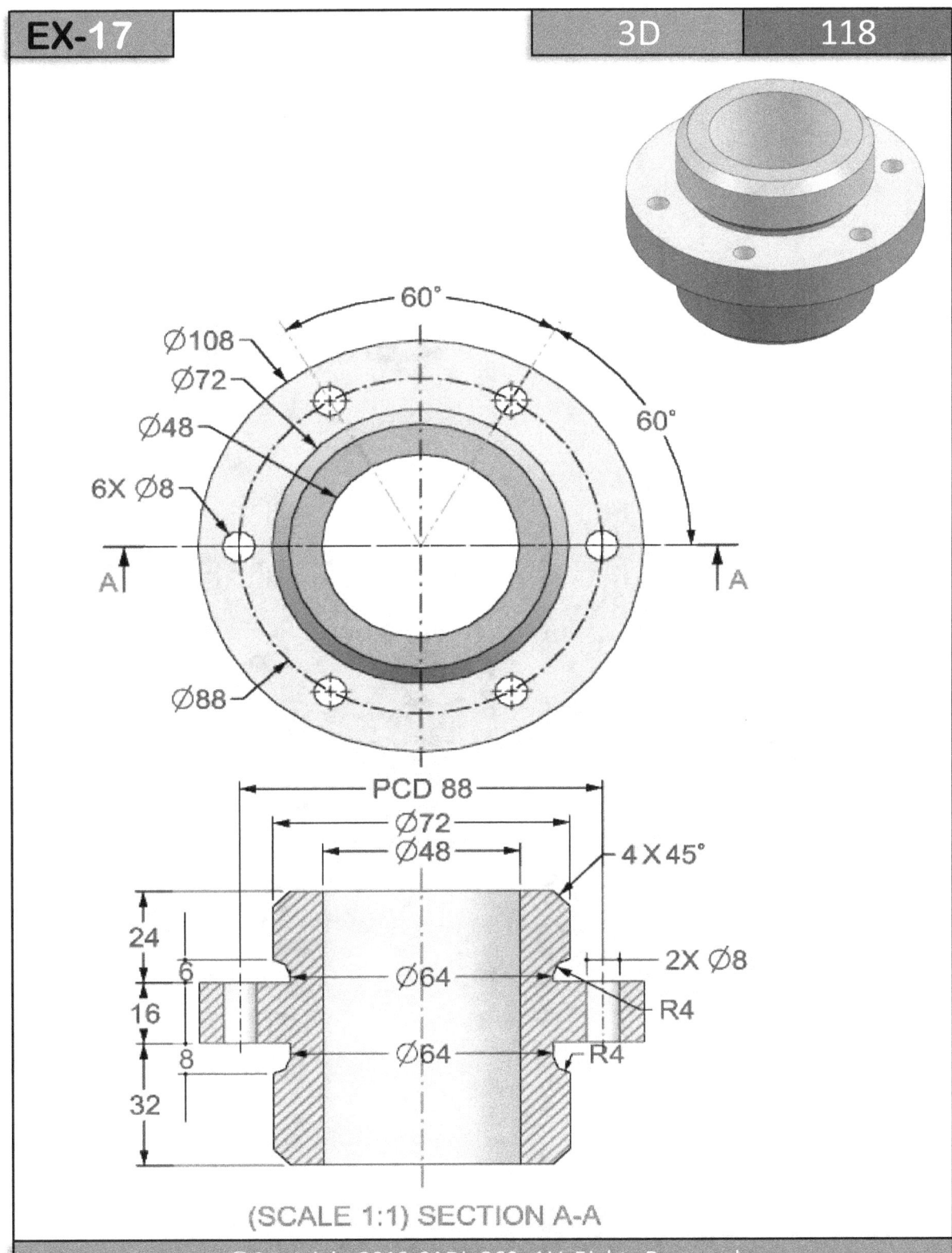

60°

Ø108
Ø72
Ø48
6X Ø8
60°
A
A
Ø88

PCD 88
Ø72
Ø48
4 X 45°

24
6
Ø64
2X Ø8
16
R4
8
Ø64
R4
32

(SCALE 1:1) SECTION A-A

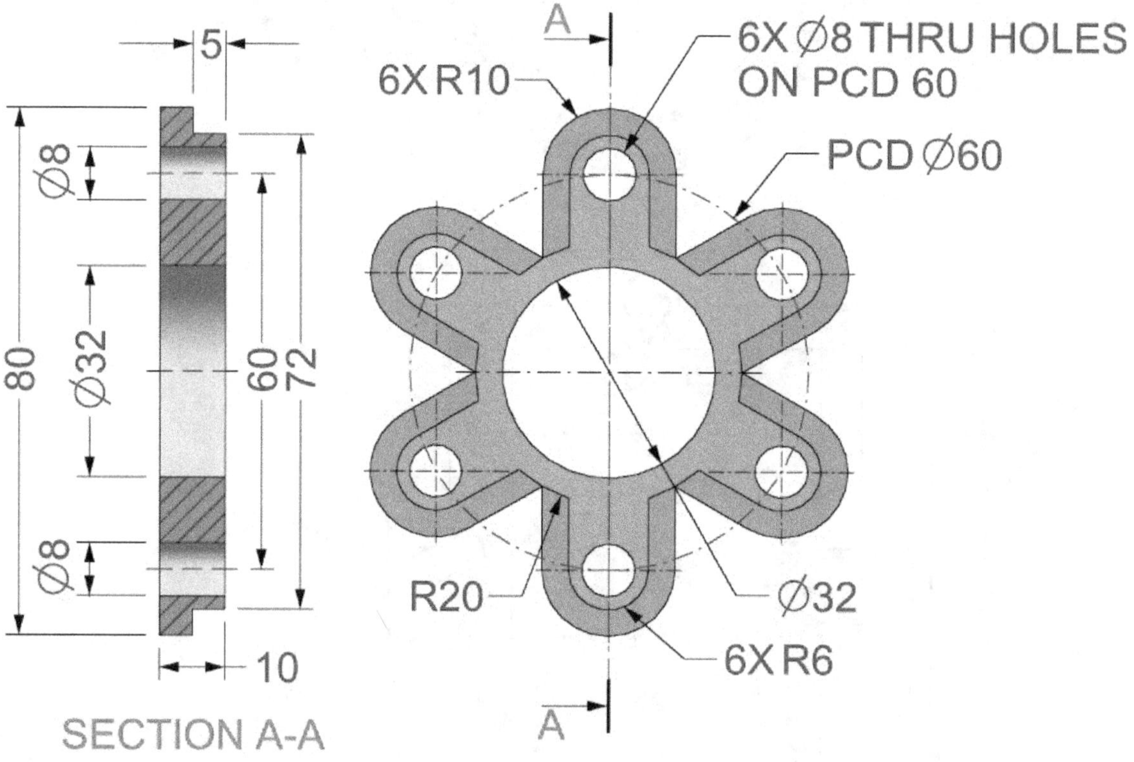

5

Ø8

Ø32

80

60

72

Ø8

10

SECTION A-A

A

6X R10

6X Ø8 THRU HOLES
ON PCD 60

PCD Ø60

R20

Ø32

6X R6

A

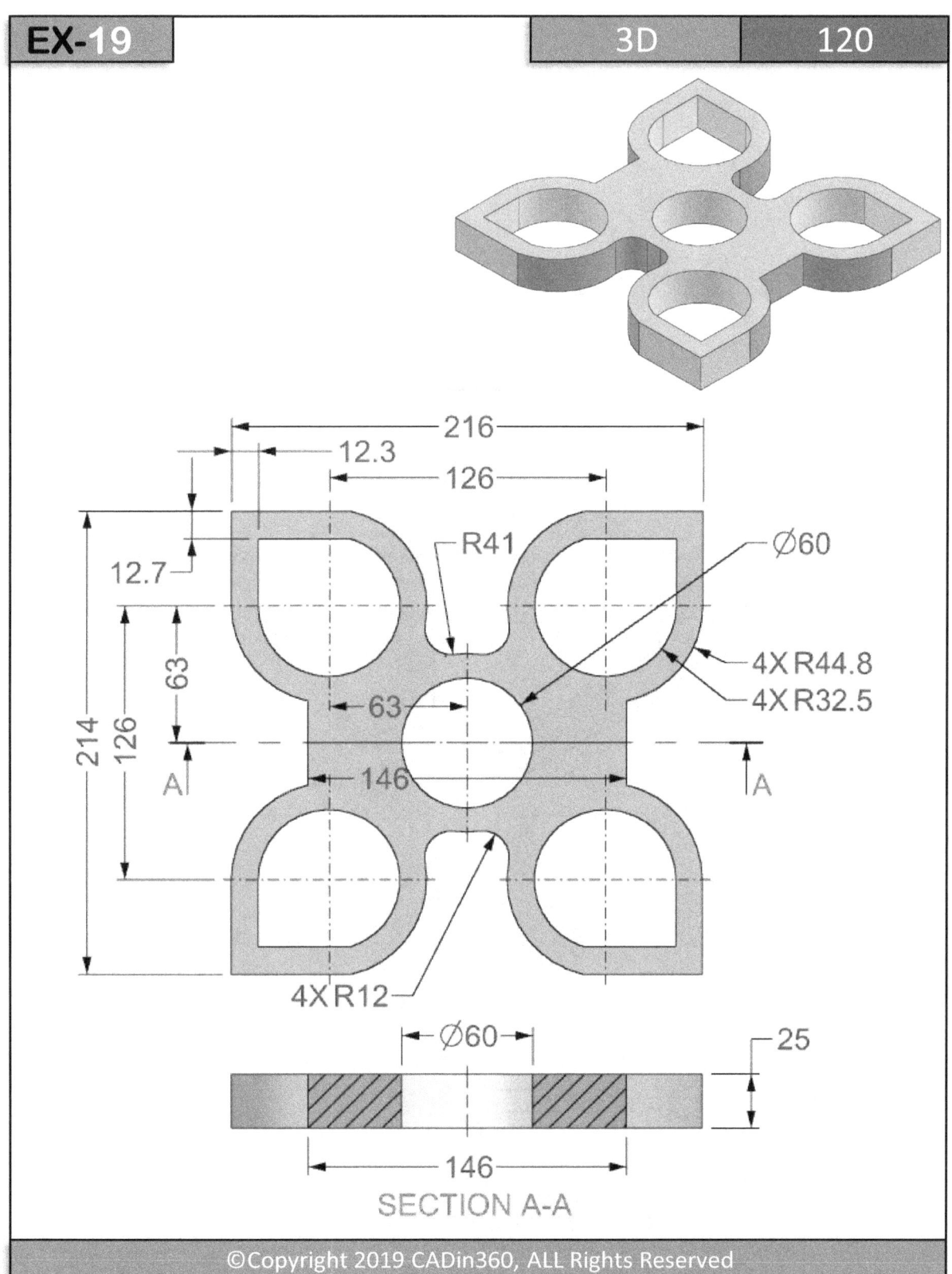

EX-19   3D   120

216
12.3
126
R41
Ø60
12.7
63
126
214
63
4X R44.8
4X R32.5
A
146
A
4X R12

Ø60
25
146
SECTION A-A

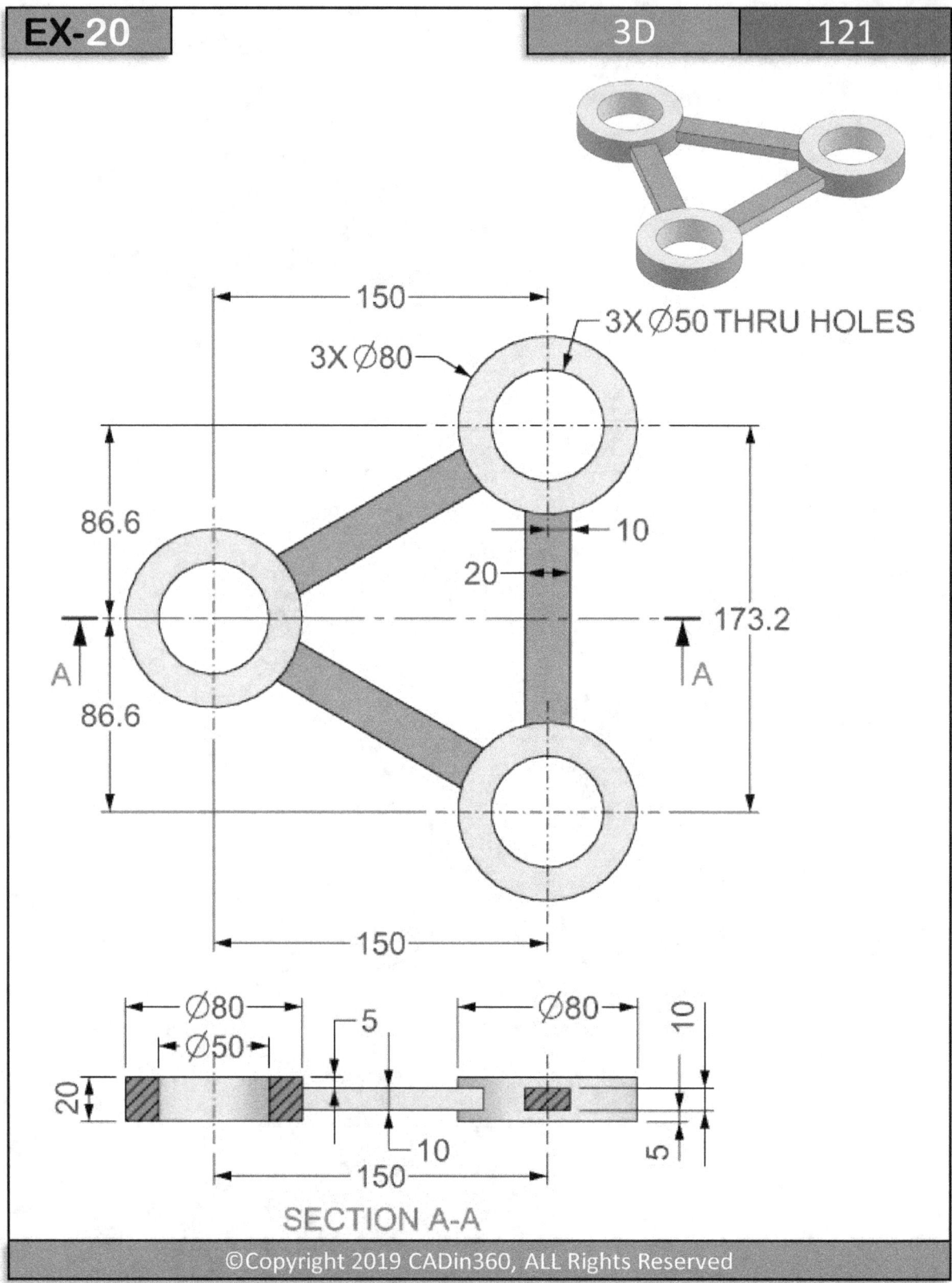

EX-20　　3D　　121

3X Ø50 THRU HOLES

3X Ø80

150

86.6

10

20

173.2

A

86.6

150

Ø80

5

Ø80

10

Ø50

20

10

5

150

SECTION A-A

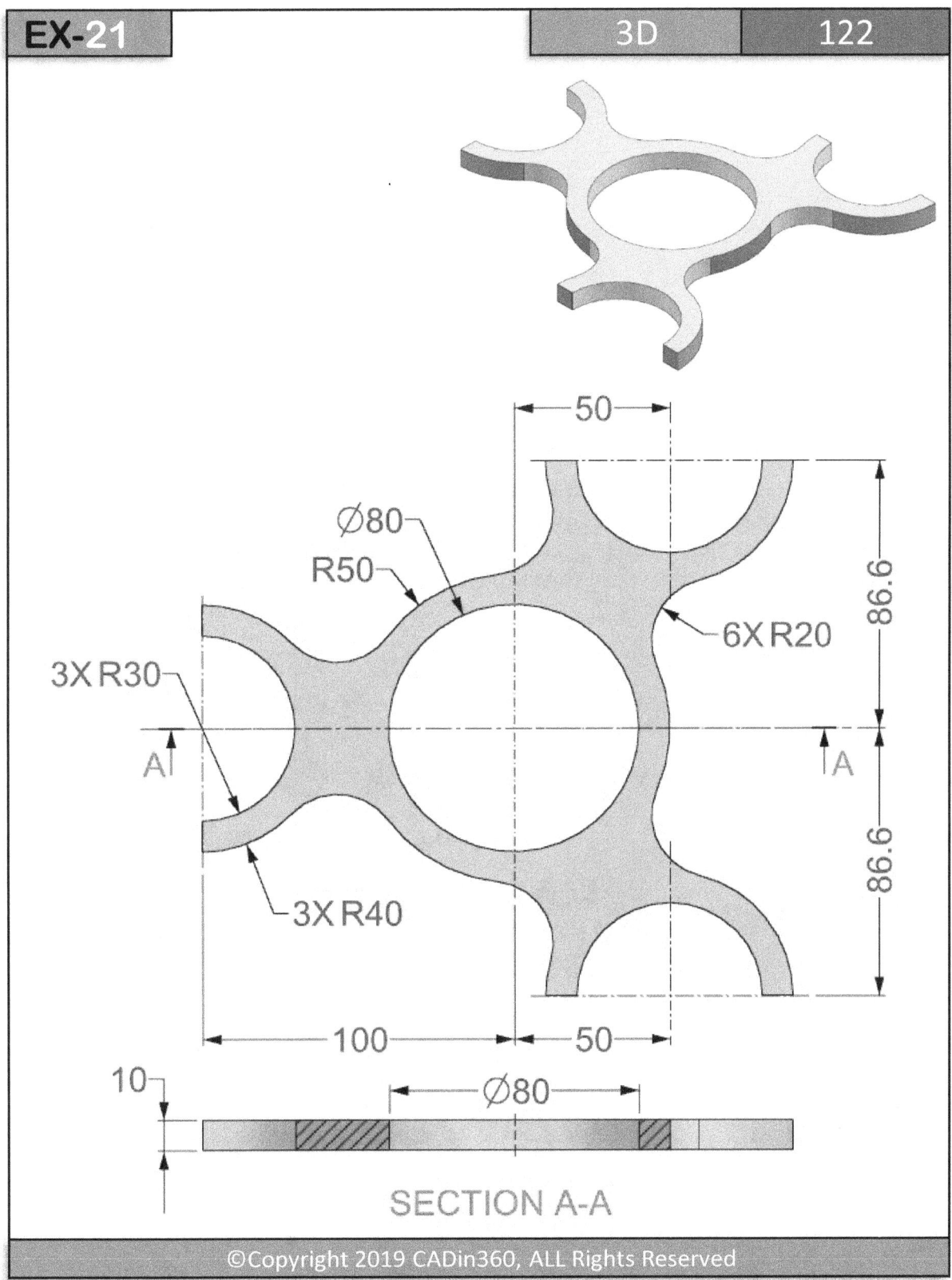

EX-21

3D 122

Ø80

R50

3X R30

3X R40

50

86.6

6X R20

86.6

100

50

10

Ø80

SECTION A-A

©Copyright 2019 CADin360, ALL Rights Reserved

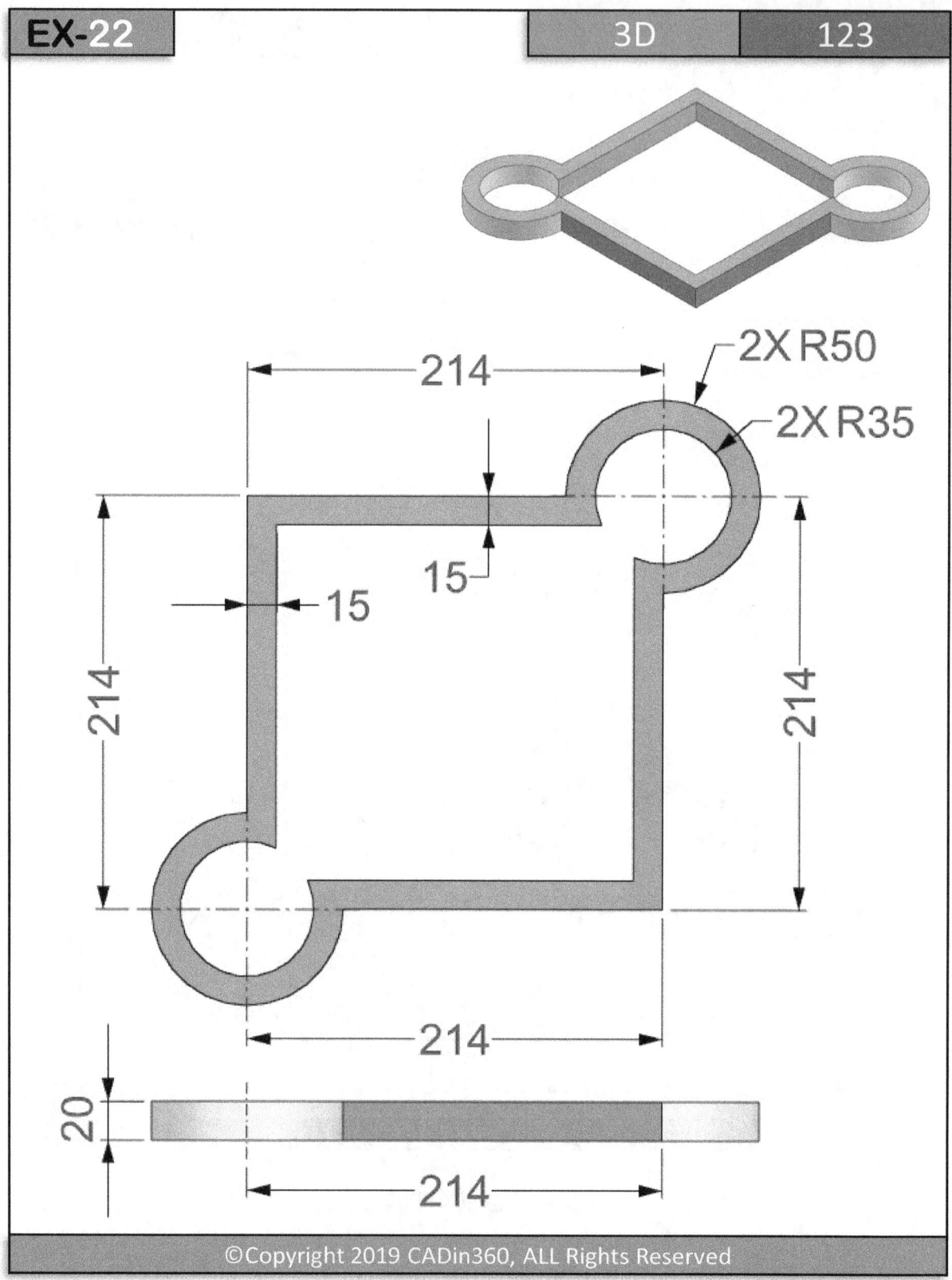

2X R50

2X R35

214

15

15

214

214

214

20

214

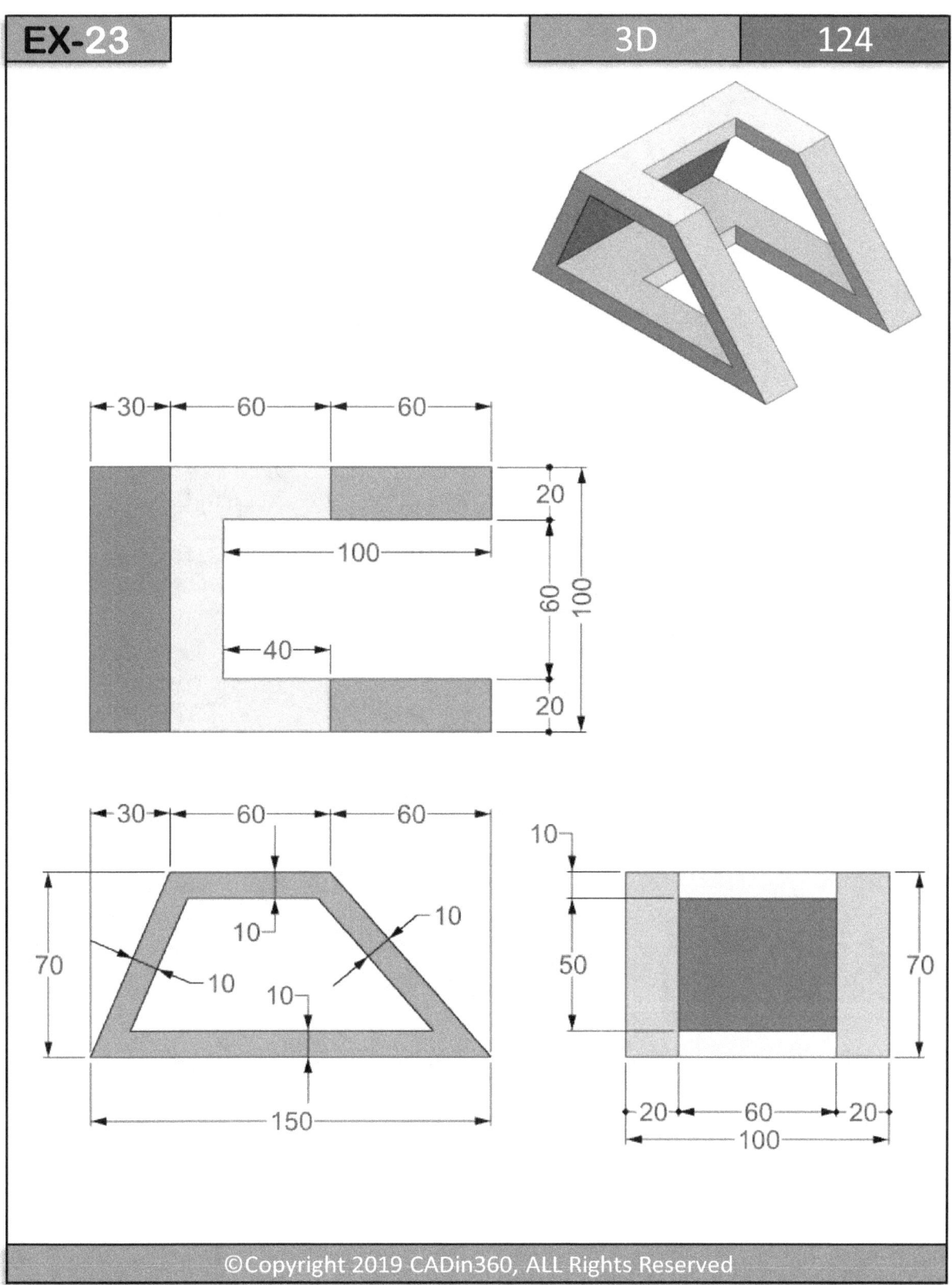

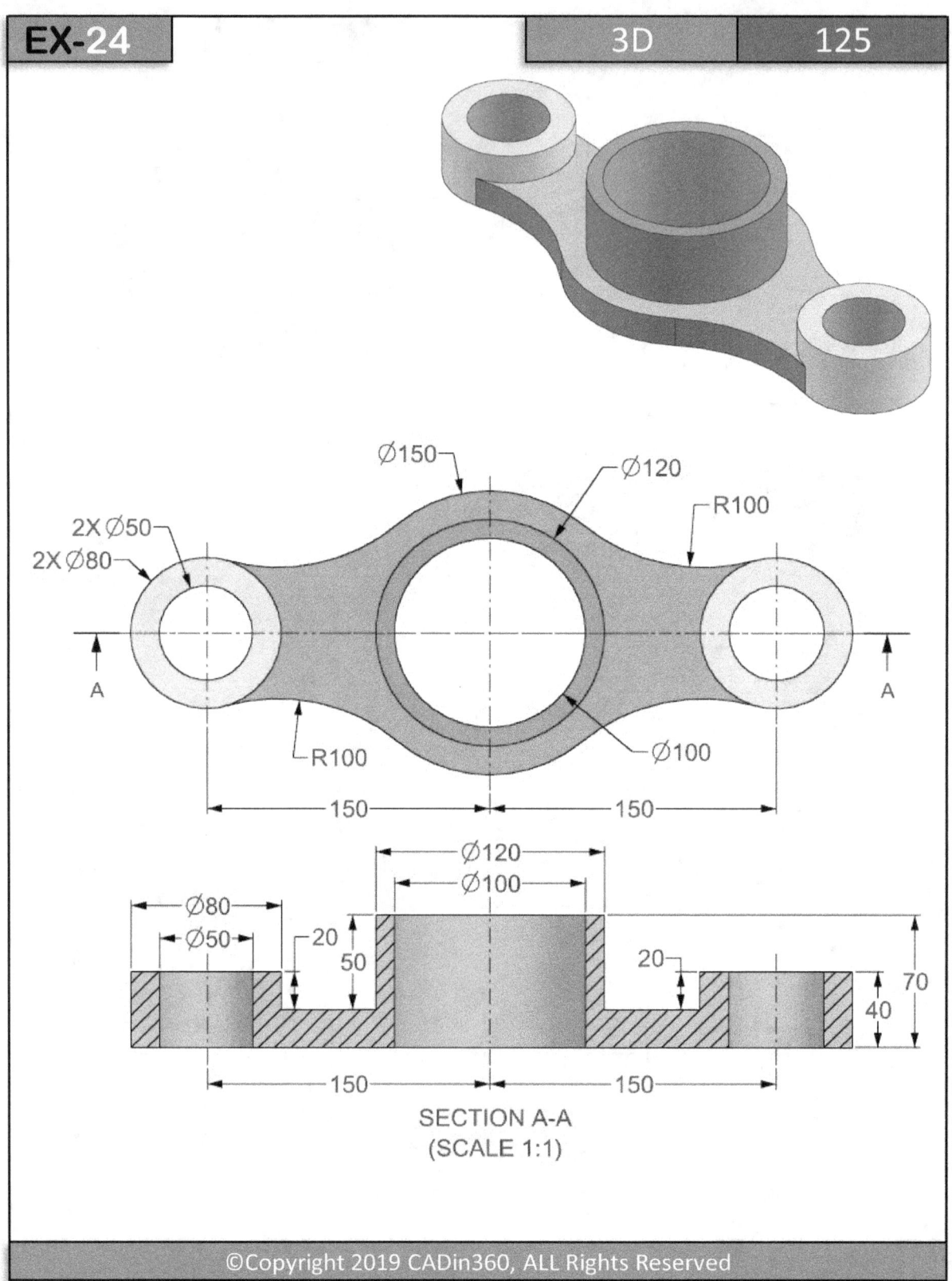

EX-24

3D   125

⌀150   ⌀120   R100

2X ⌀50
2X ⌀80

⌀100

R100

150   150

⌀120
⌀100

⌀80
⌀50
20
50
20
70
40

150   150

SECTION A-A
(SCALE 1:1)

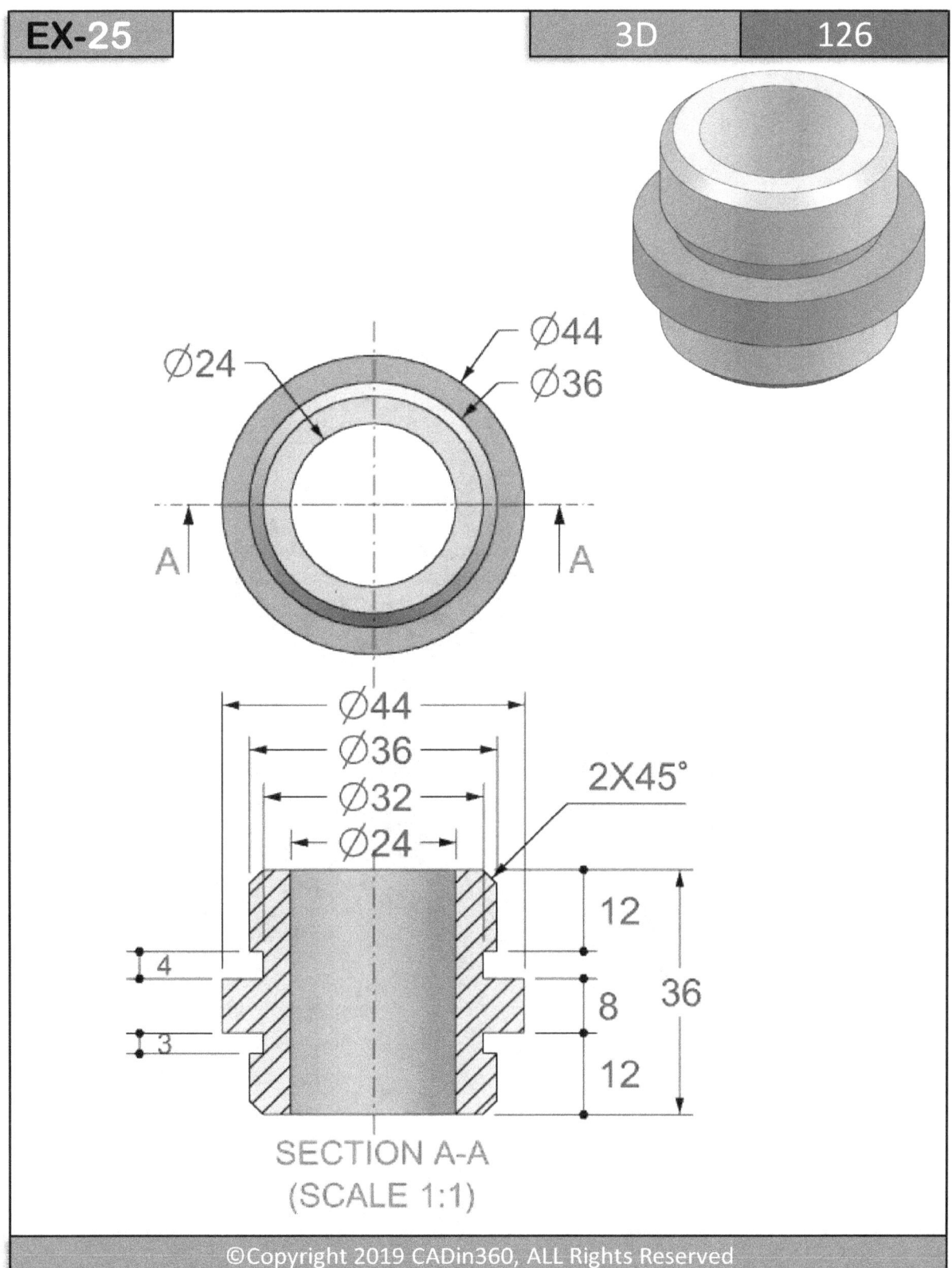

EX-25　　3D　　126

Ø24
Ø44
Ø36

A   A

Ø44
Ø36
Ø32
Ø24
2X45°

12

4

8   36

3

12

SECTION A-A
(SCALE 1:1)

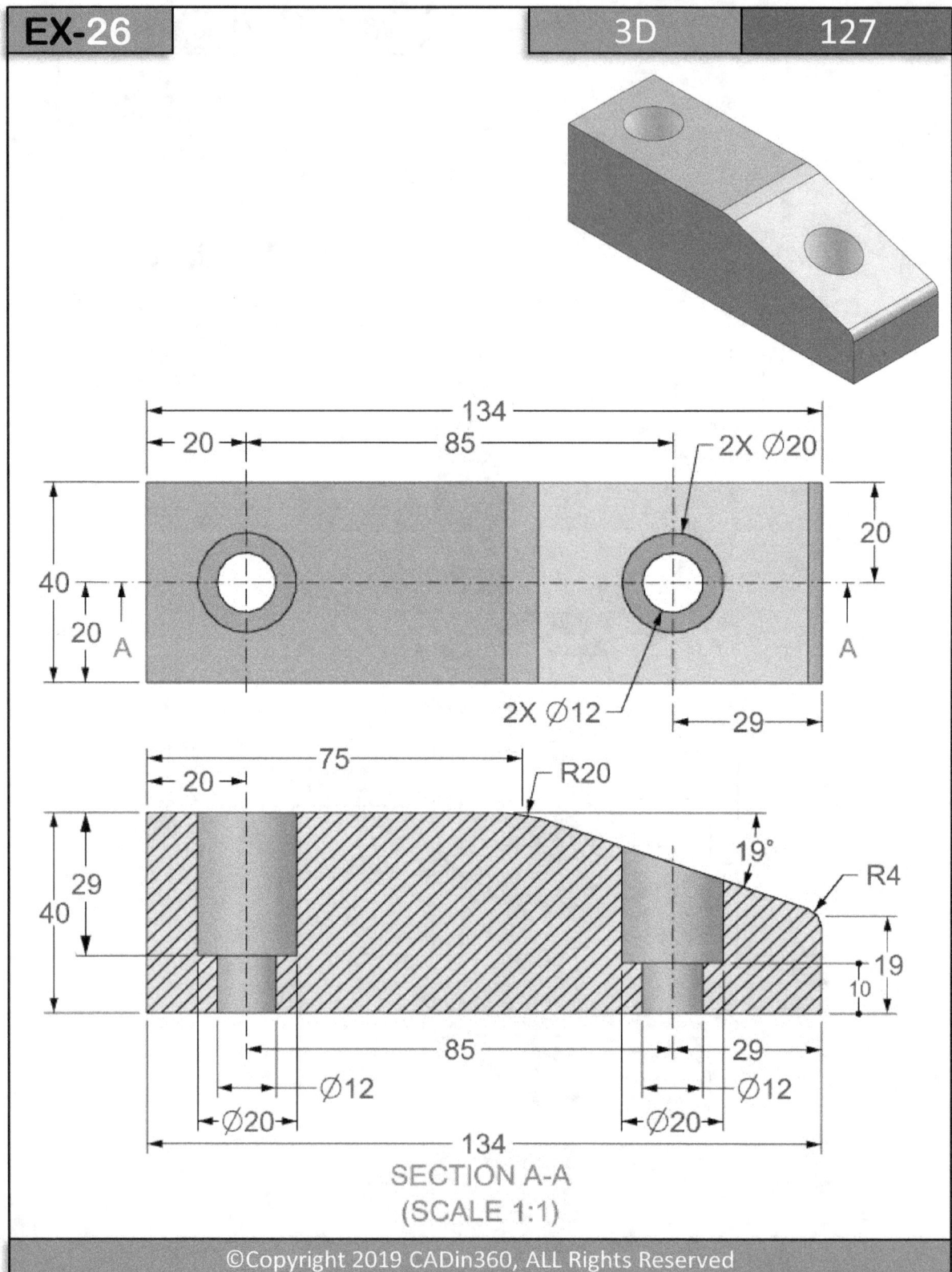

134

20 85 2X ∅20

40 20

20 A A

2X ∅12 29

75 R20

20 19°

29 R4

40 19

10

85 29

∅12 ∅12

∅20 ∅20

134

SECTION A-A
(SCALE 1:1)

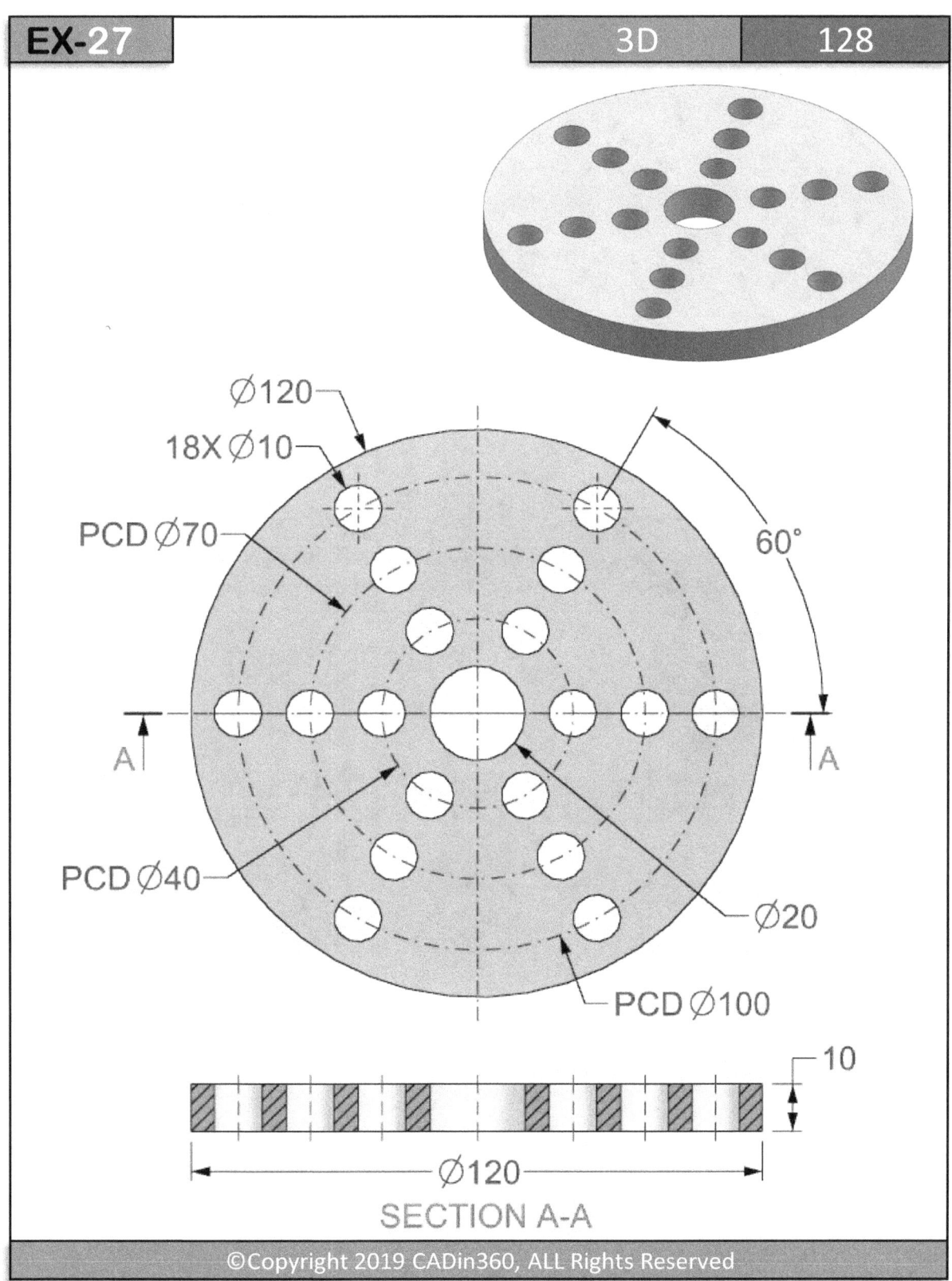

Ø120

18X Ø10

PCD Ø70

60°

PCD Ø40

Ø20

PCD Ø100

10

Ø120

SECTION A-A

A

A

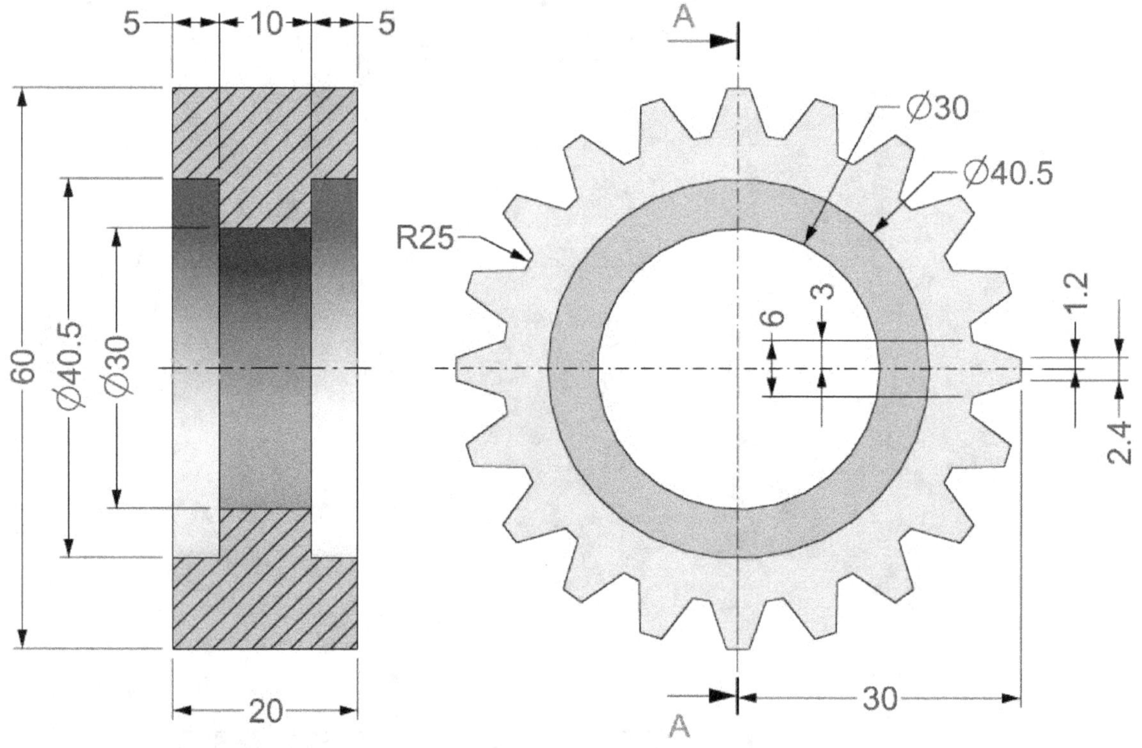

5 | 10 | 5

60

Ø40.5

Ø30

20

SECTION A-A

A

Ø30

Ø40.5

R25

6

3

1.2

2.4

30

A

A

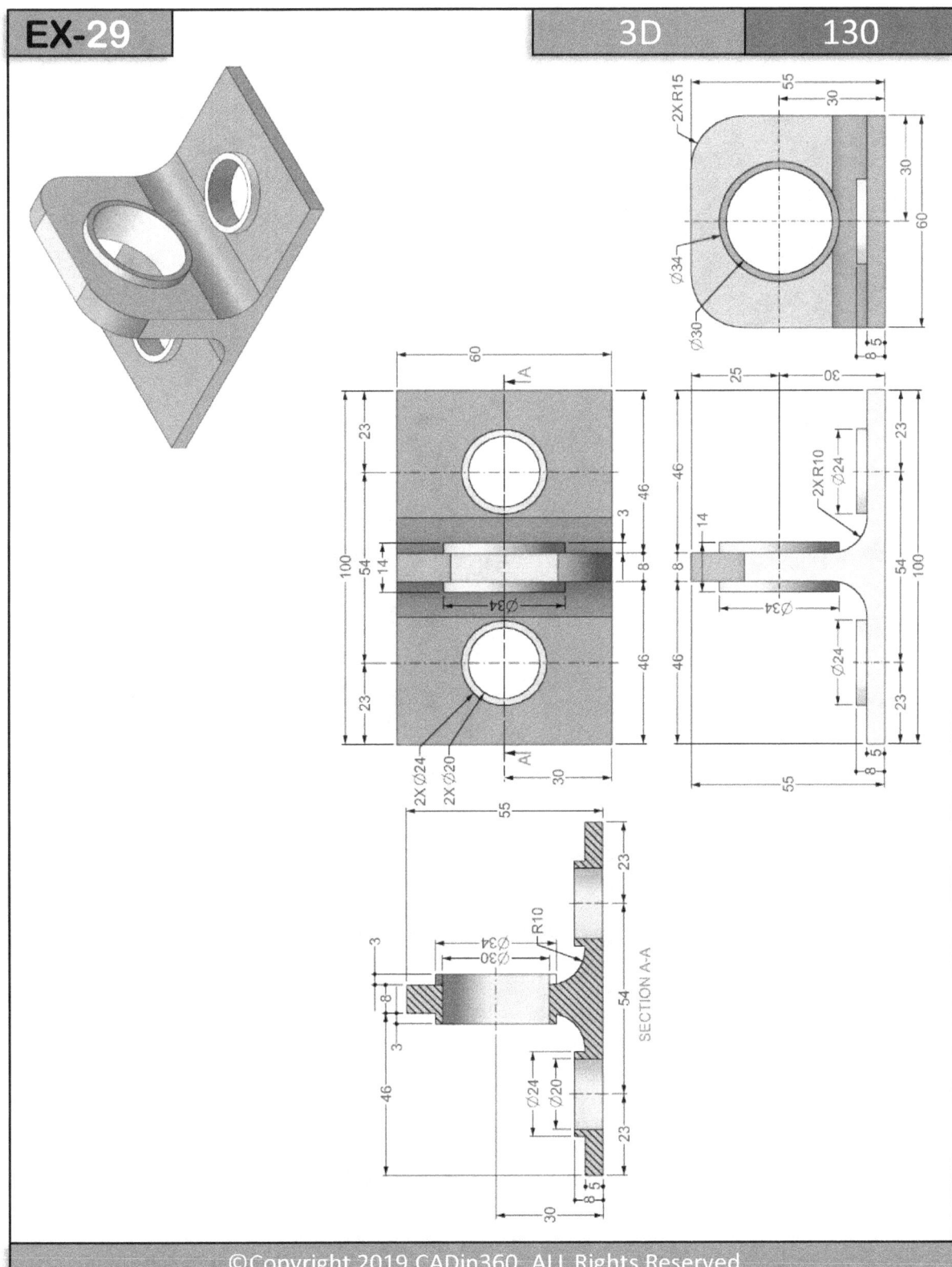

EX-29

3D

130

SECTION A-A

©Copyright 2019 CADin360, ALL Rights Reserved

SECTION A-A

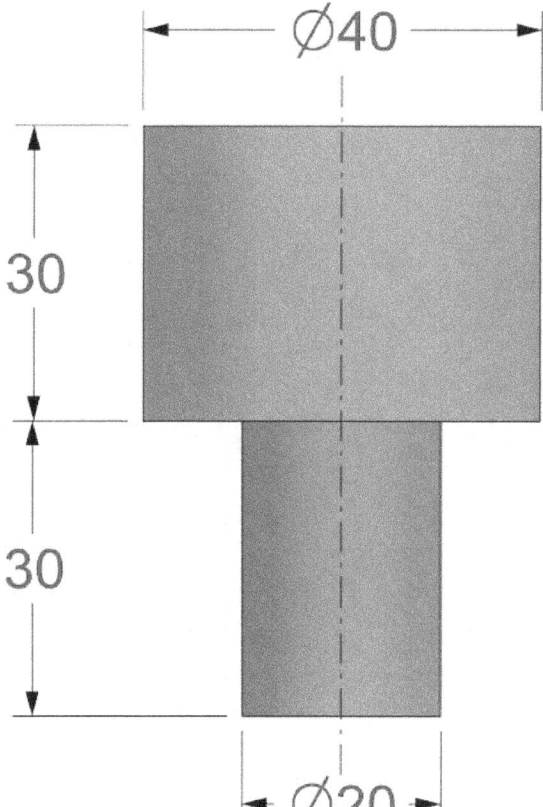

∅40

30

30

∅20

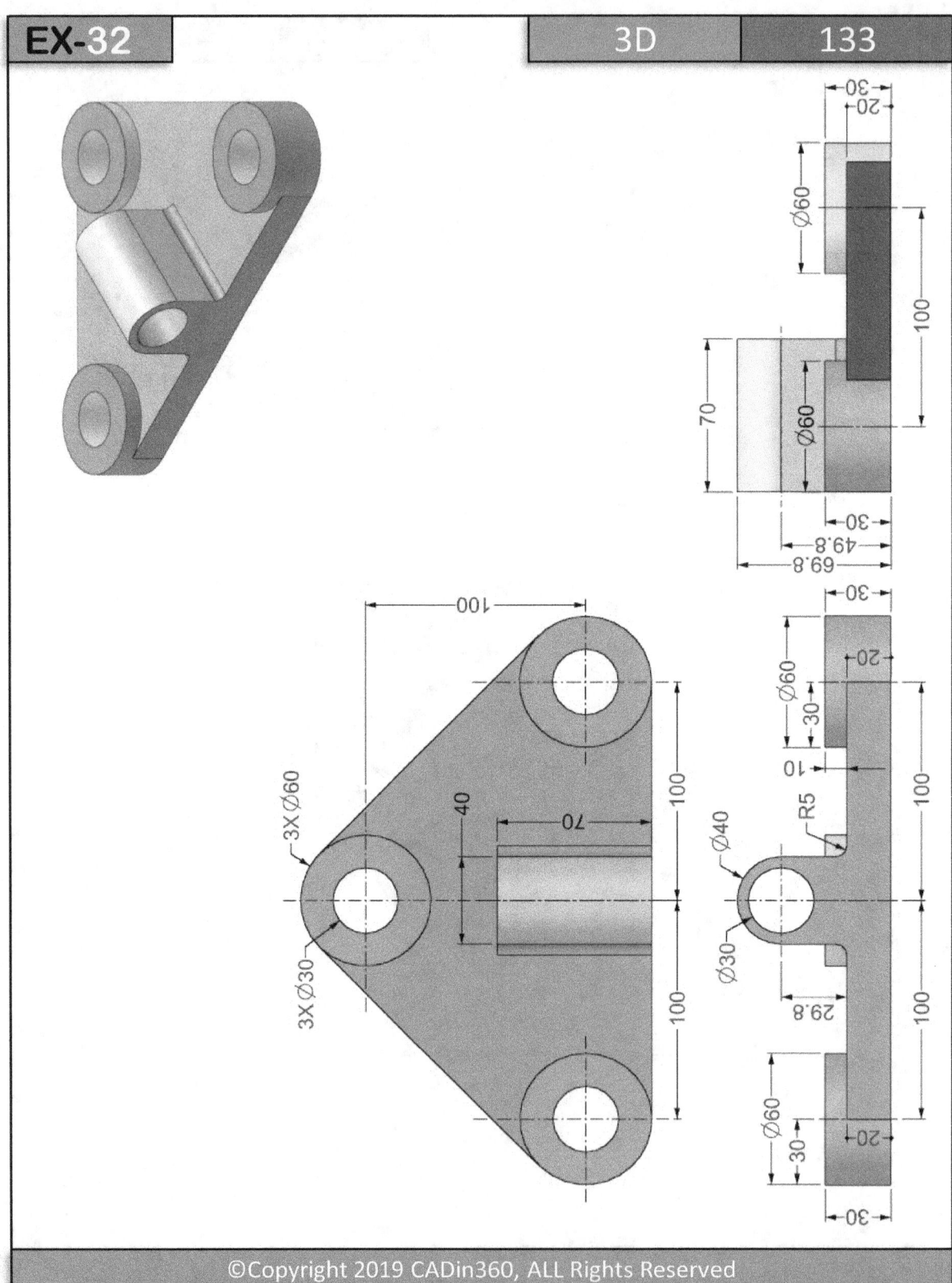

EX-32

3D

133

©Copyright 2019 CADin360, ALL Rights Reserved

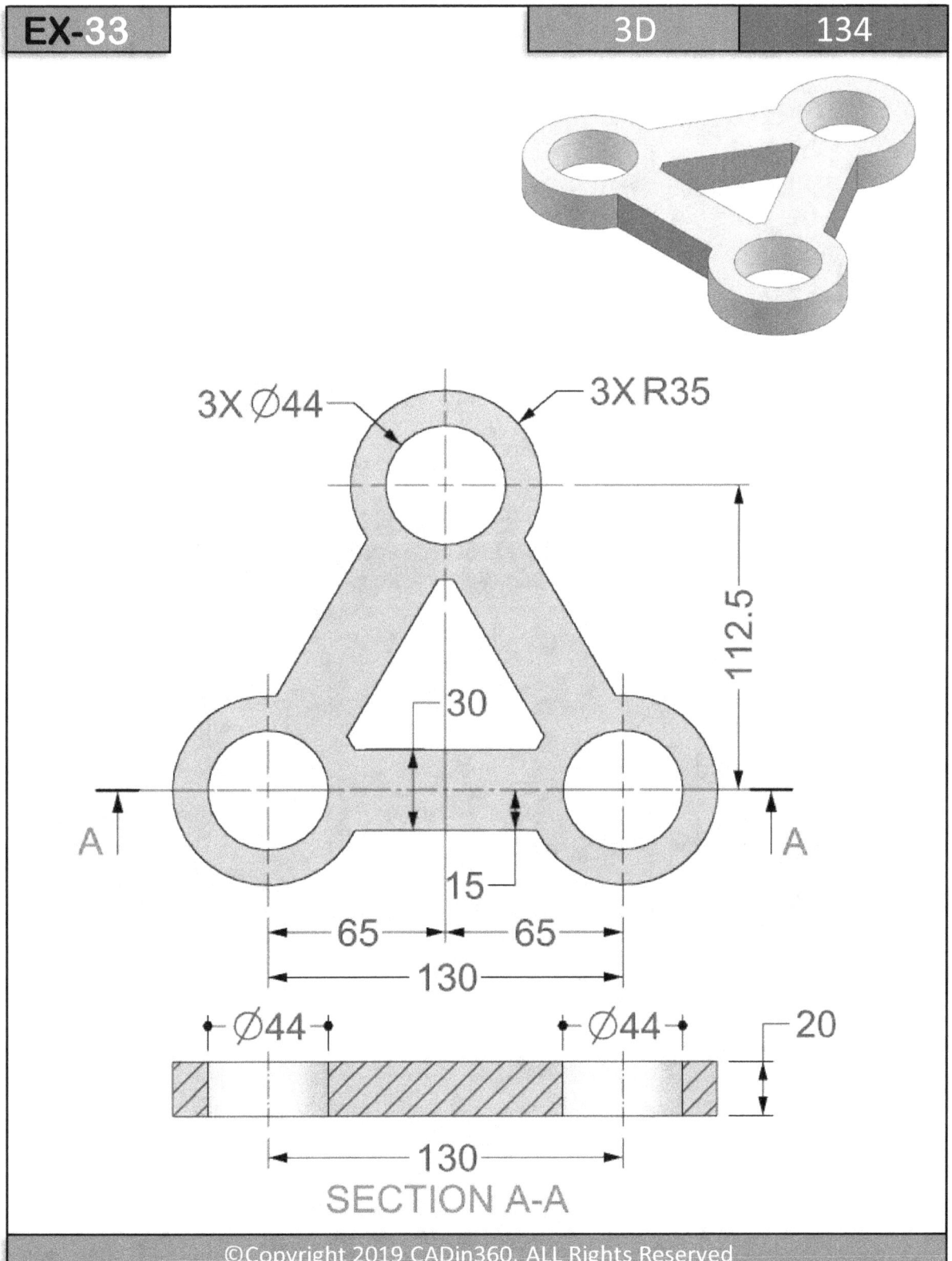

3X Ø44   3X R35

112.5

30

15

65   65

130

Ø44   Ø44

20

130

SECTION A-A

A   A

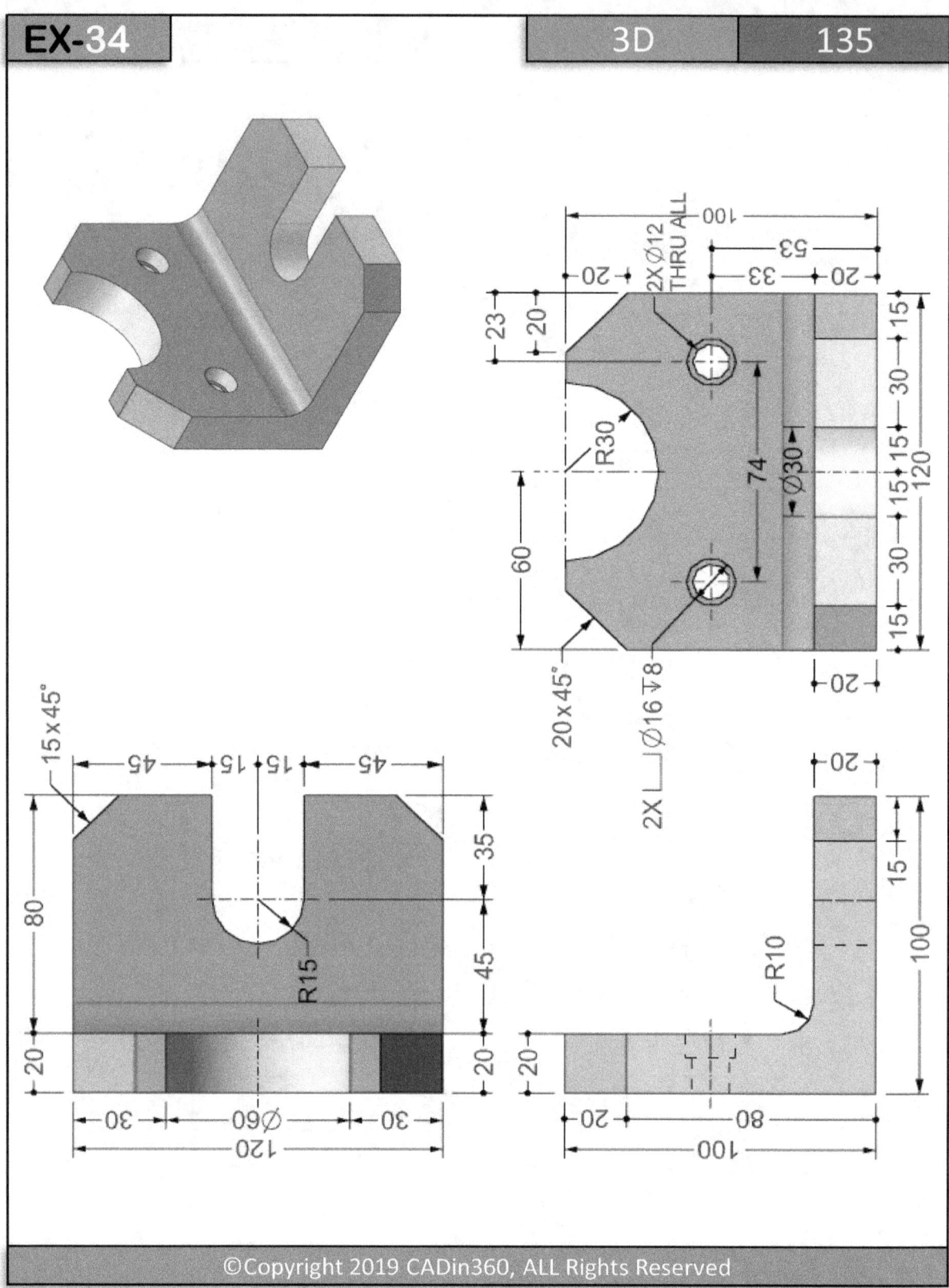

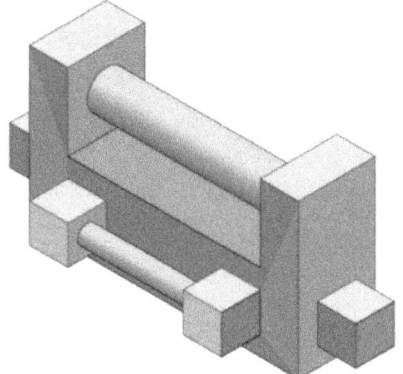

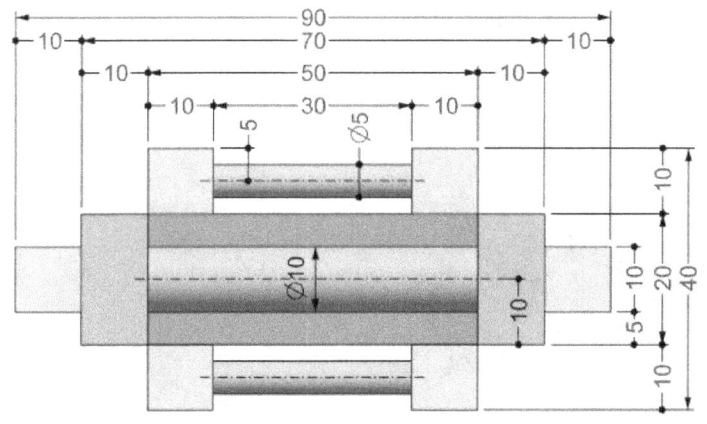

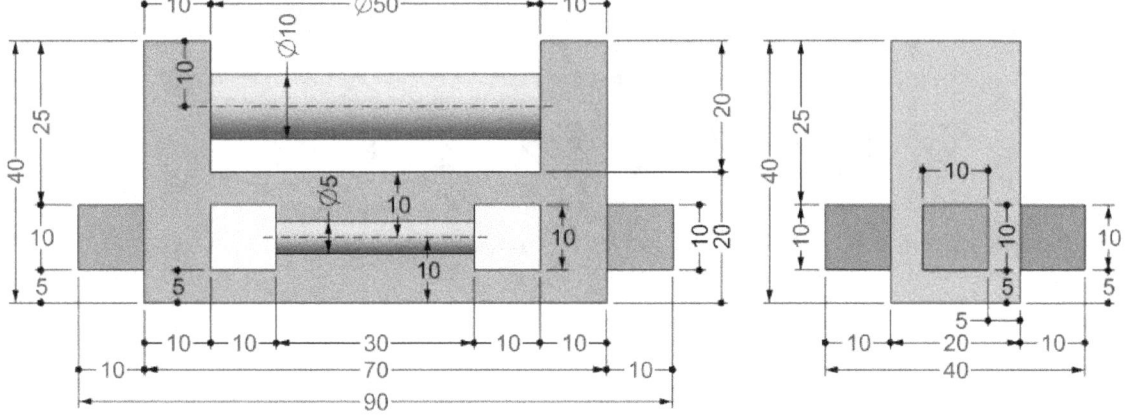

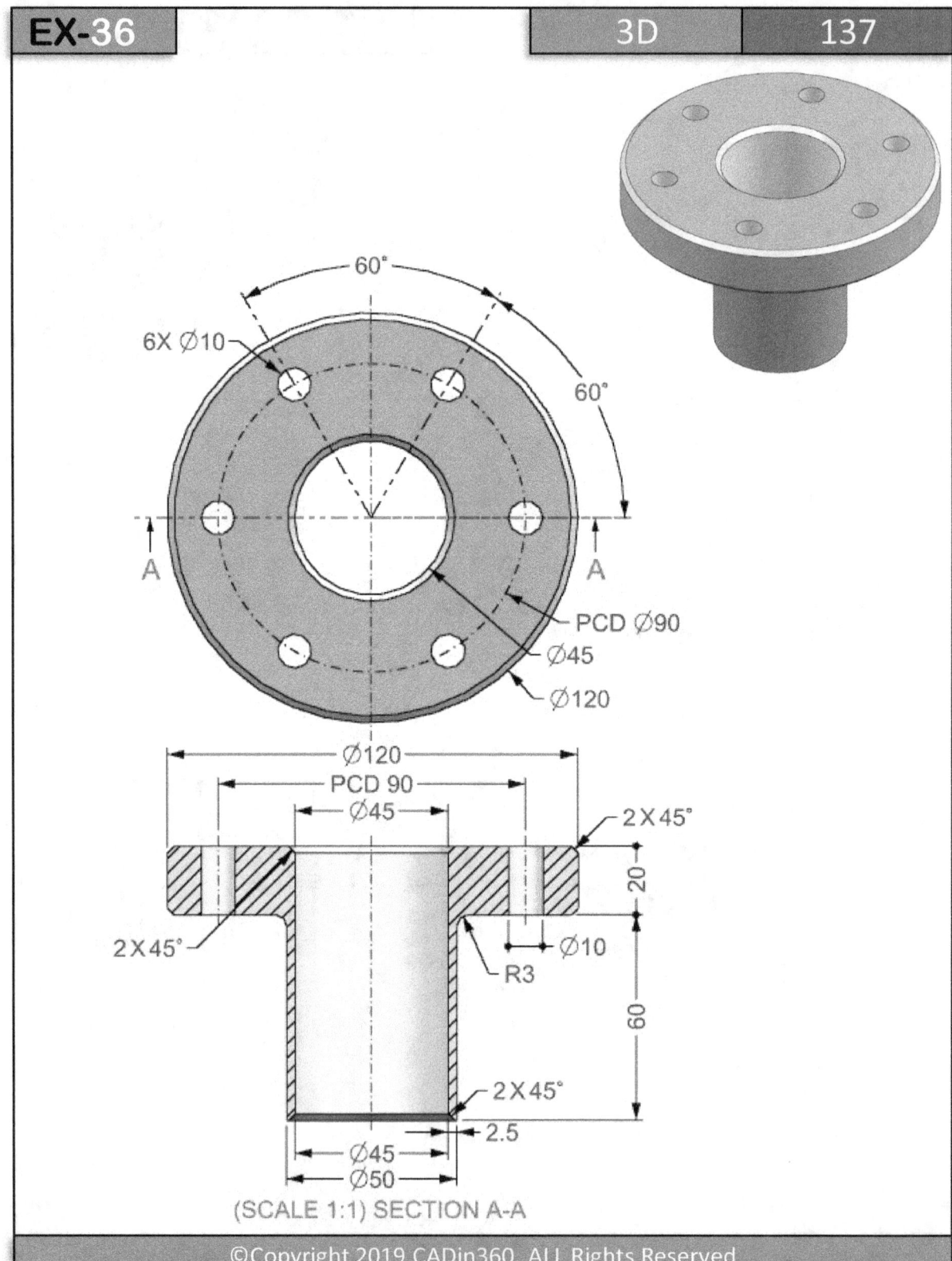

60°

6X Ø10

60°

A　　　　A

PCD Ø90

Ø45

Ø120

Ø120

PCD 90

Ø45

2 X 45°

2 X 45°

20

Ø10

R3

60

2 X 45°

2.5

Ø45

Ø50

(SCALE 1:1) SECTION A-A

SECTION A-A

∅12
∅20
∅12
∅6

25
34
26
120
26
34
70
22
22
10
25
5

16.7 16.6 16.7
16.7 16.6 11.7
50
22
10

5X45°
50
10 15 15 10
A
∅12
∅6
∅20
4X ∅10
10 15 15 10
11.7 11.7
8.3 8.3
11.7 8.3 8.3
11.7
60 26 26 60 9
A

25
15
11.7
16.6
11.7
15
25
70
120
22
10

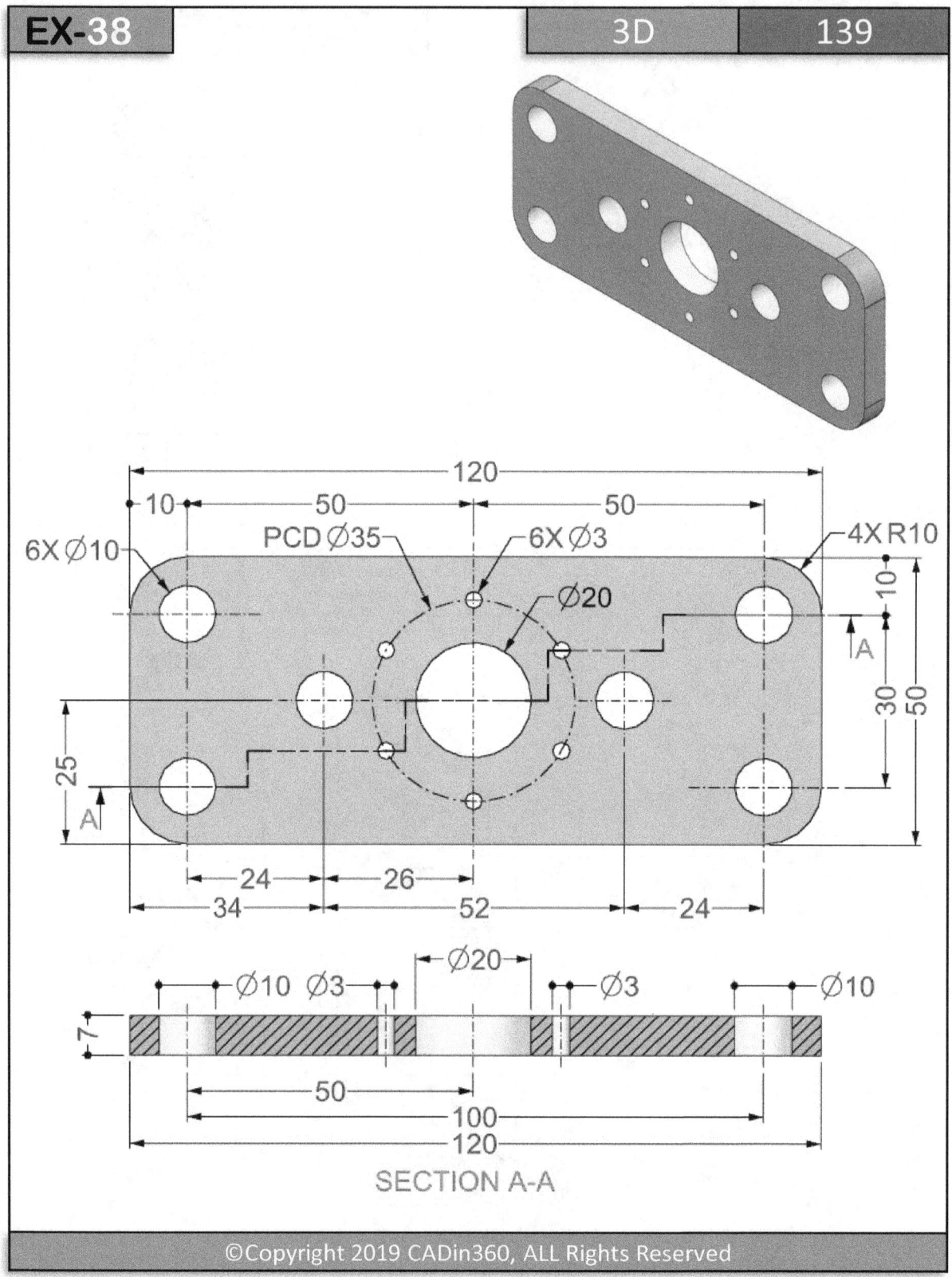

120

10   50   50

4X R10

6X Ø10

PCD Ø35

6X Ø3

Ø20

10

A

30

50

25

A

24   26

34   52   24

Ø20

Ø10   Ø3   Ø3   Ø10

7

50

100

120

SECTION A-A

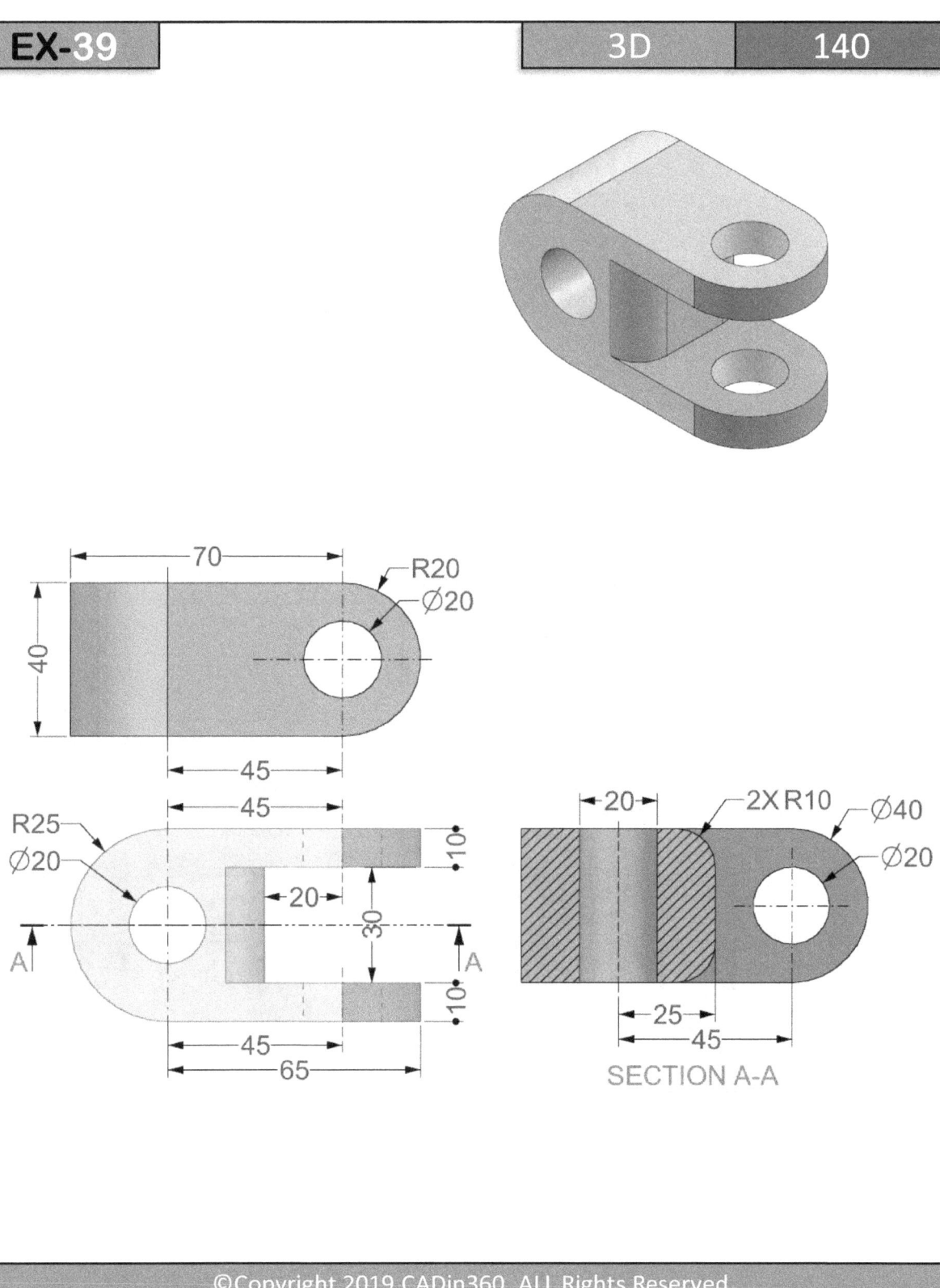

R20
Ø20
70
40
45

R25
Ø20
45
20
30
10
10
45
65

20
2X R10
Ø40
Ø20
25
45

SECTION A-A

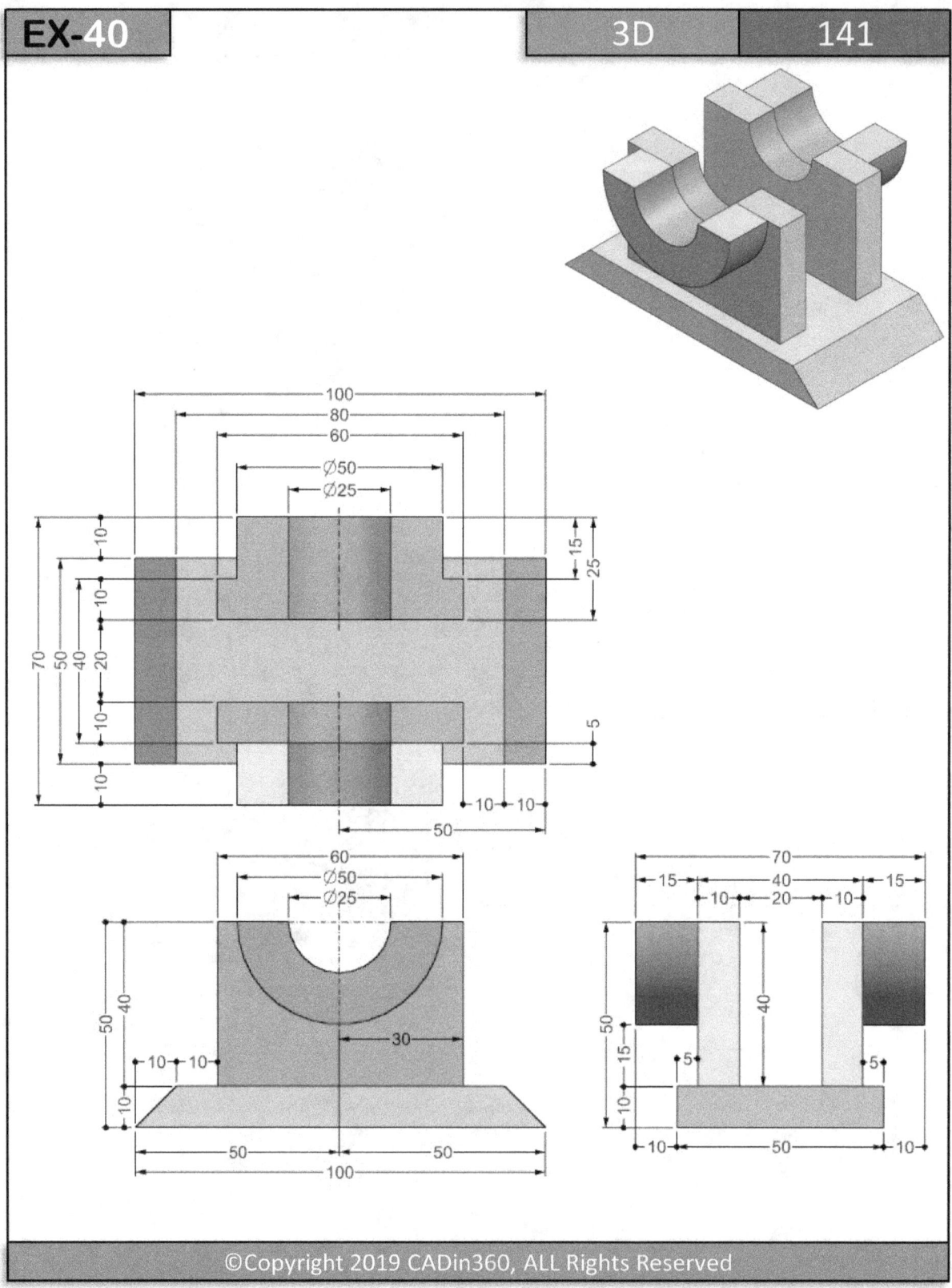

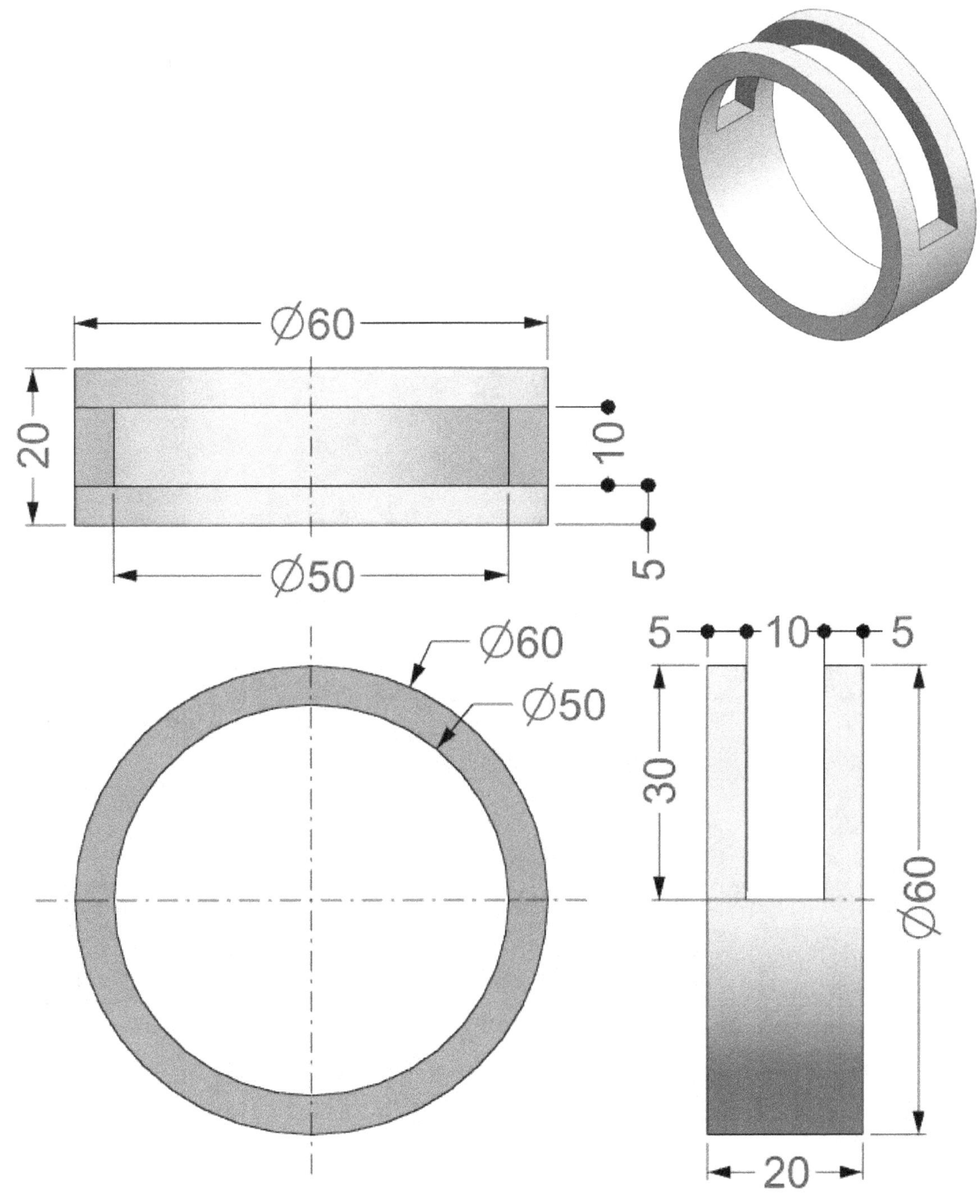

Ø60

20

Ø50

10

5

Ø60

Ø50

5   10   5

30

Ø60

20

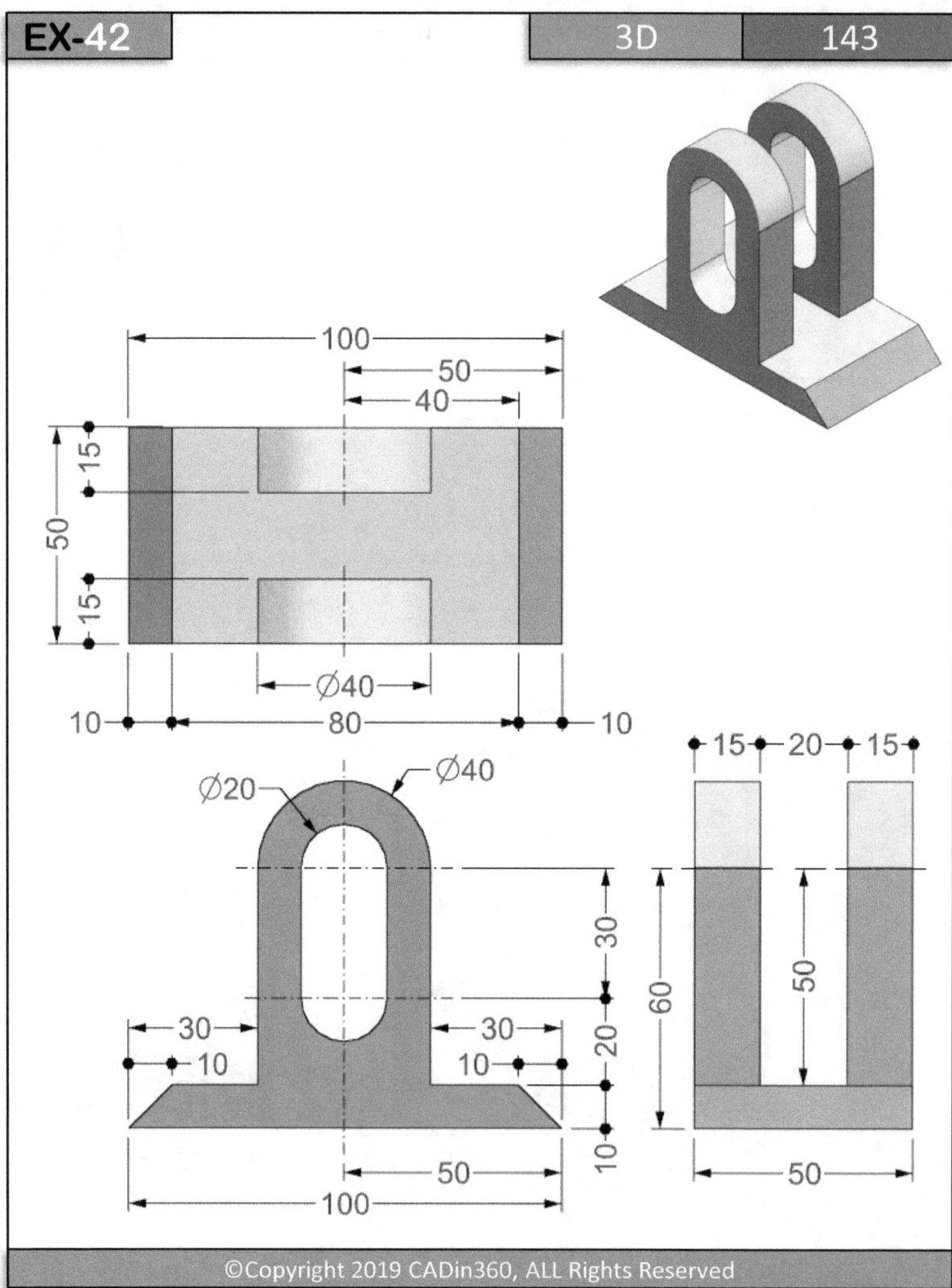

Ø30
Ø24

30 15 25 10

10

50

10

10 10 30

100

25

35

80

10 30

25

25

Ø30

R2

45

10

10

10

5

10

10

10

100

10 30 10

25

45

Ø30

R2

10

10

5

50

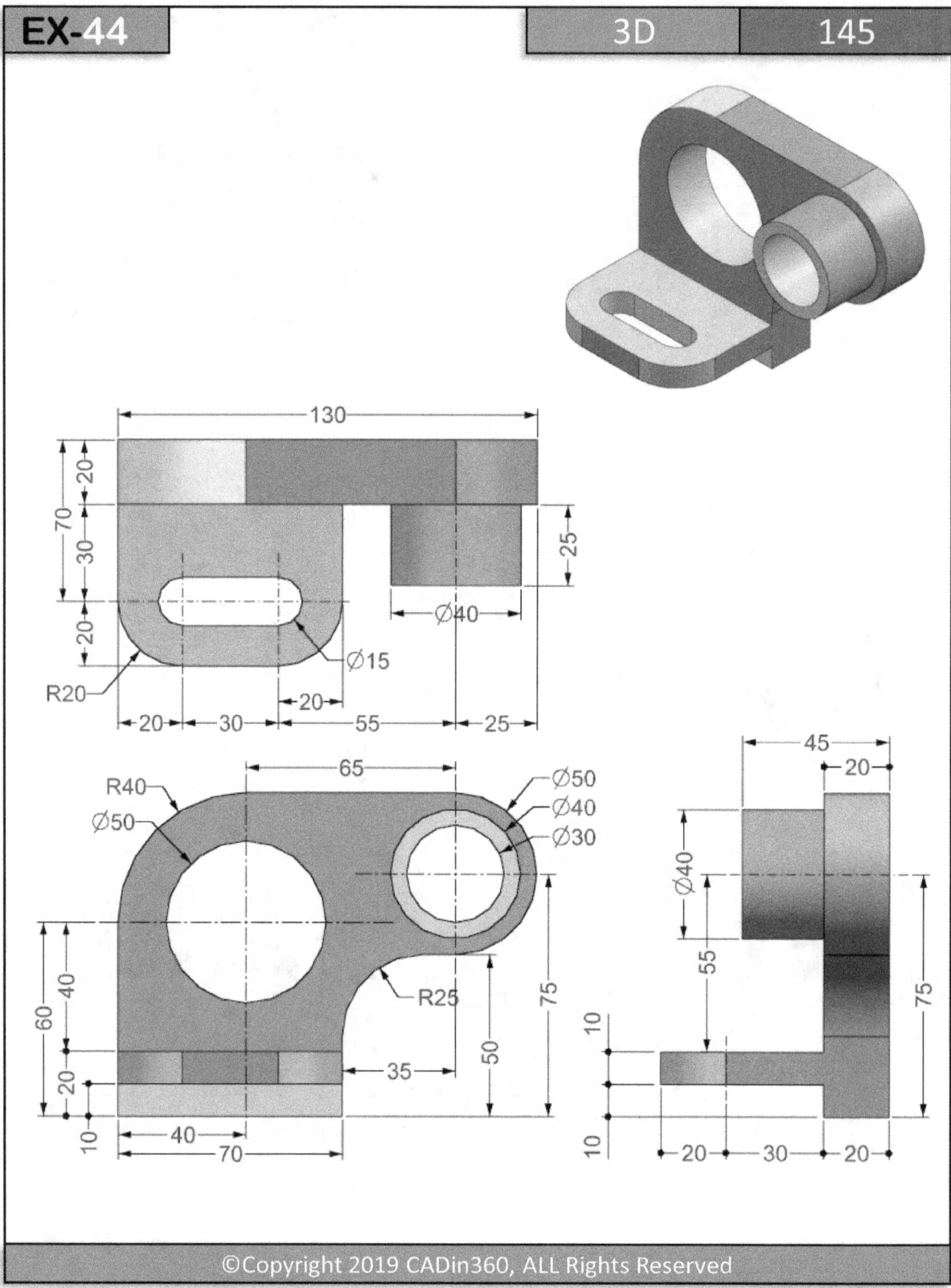

EX-44

3D   145

130

20

70

30

20

25

⌀40

R20

⌀15

20   30   20   55   25

R40

⌀50

65

⌀50
⌀40
⌀30

R25

60

40

20

10

40

70

50

75

35

45

20

⌀40

55

75

10

10

20   30   20

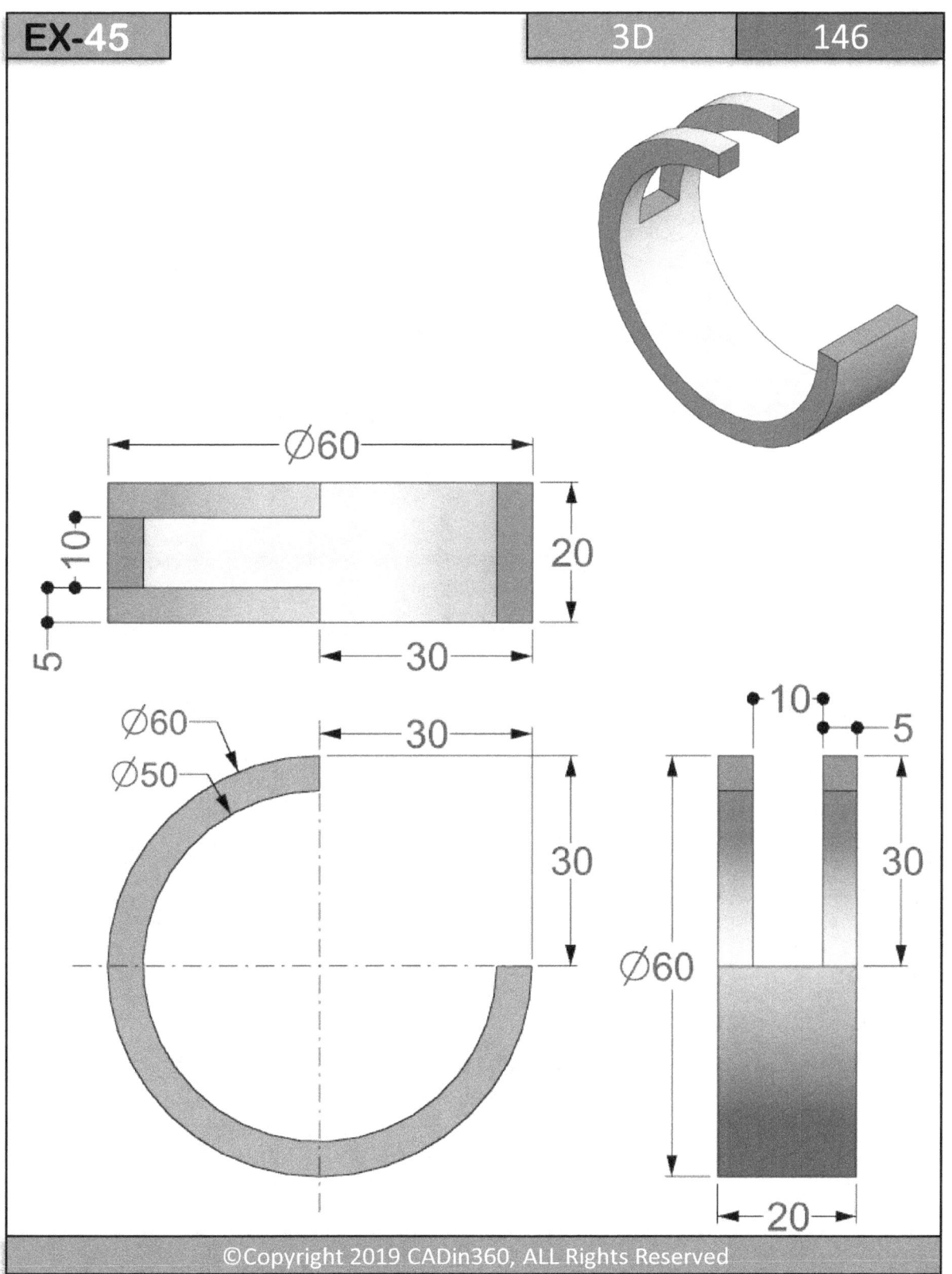

EX-45   3D   146

⌀60

20

10

5

30

⌀60
⌀50

30

30

10

5

30

⌀60

20

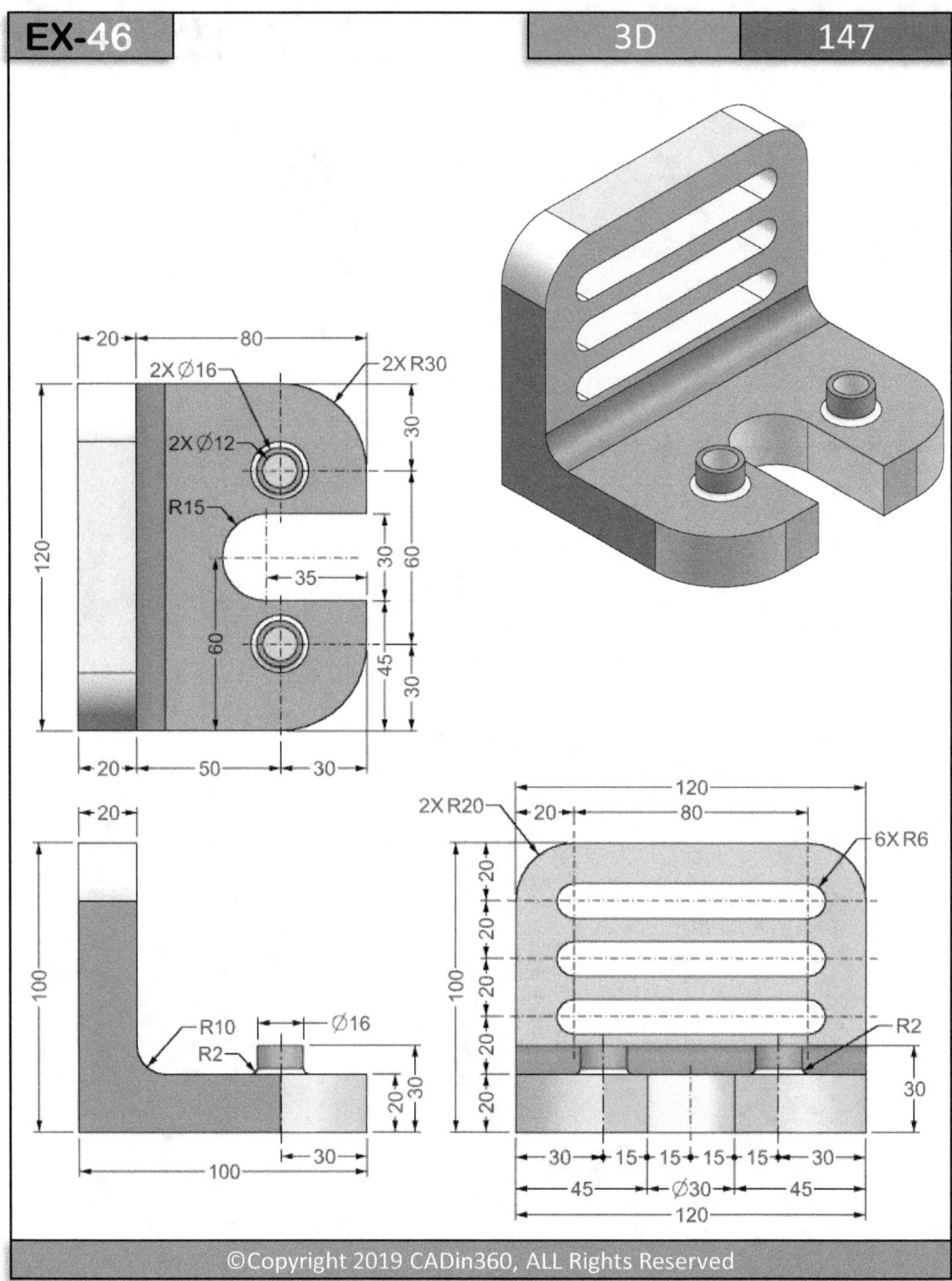

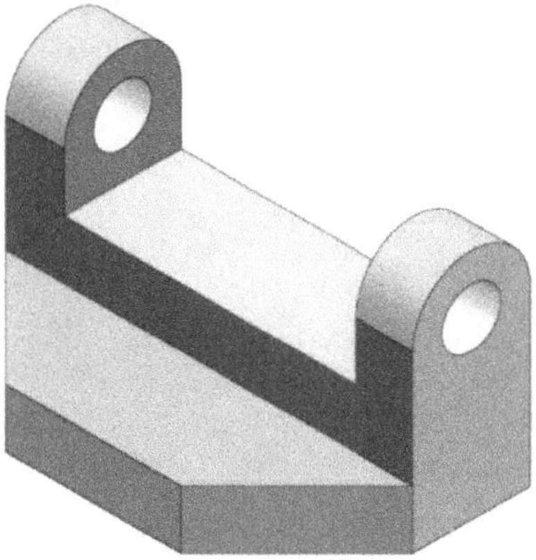

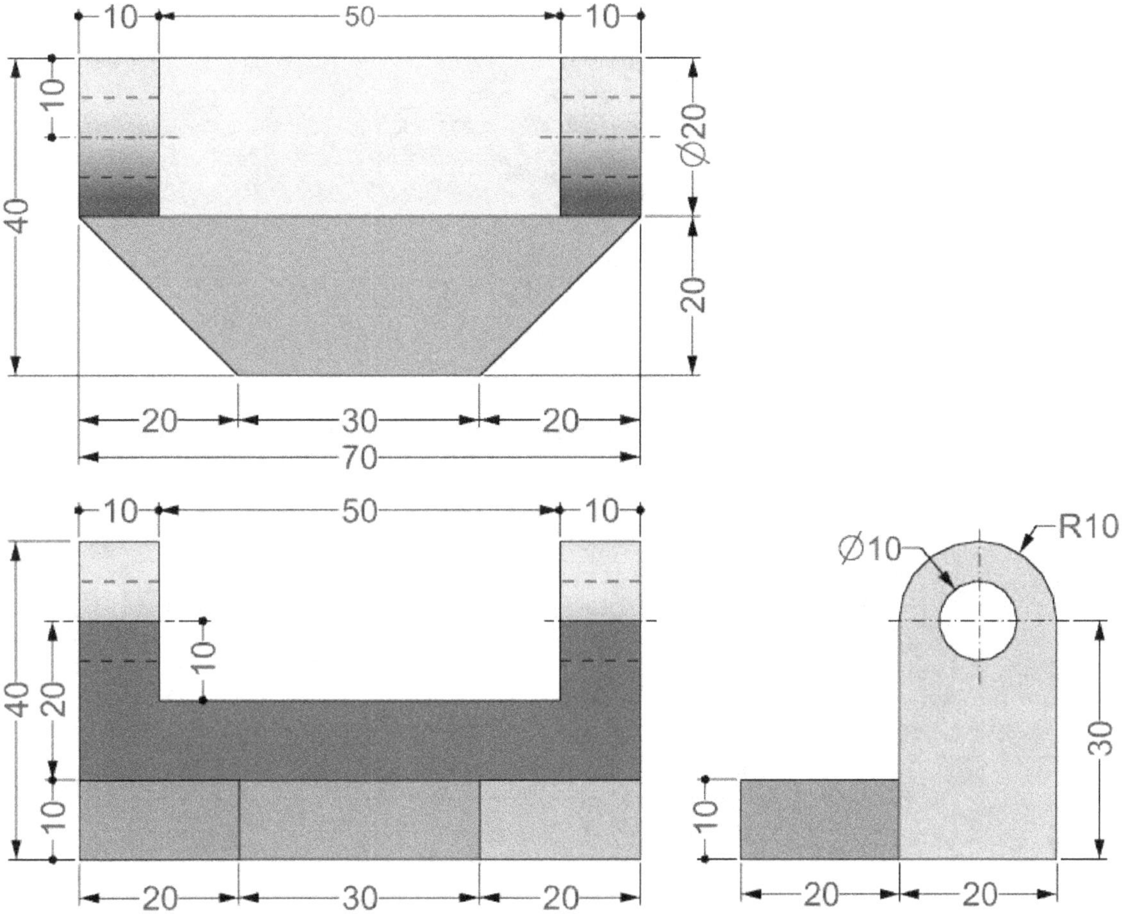

10 — 50 — 10

10
40
Ø20
20

20 — 30 — 20
70

10 — 50 — 10
40
20
10
10
20 — 30 — 20

Ø10 — R10
30
10
20 — 20

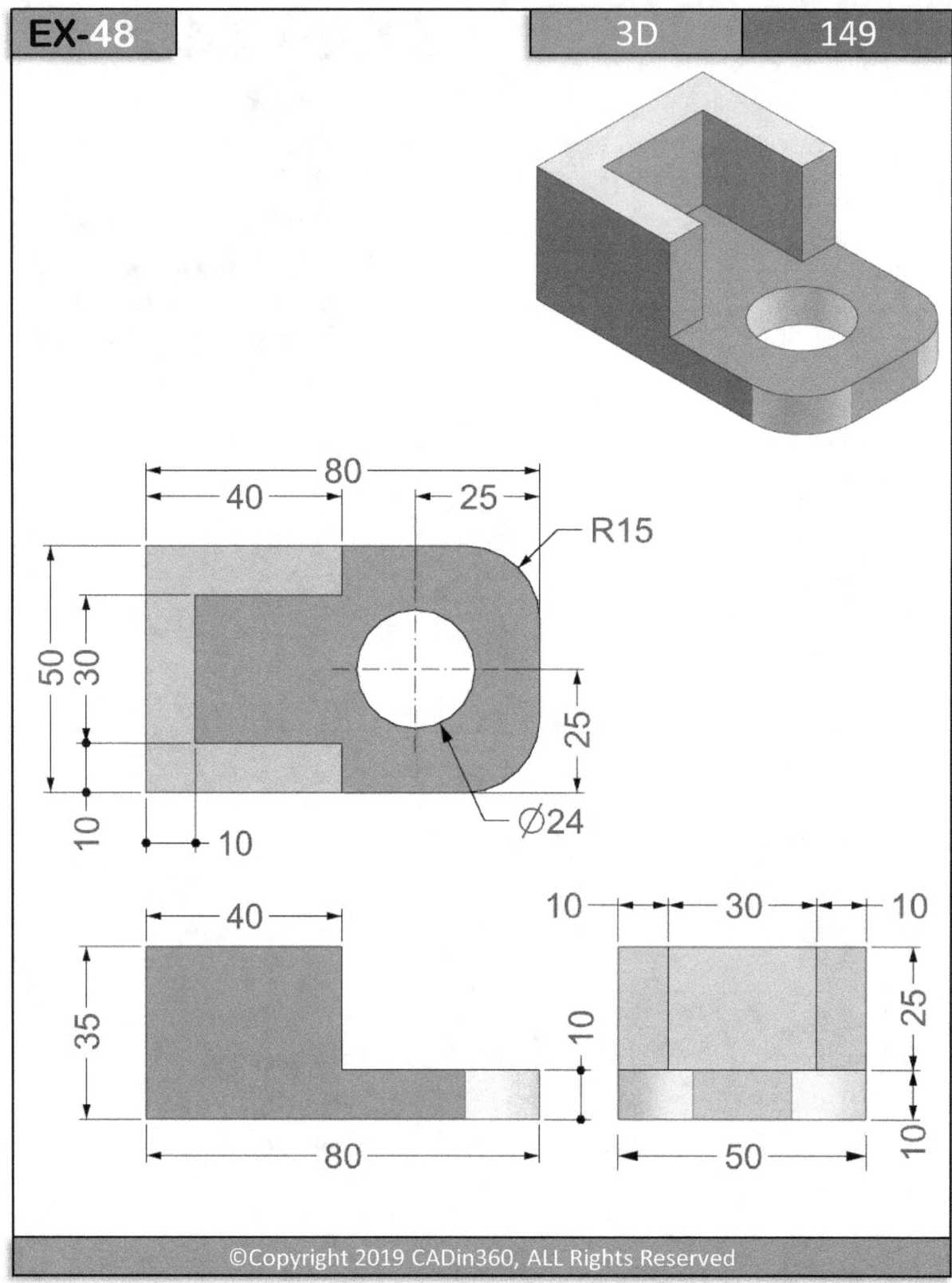

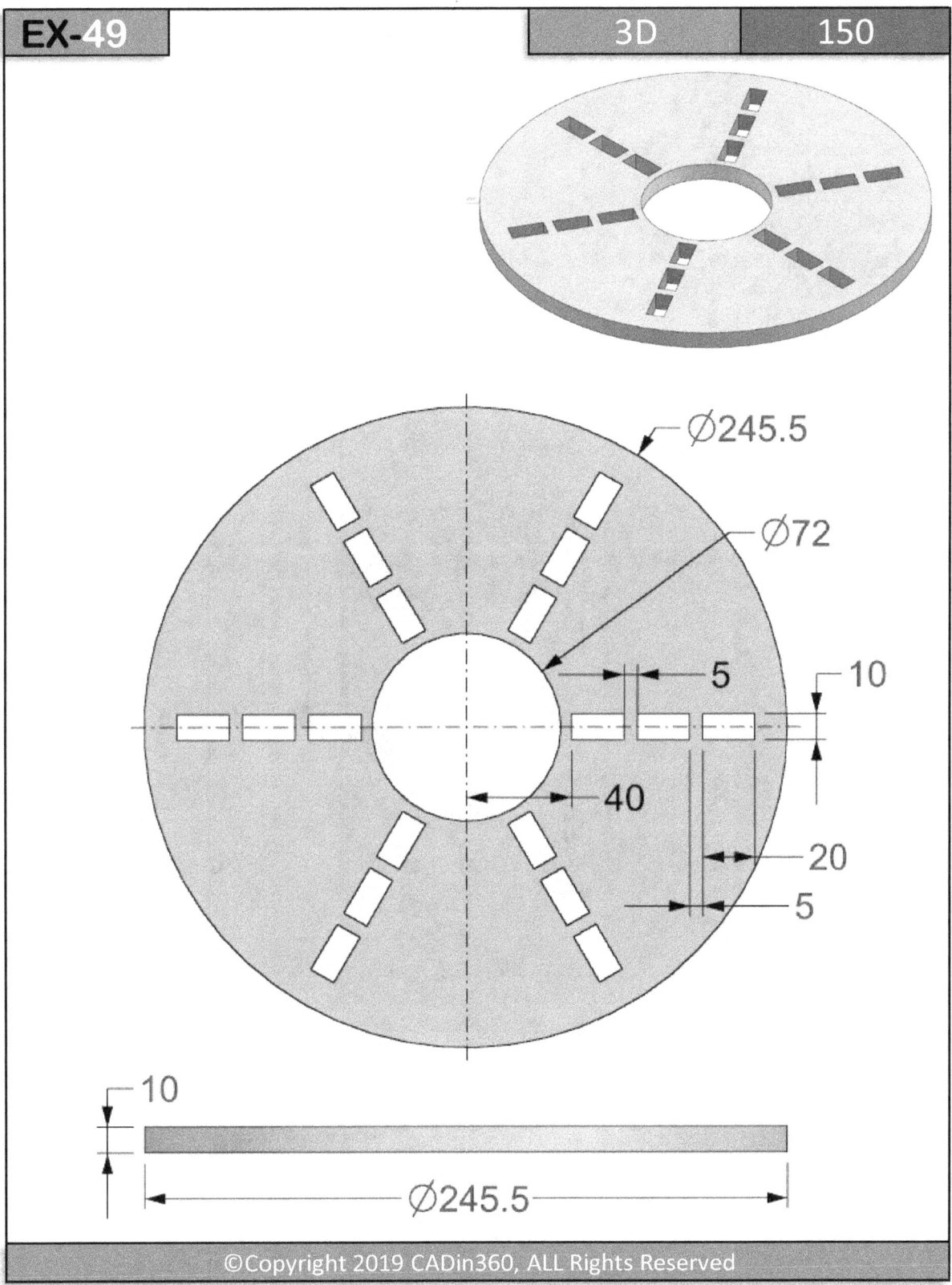

Ø245.5

Ø72

5

10

40

20

5

10

Ø245.5

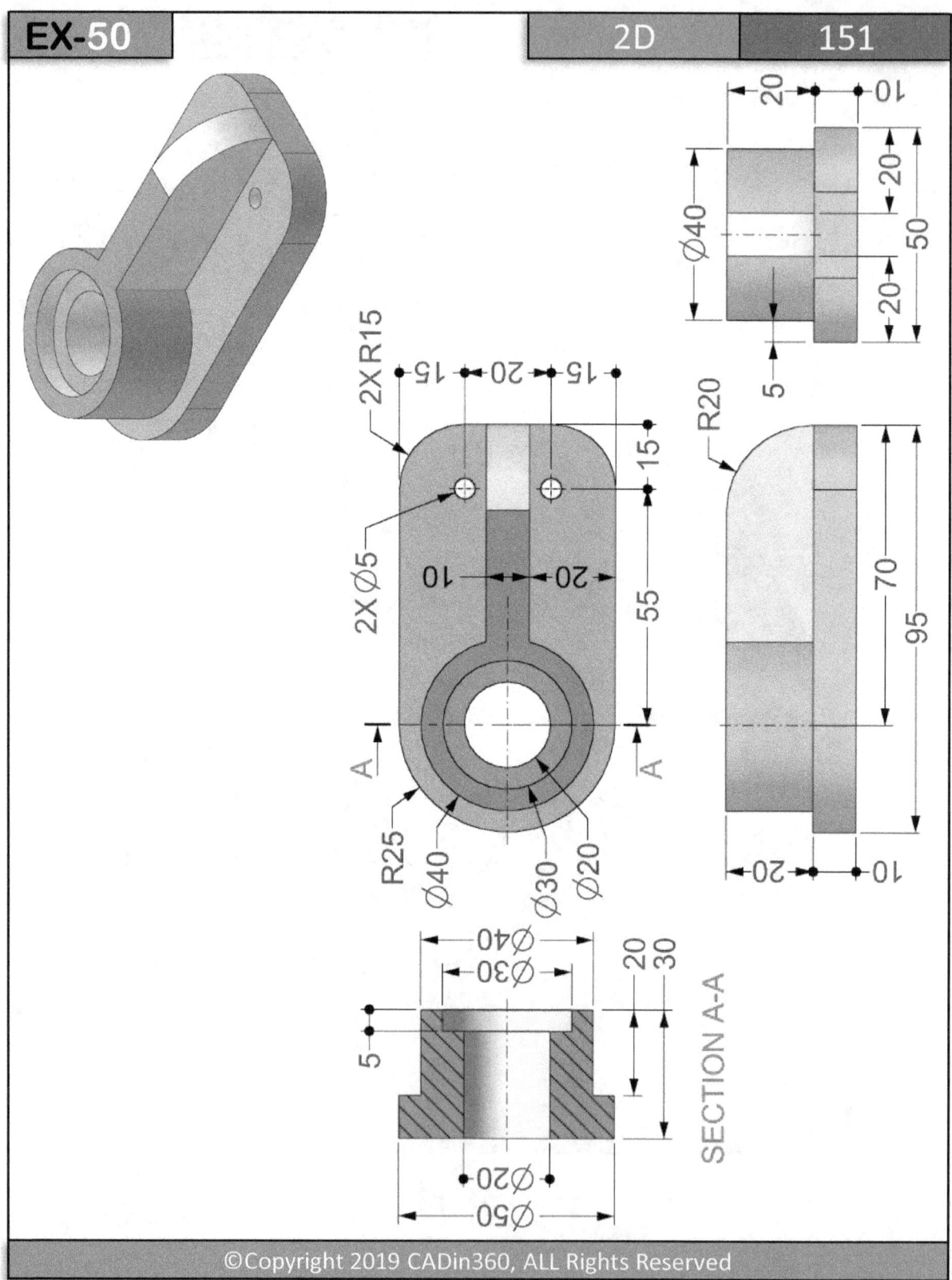

EX-50

2D

151

2×R15

2×∅5

15 20 15

15

10 20

55

A A

R25 ∅40 ∅30 ∅20

∅40

50

20

20

20

5

R20

70

95

20 10

∅40
∅30

20
30

5

∅20
∅50

SECTION A-A

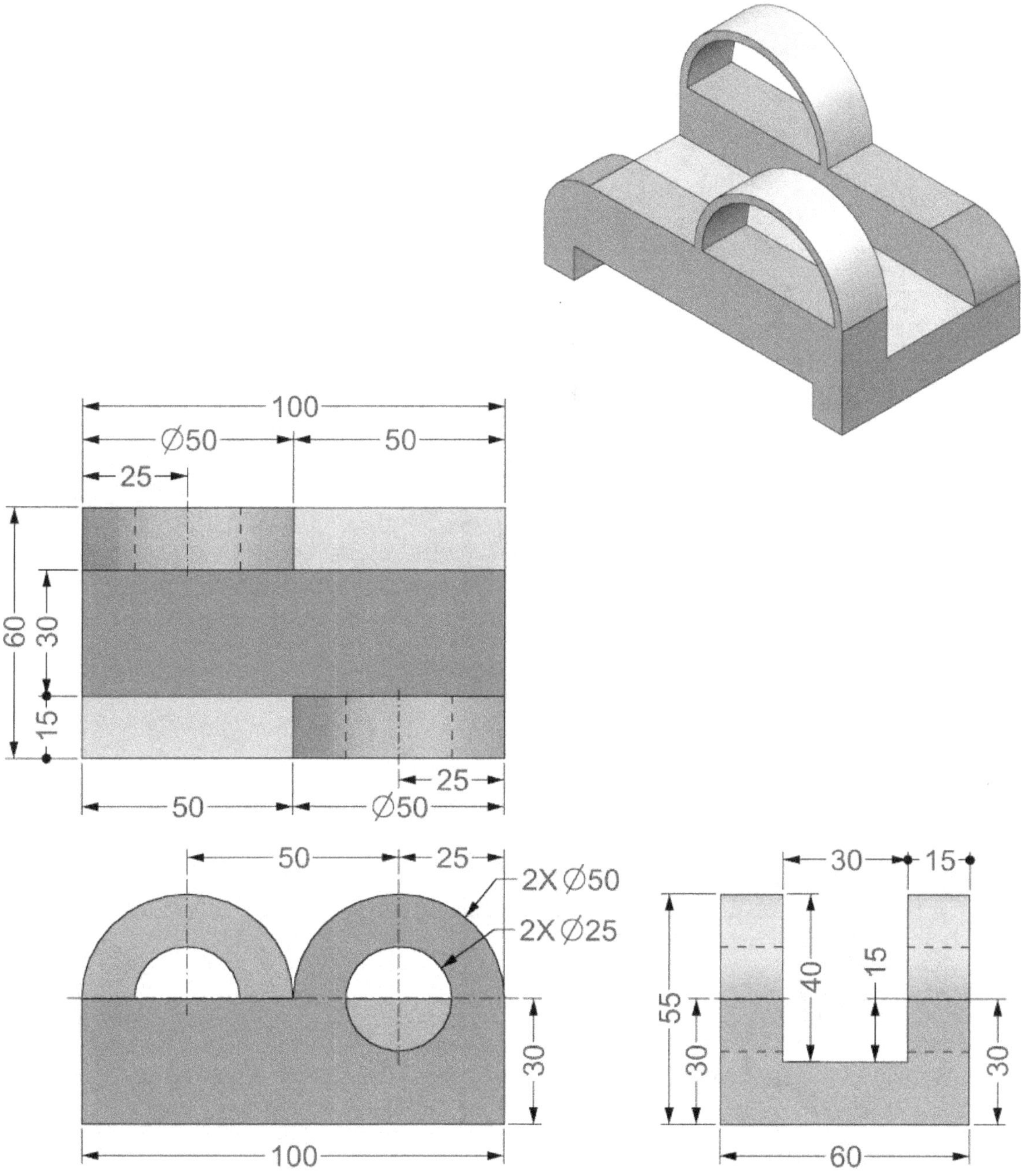

100
Ø50   50
25
60   30
15
50   Ø50
25

50   25
2X Ø50
2X Ø25
30
100

30   15
55   40   15
30   30
60

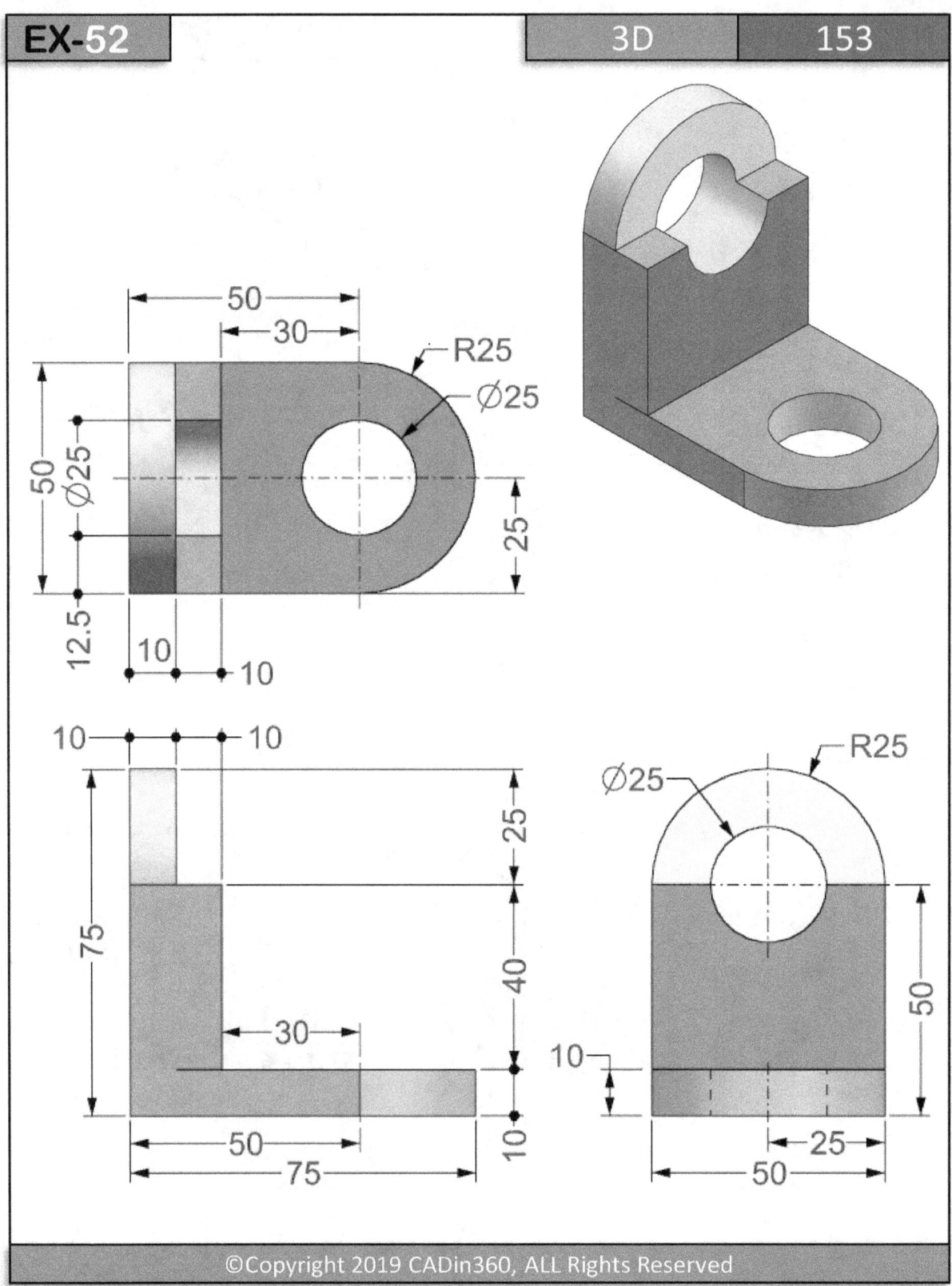

EX-52

3D   153

50

30

R25

Ø25

50

Ø25

25

12.5

10

10

10

10

75

25

40

30

50

10

75

Ø25

R25

50

10

25

50

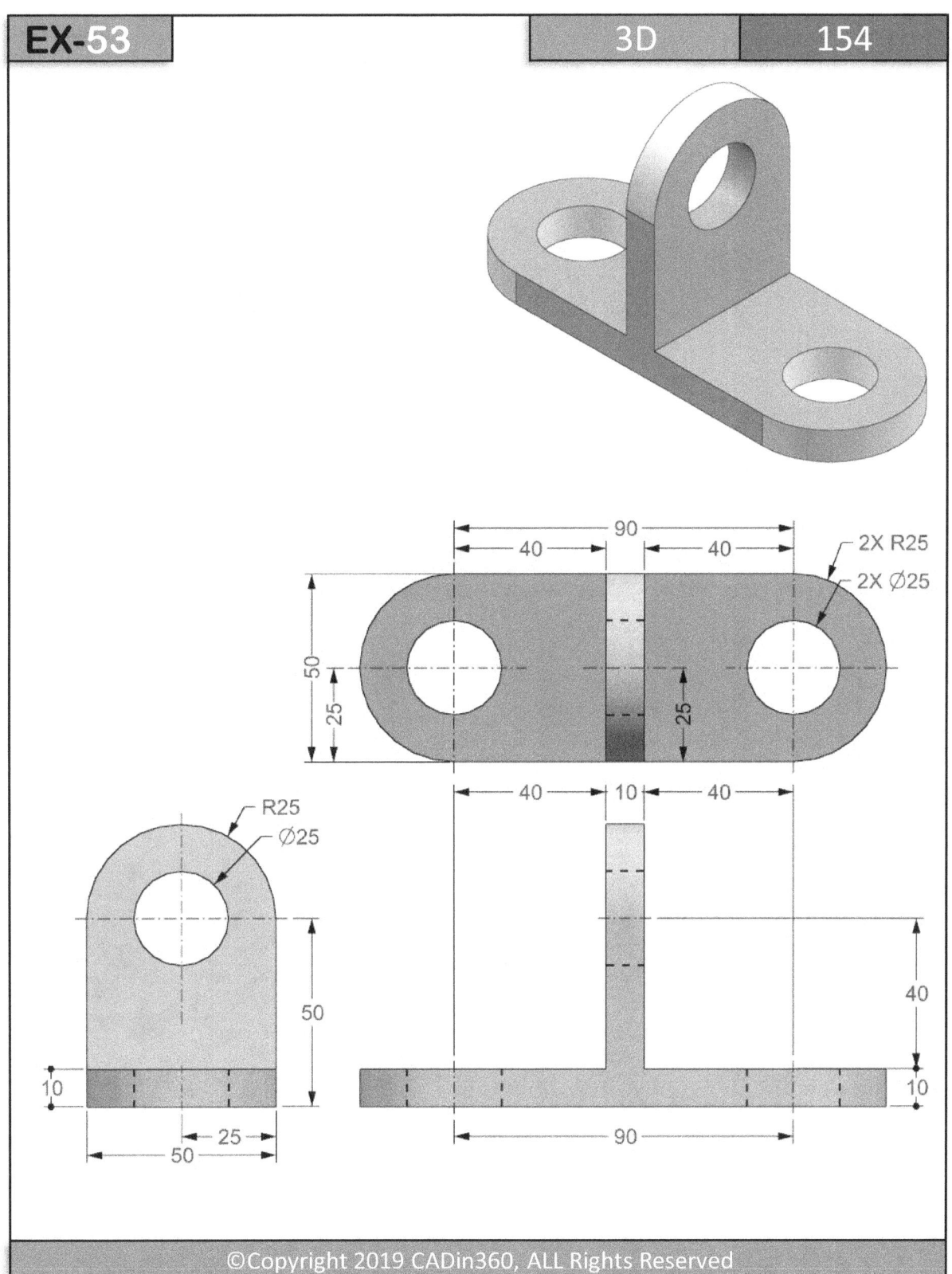

2X R25

2X Ø25

90

40

40

50

25

25

40

10

40

R25

Ø25

50

40

10

10

25

50

90

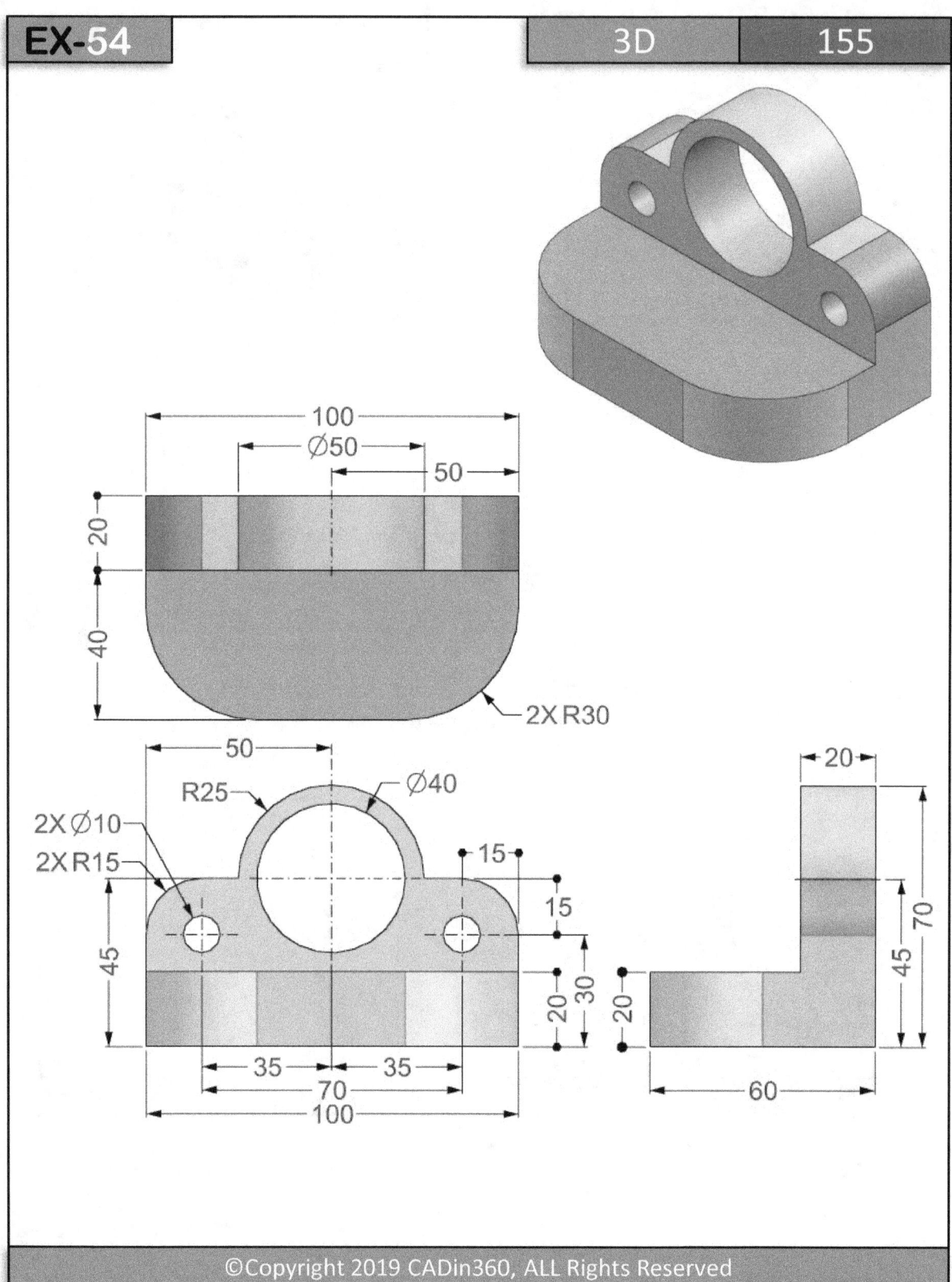

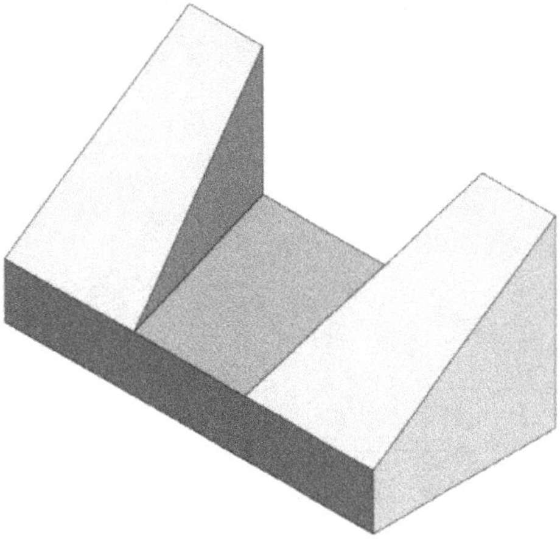

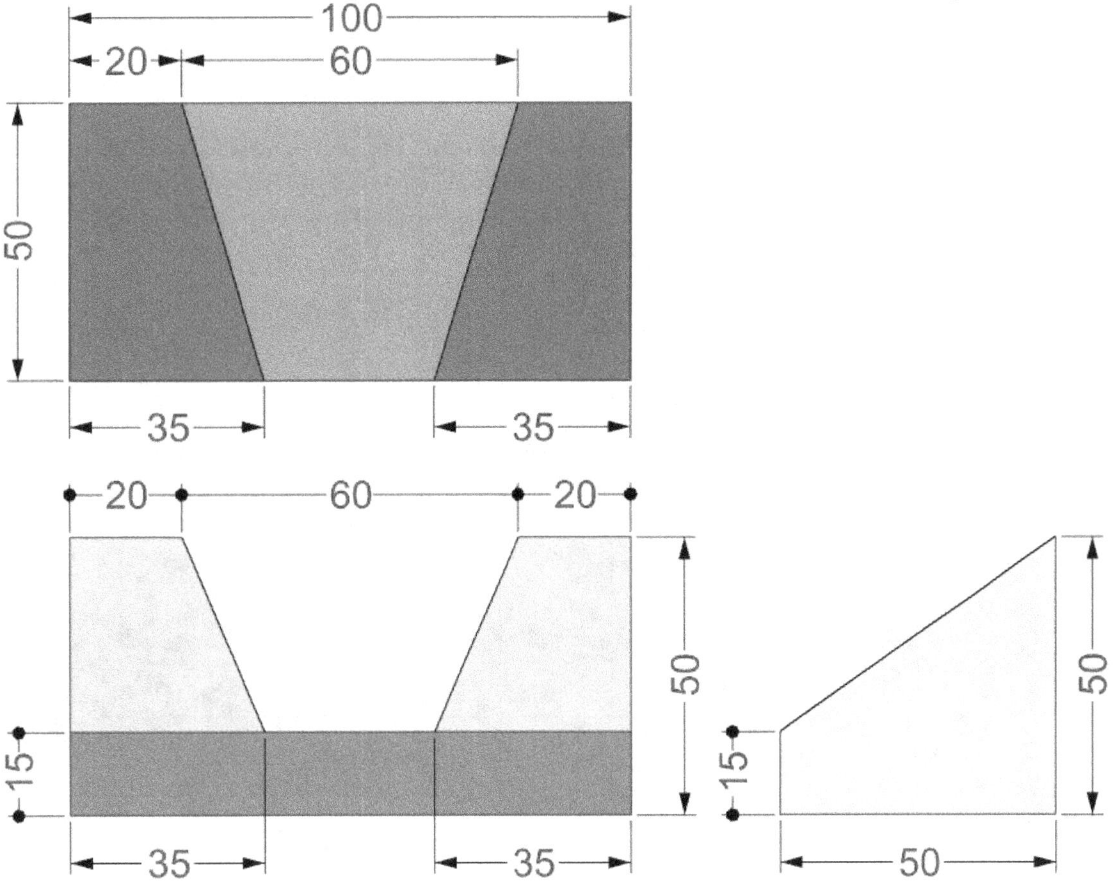

100

20 | 60

50

35 | 35

20 | 60 | 20

50

15

35 | 35

15

50 | 50

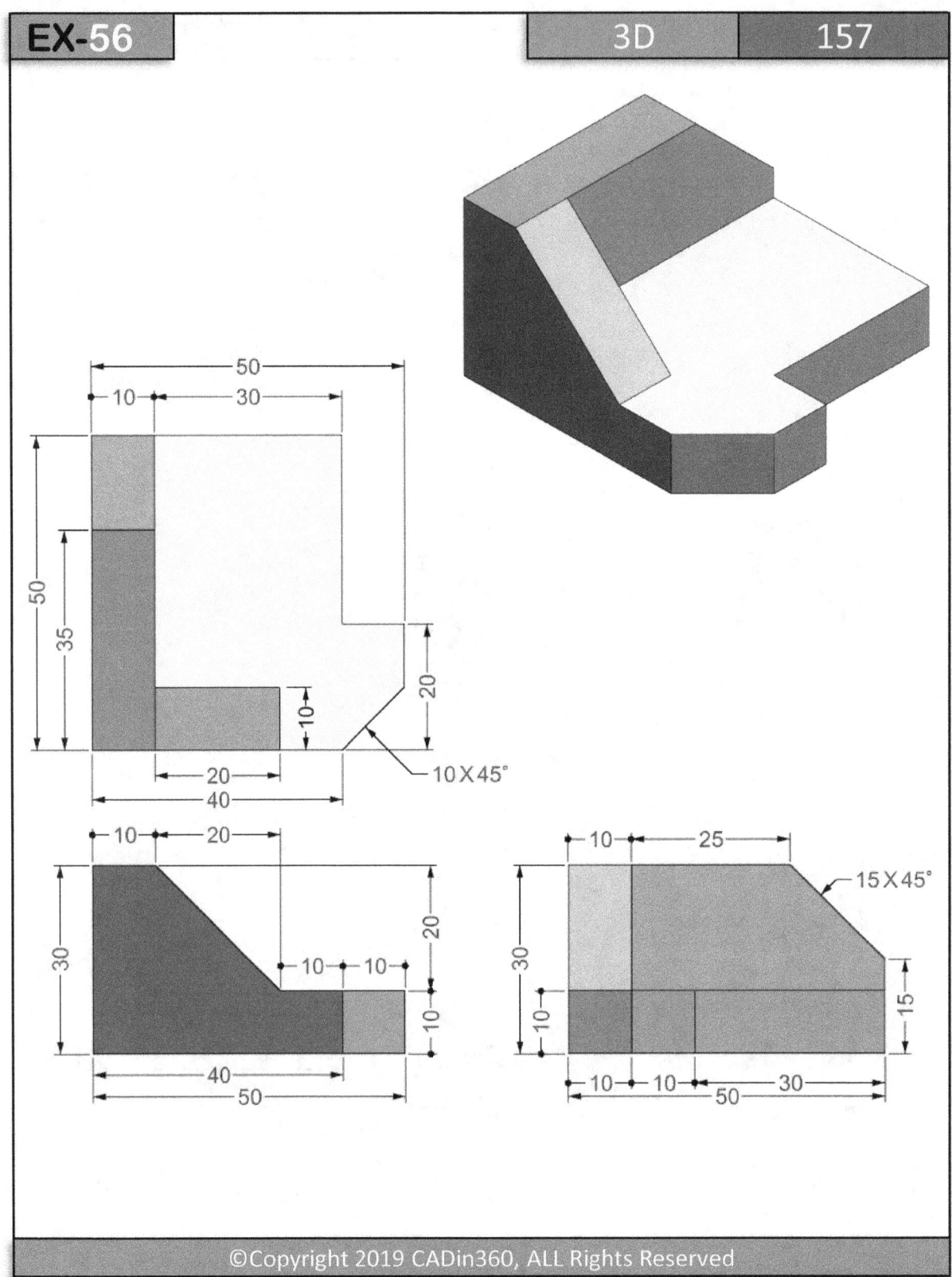

50
10 30
50
35
20
10
20
40
10 X 45°

10 20
30
10 10
20
10
40
50

10 25
15 X 45°
30
10
10 10 30
50
15

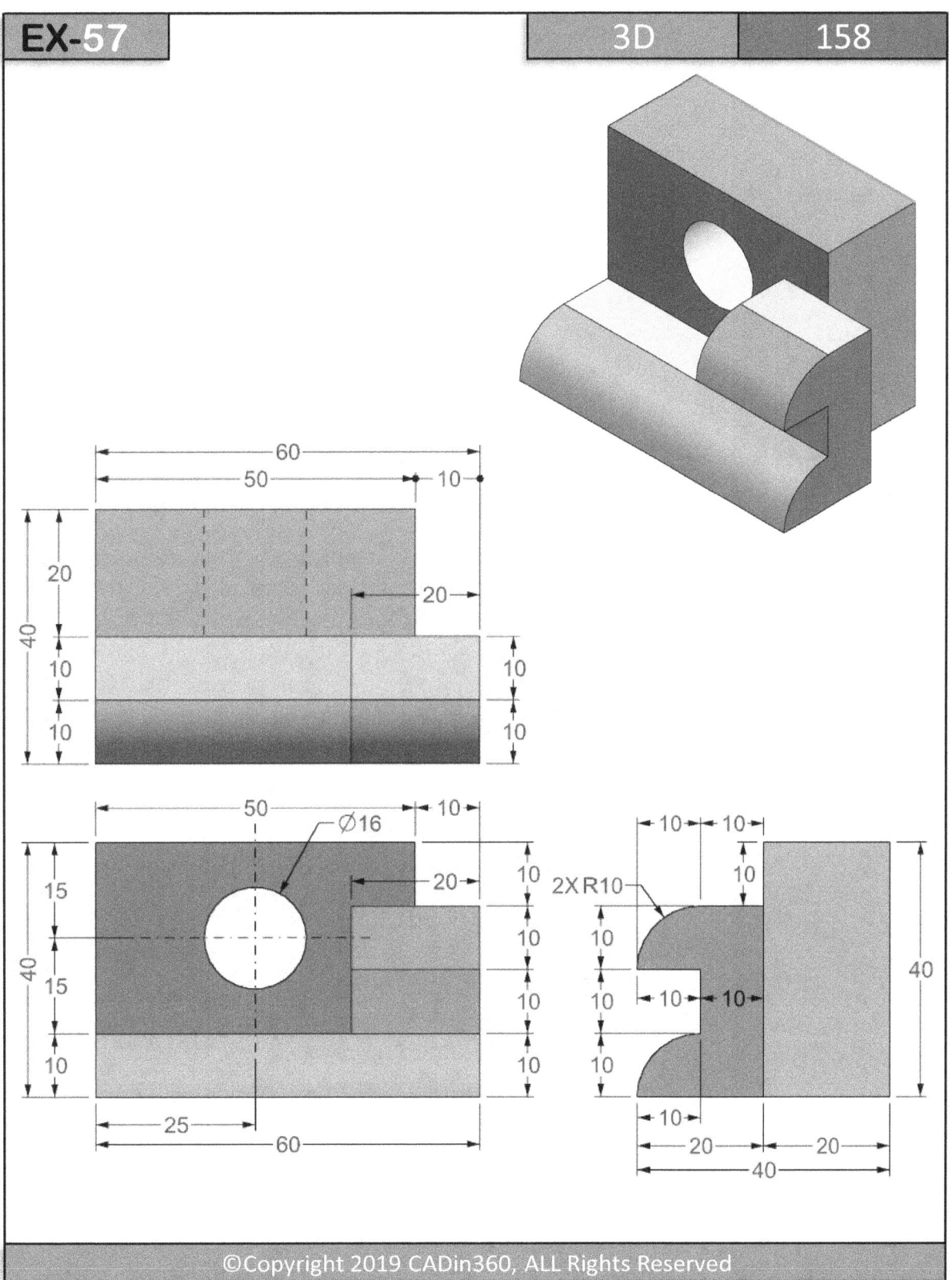

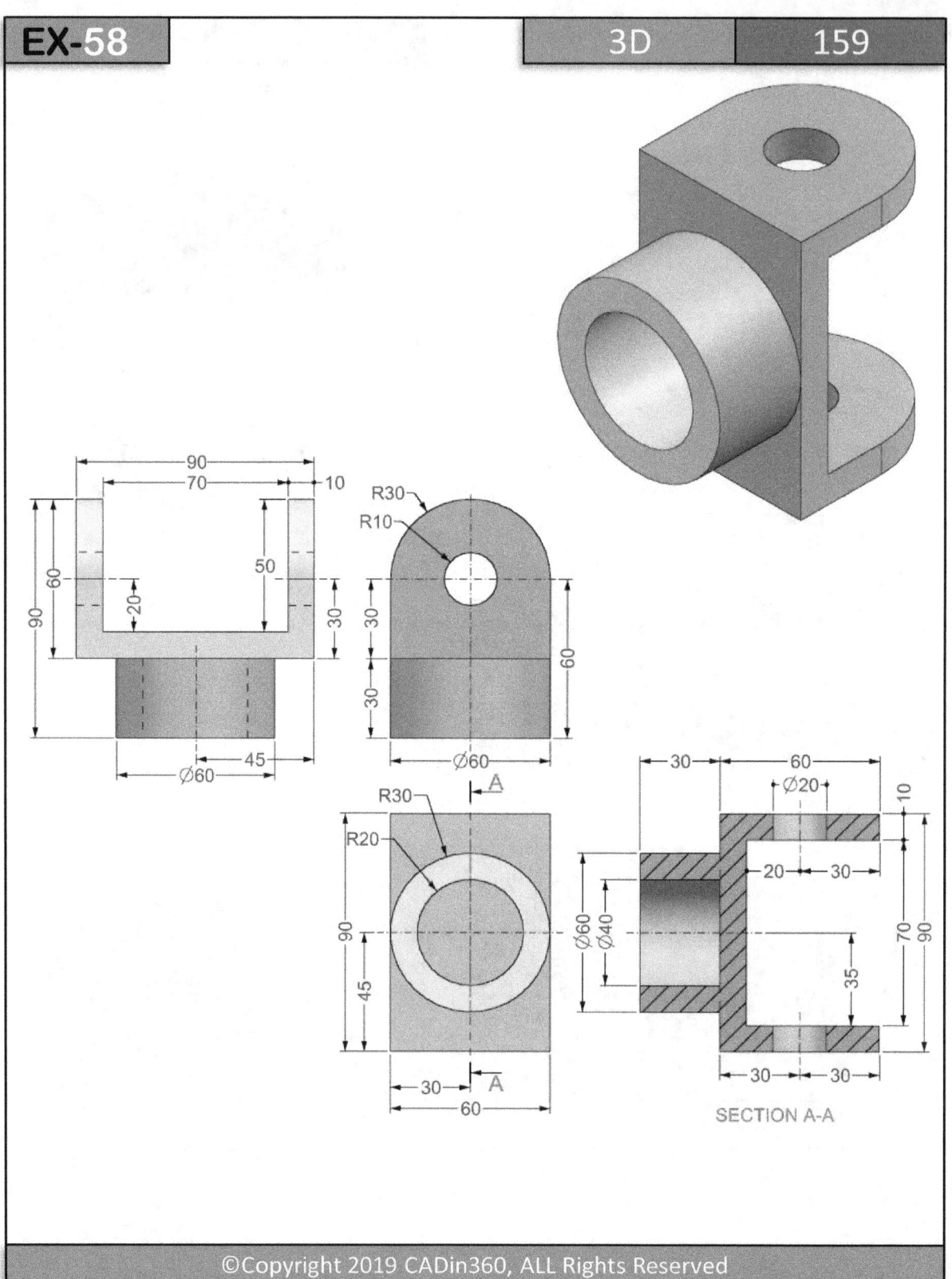

90
70
10
60
50
90
20
30
45
Ø60

R30
R10
30
60
30
Ø60
A

R30
R20
90
45
30
A
60

30
60
Ø20
10
20
30
Ø60
Ø40
70
90
35
30
30
SECTION A-A

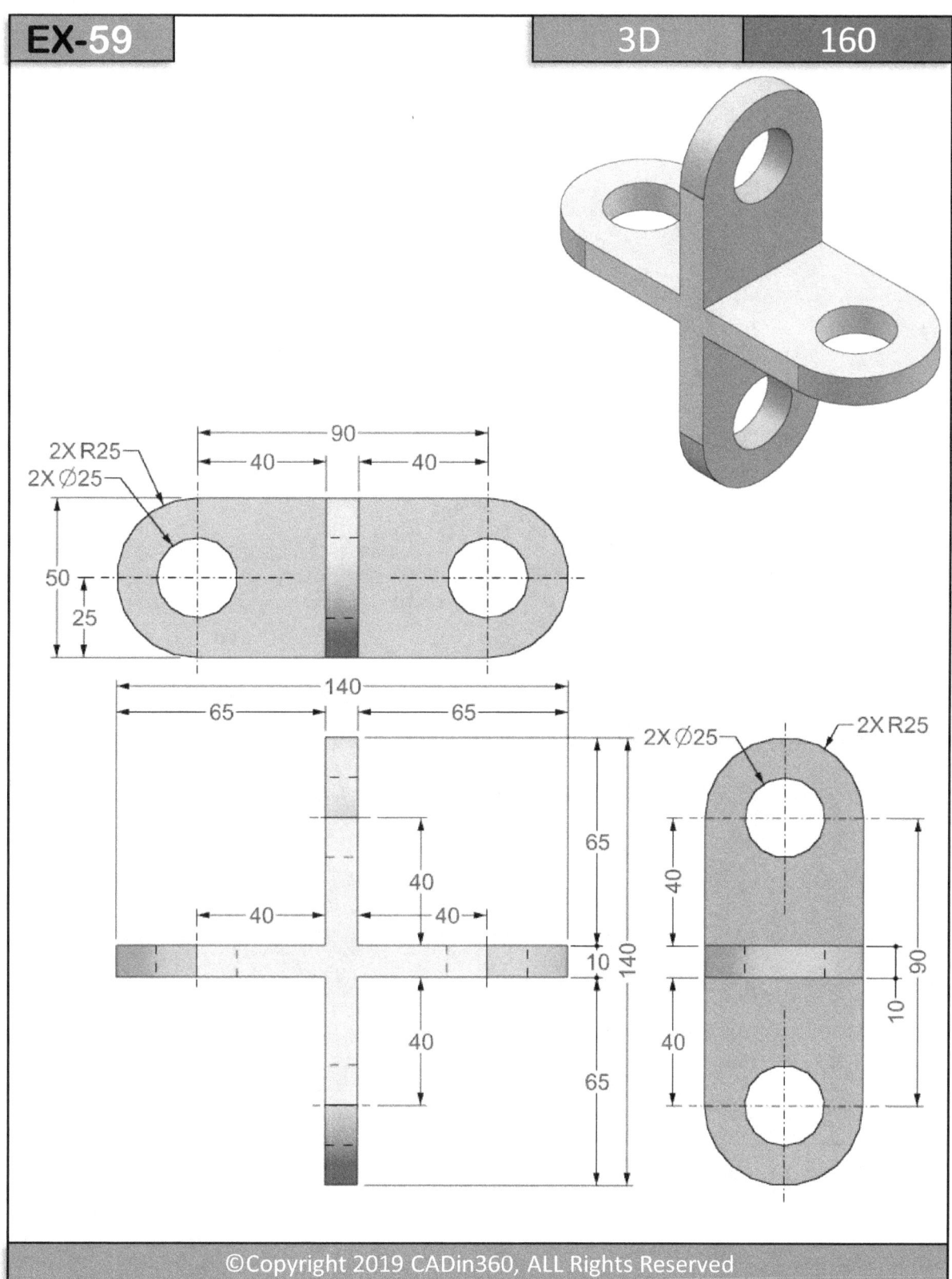

2X R25
2X Ø25

90
40
40
50
25

140
65
65
65
40
40
40
10
140
65

2X Ø25
2X R25
40
90
10
40

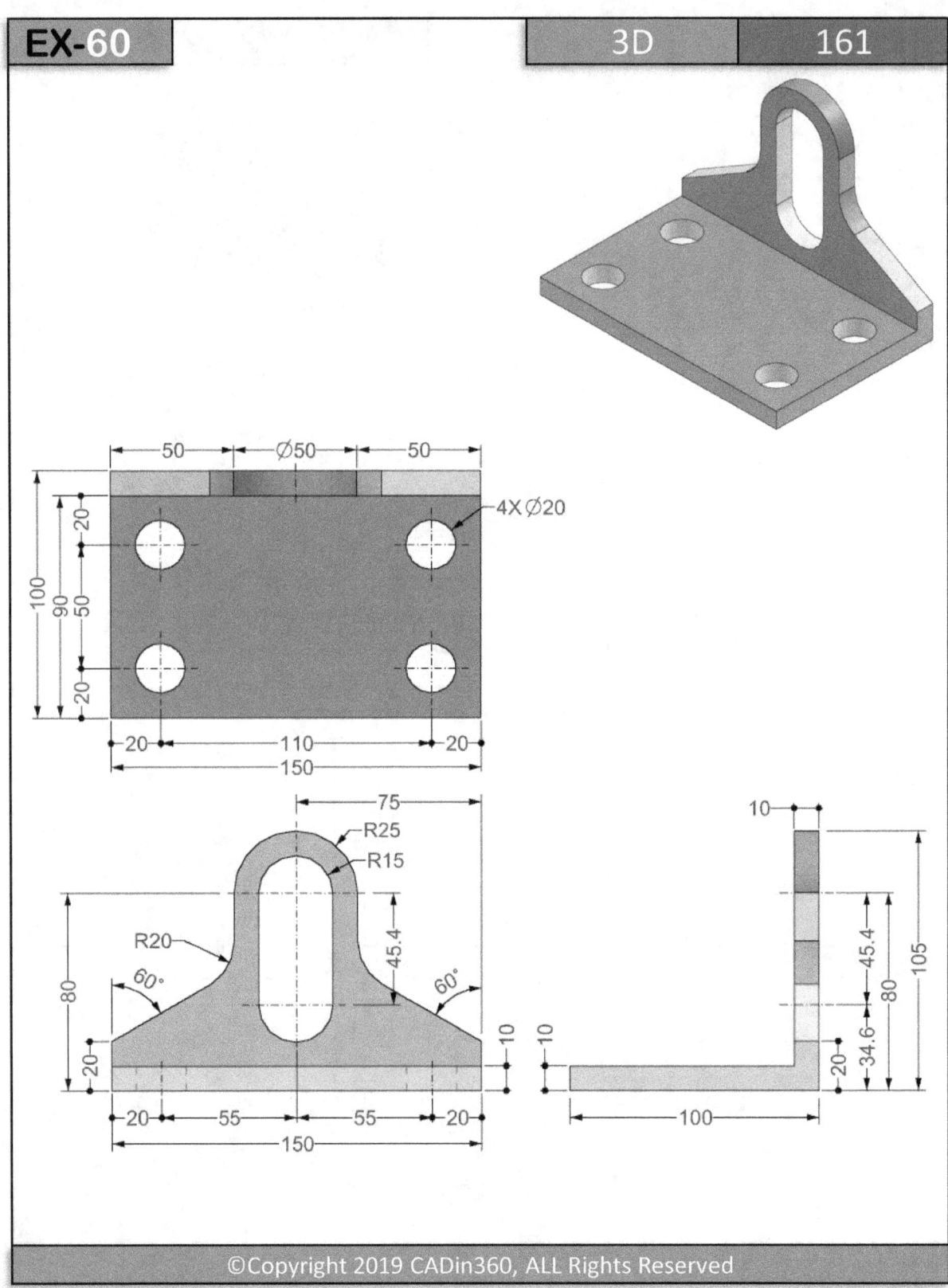

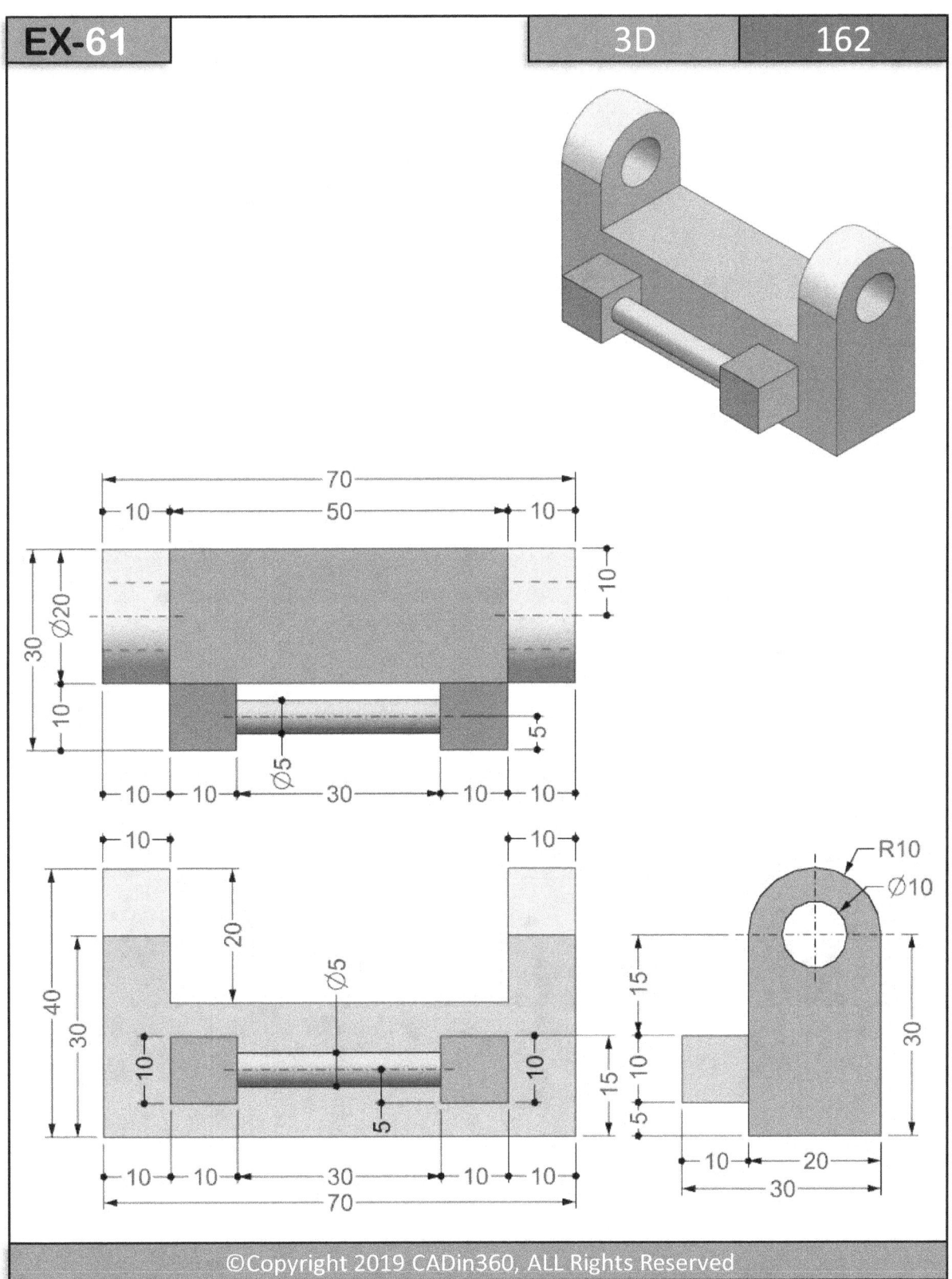

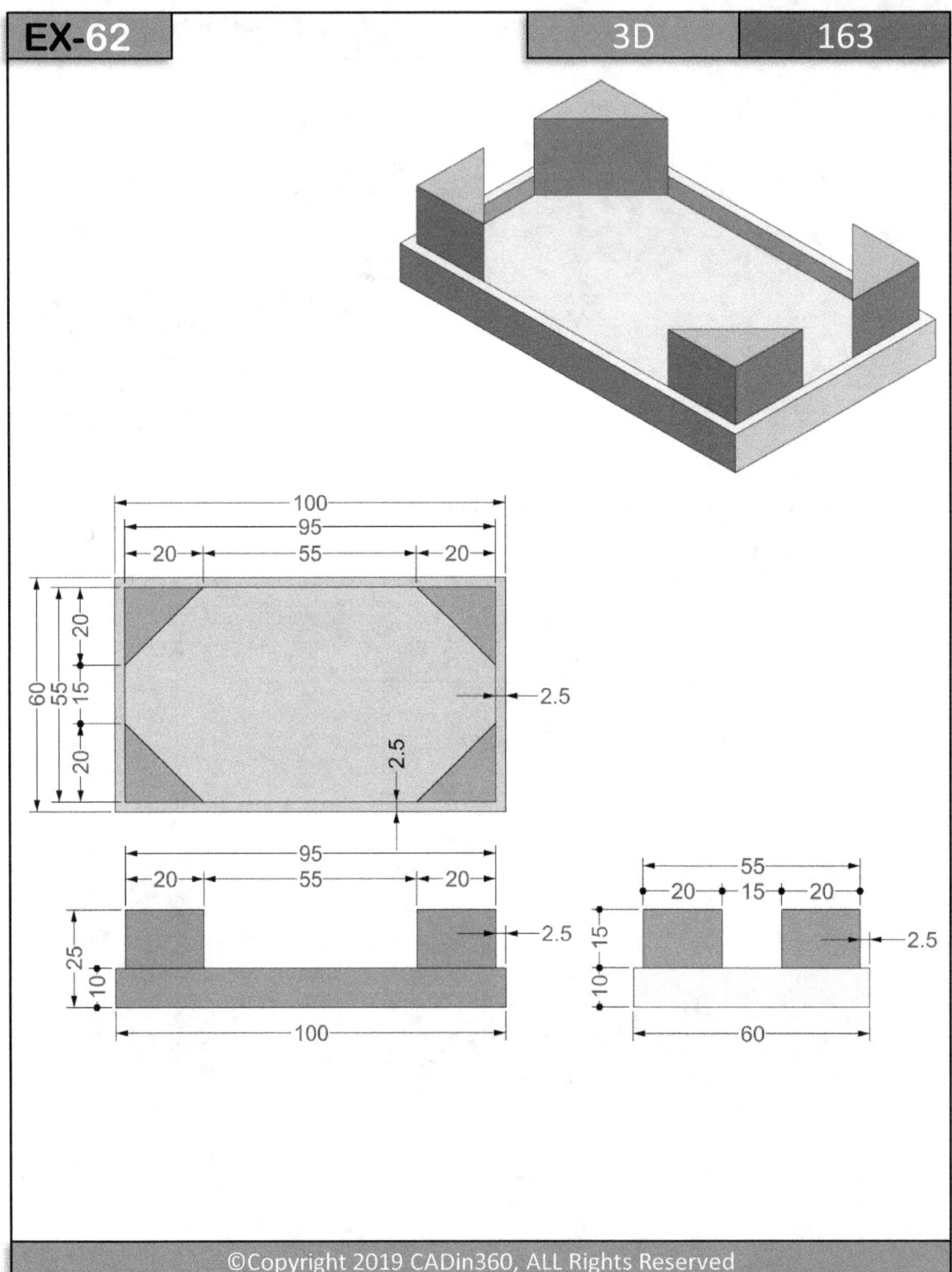

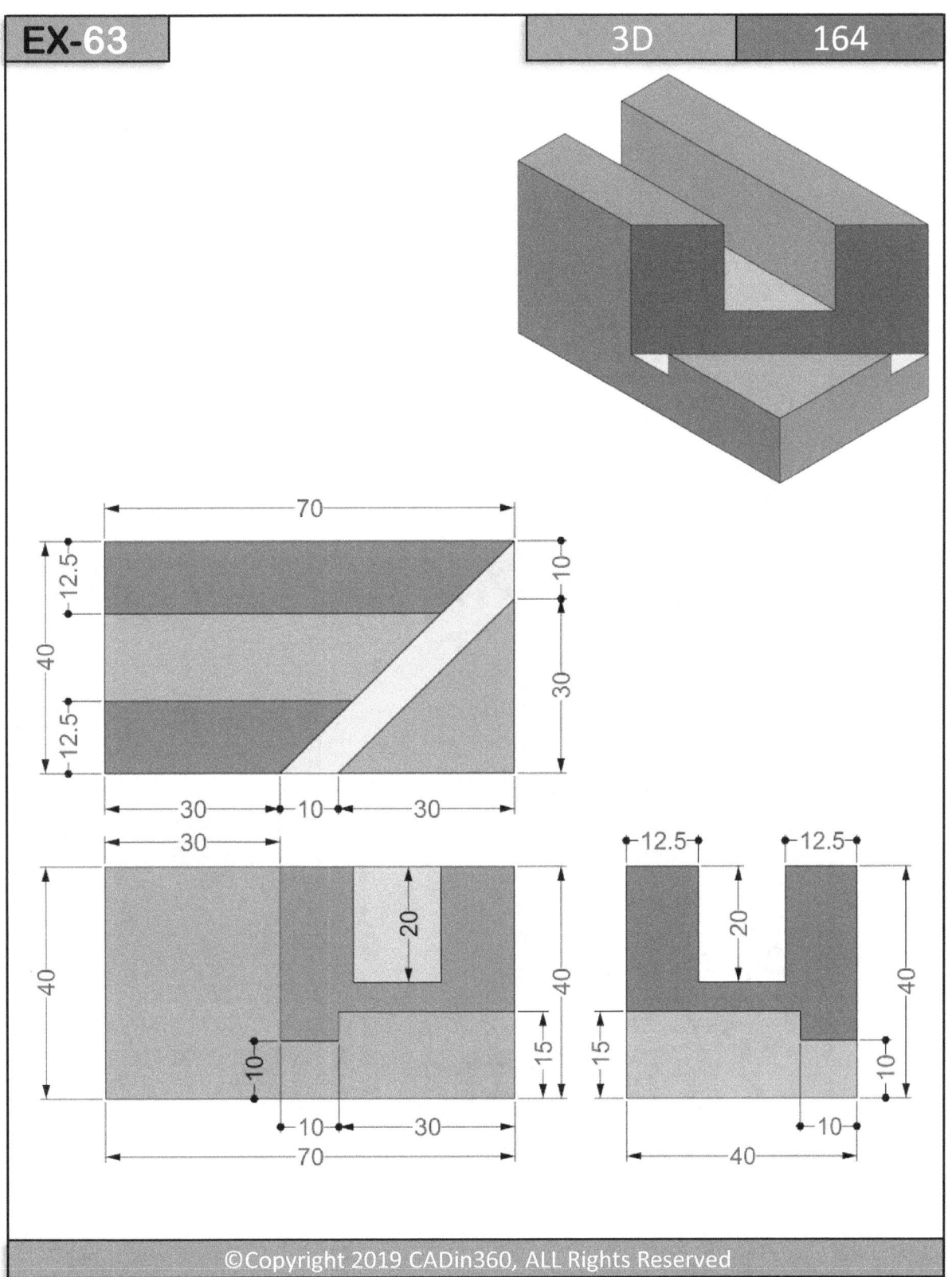

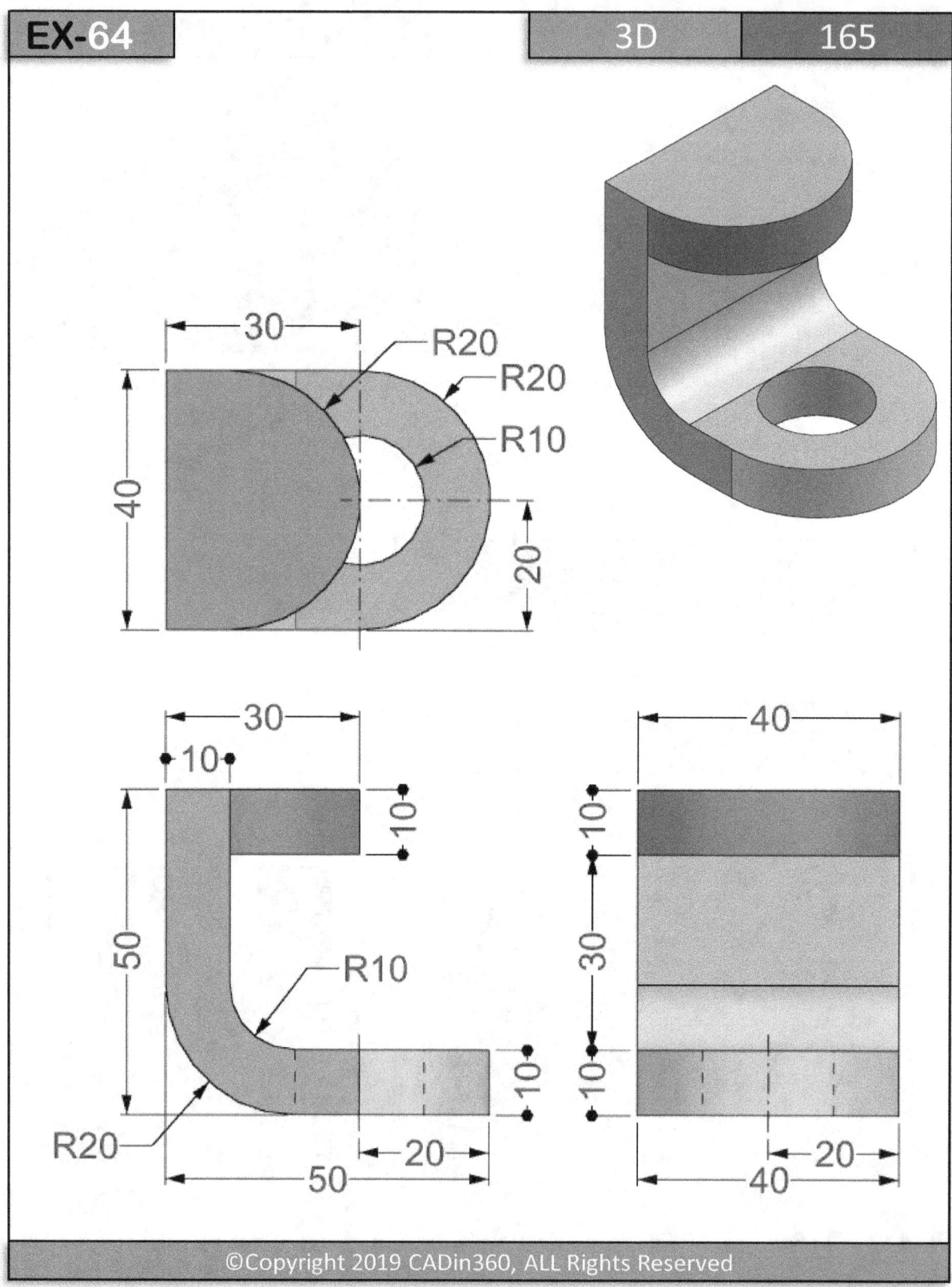

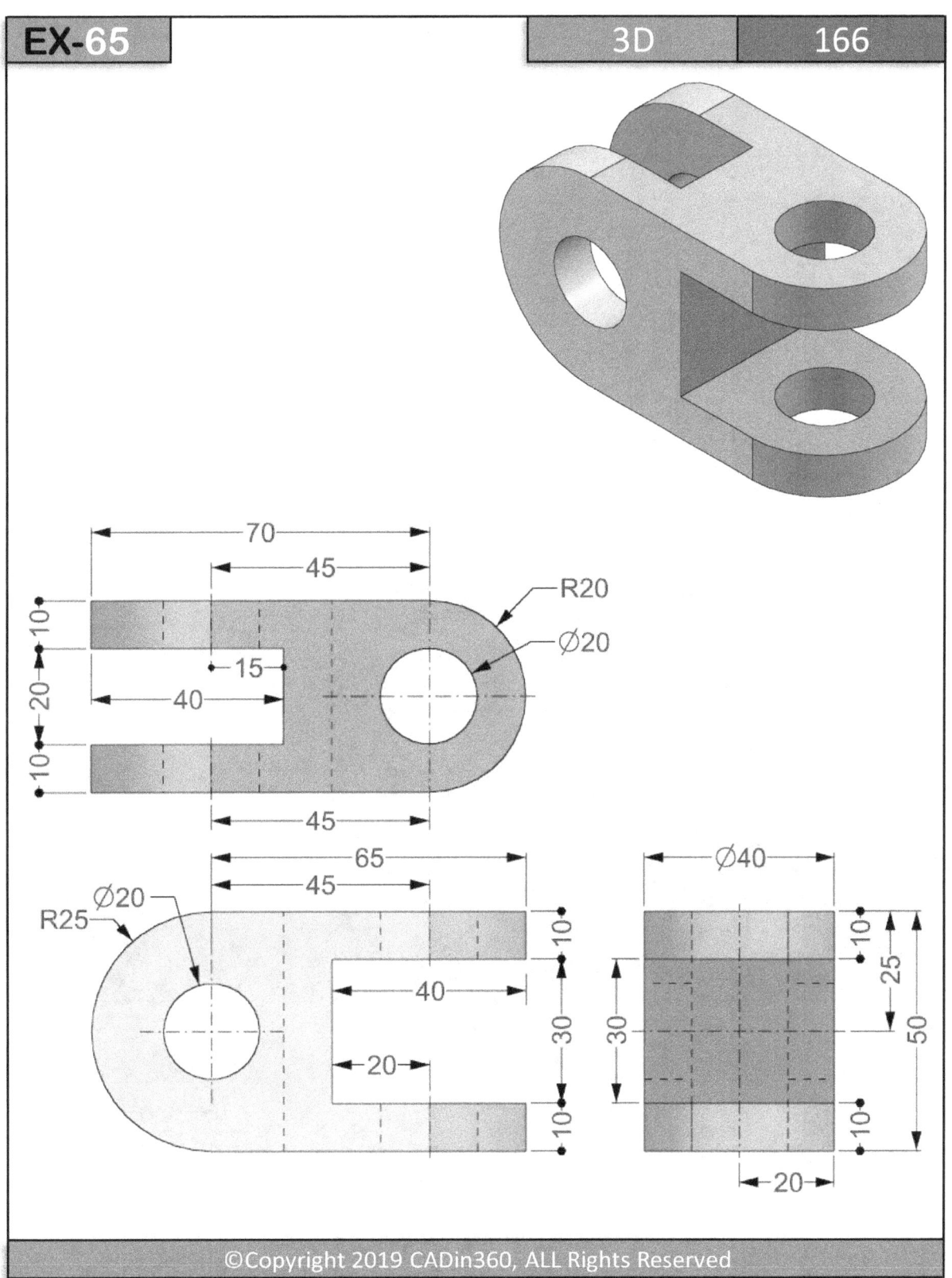

R20
Ø20

70
45
15
40
45
10
20
10

R25
Ø20
65
45
40
20
10
30
10

Ø40
10
25
50
30
10
20

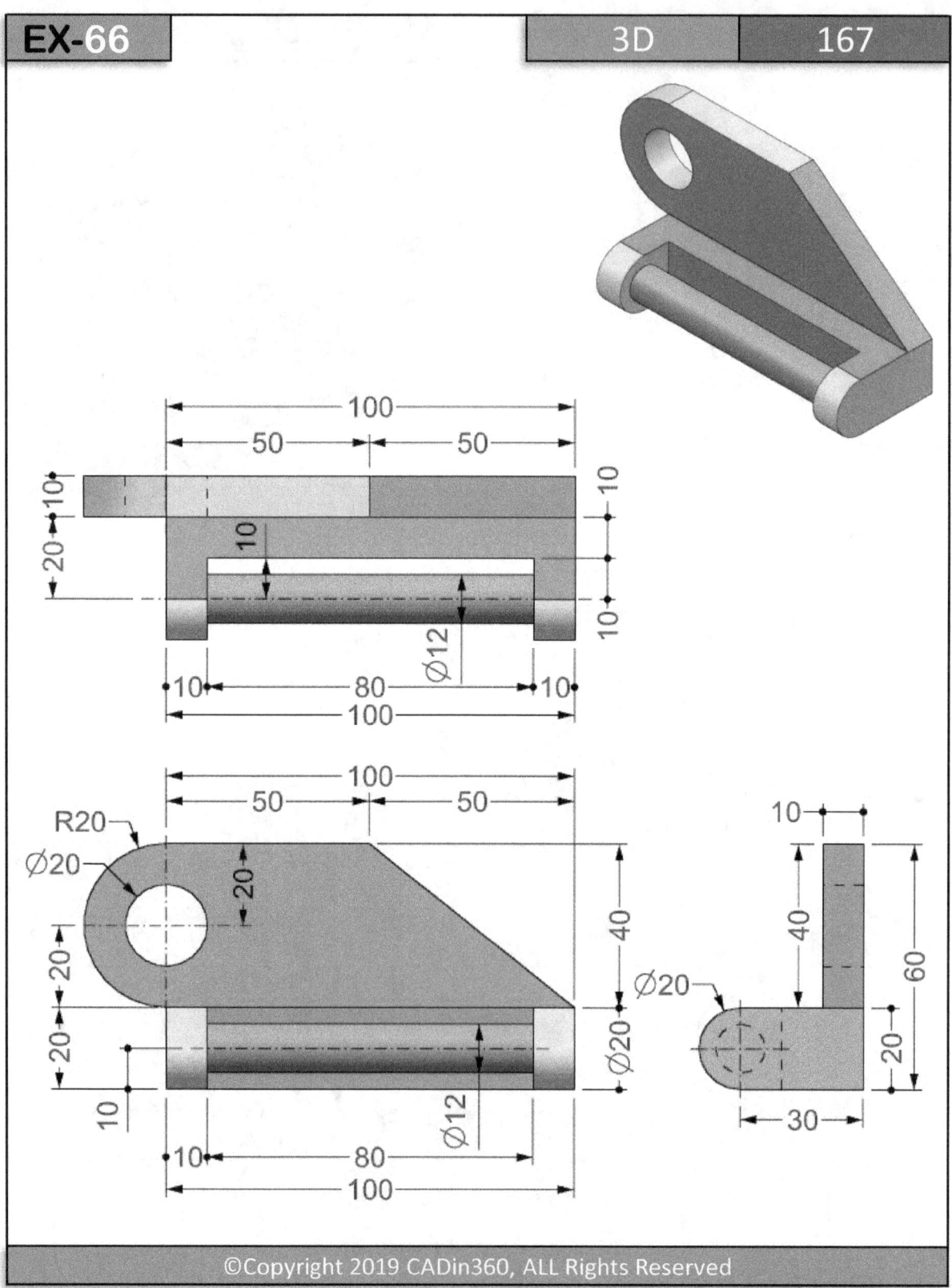

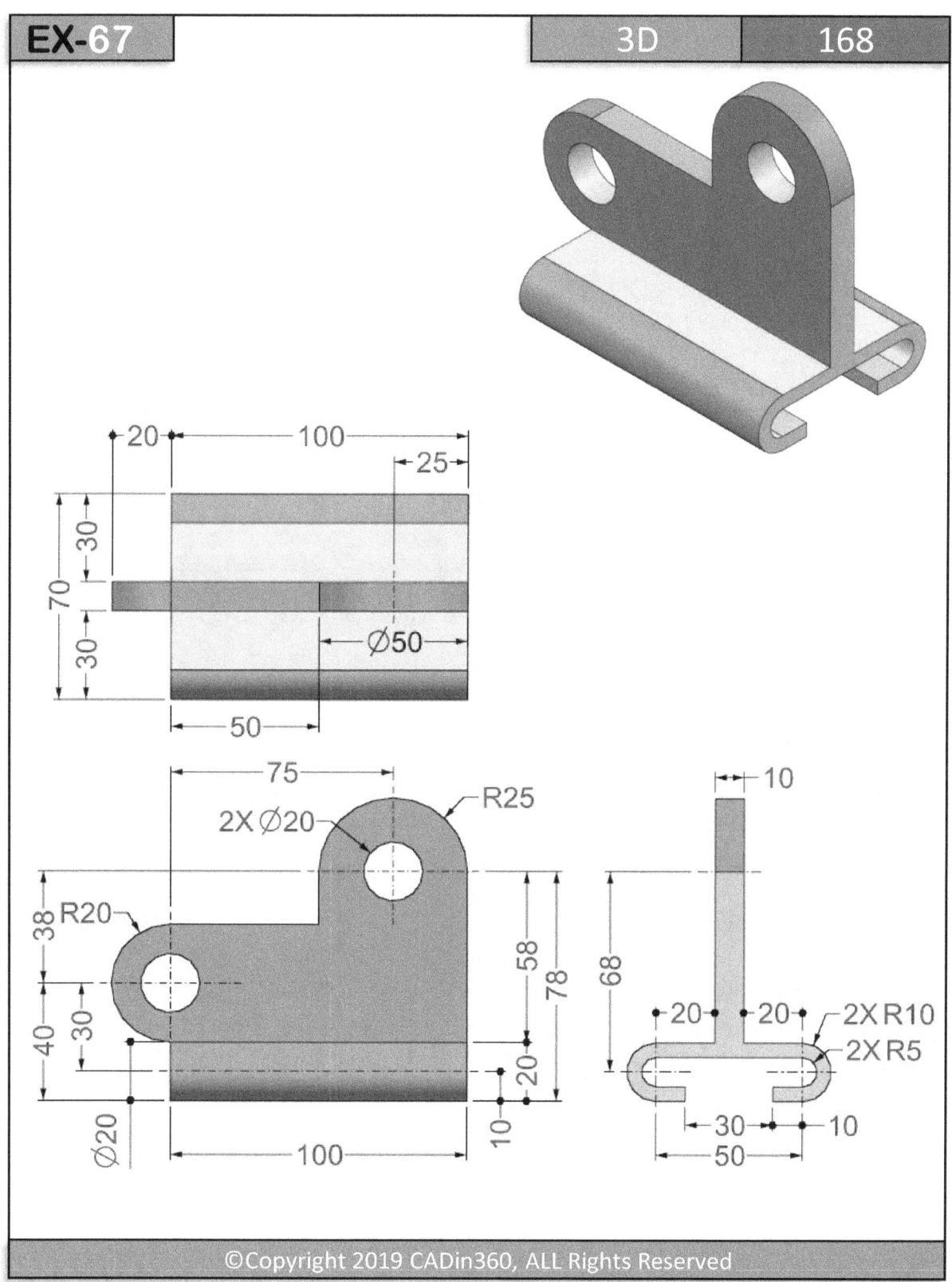

20 100 25
30
70
30
Ø50
50

75 R25
2X Ø20
R20
38
58
78
30
40
20
Ø20 10
100

10
68
20 20 2X R10
2X R5
30 10
50

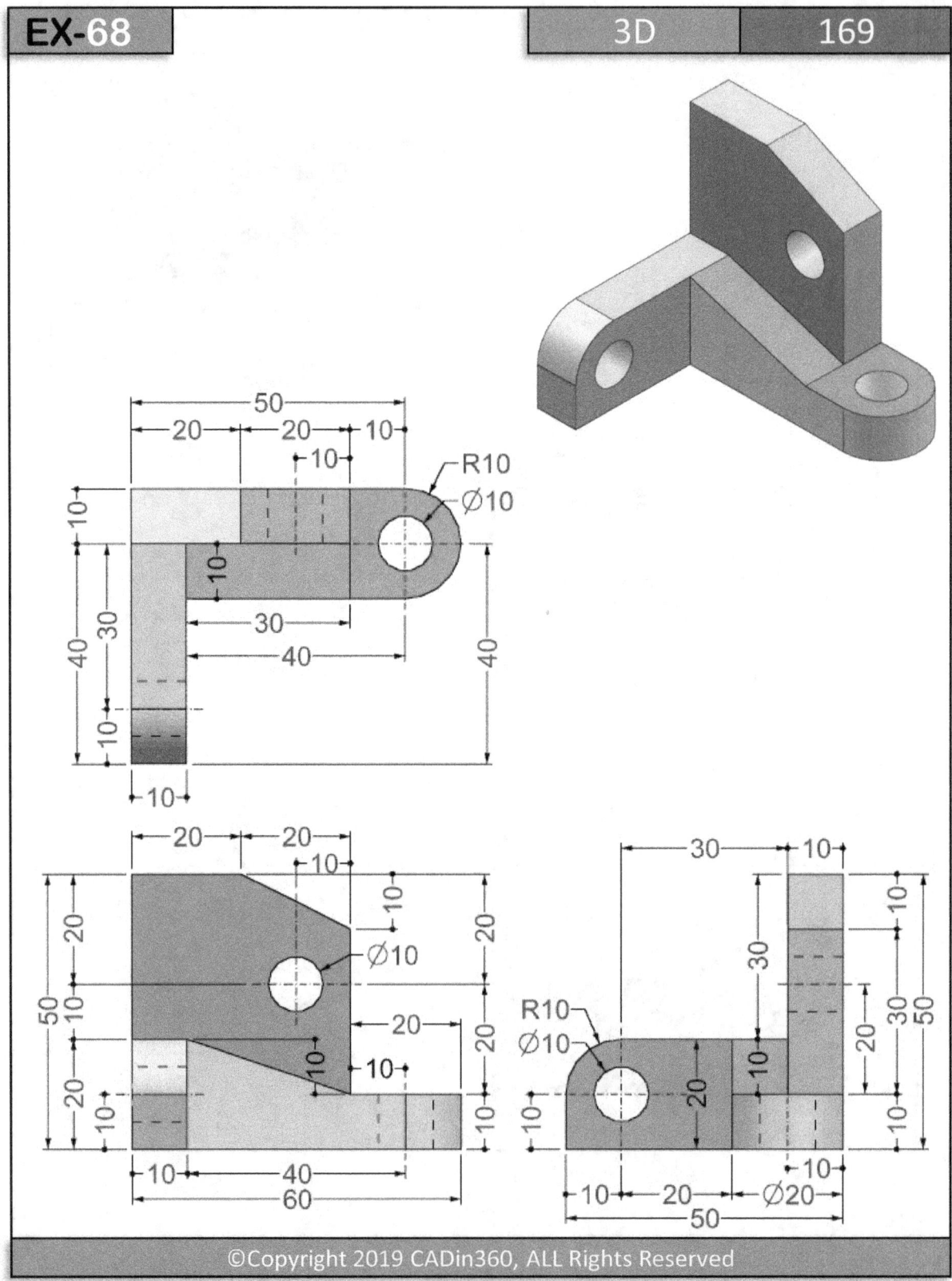

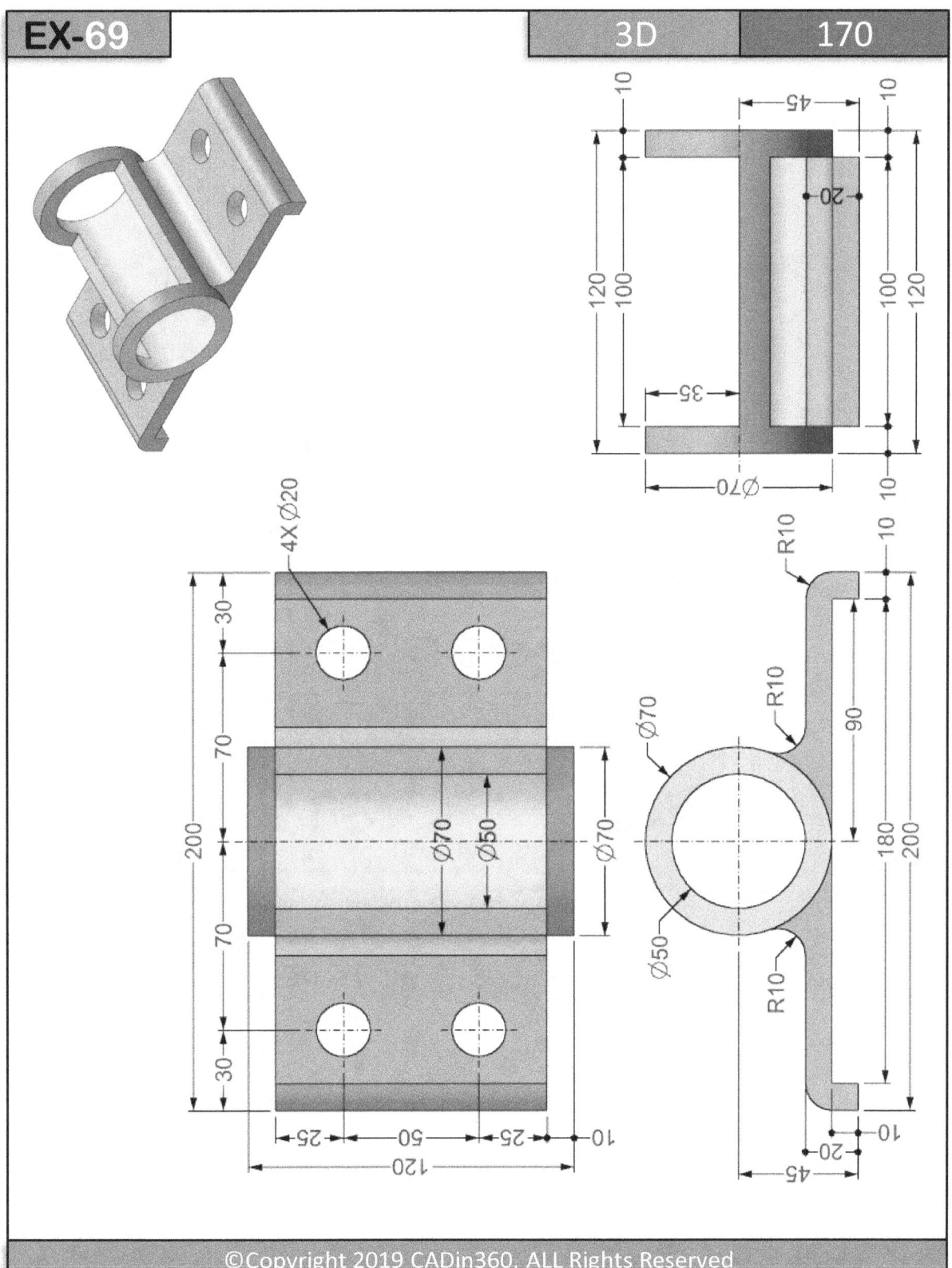

EX-69   3D   170

4X Ø20

30
70
200
70
30

Ø70
Ø50
Ø70

25
50
25
10
120

10
45
120
100
35
20
Ø70
10

R10
R10
R10
Ø70
Ø50
90
180
200
10
20
45

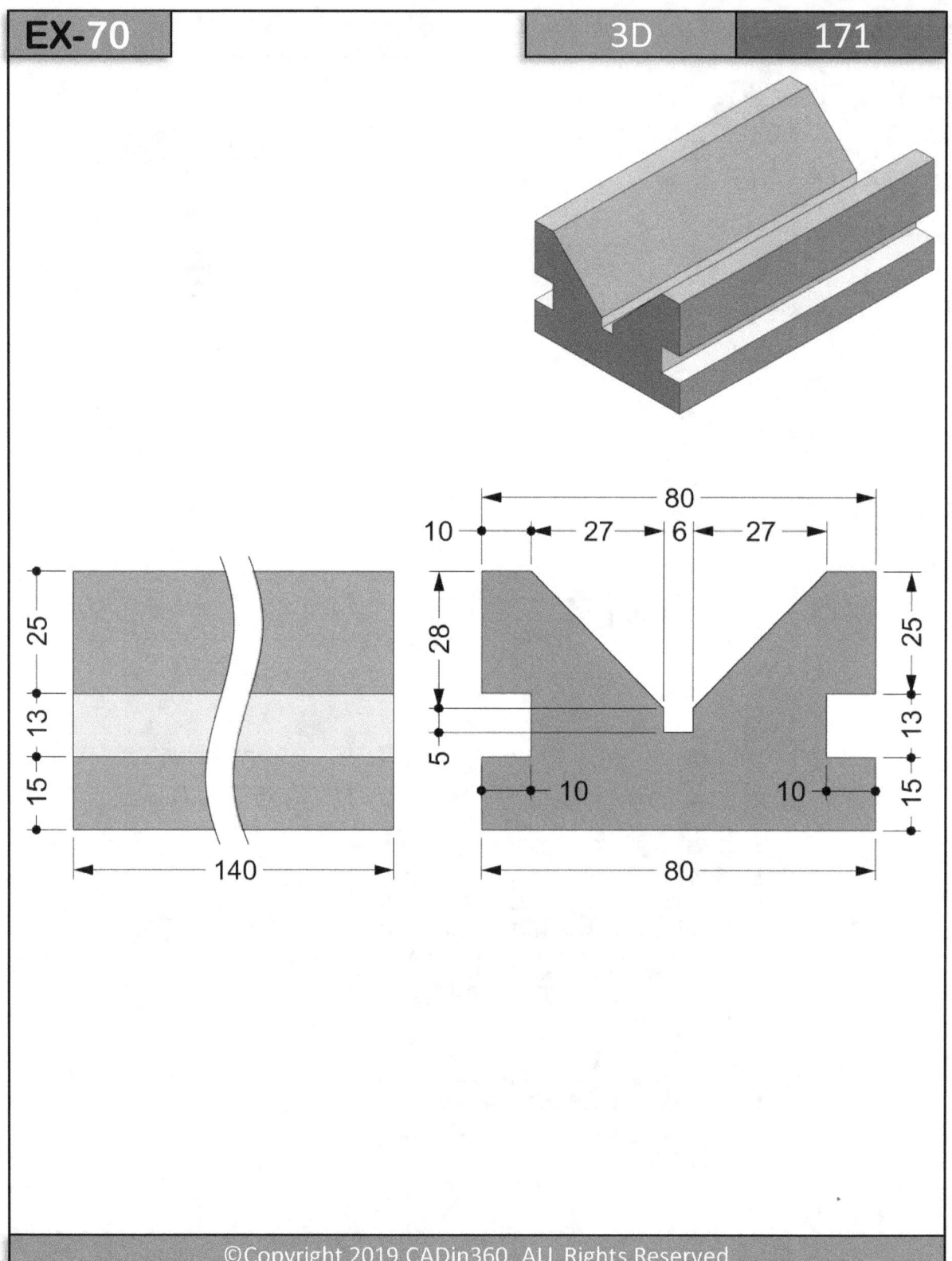

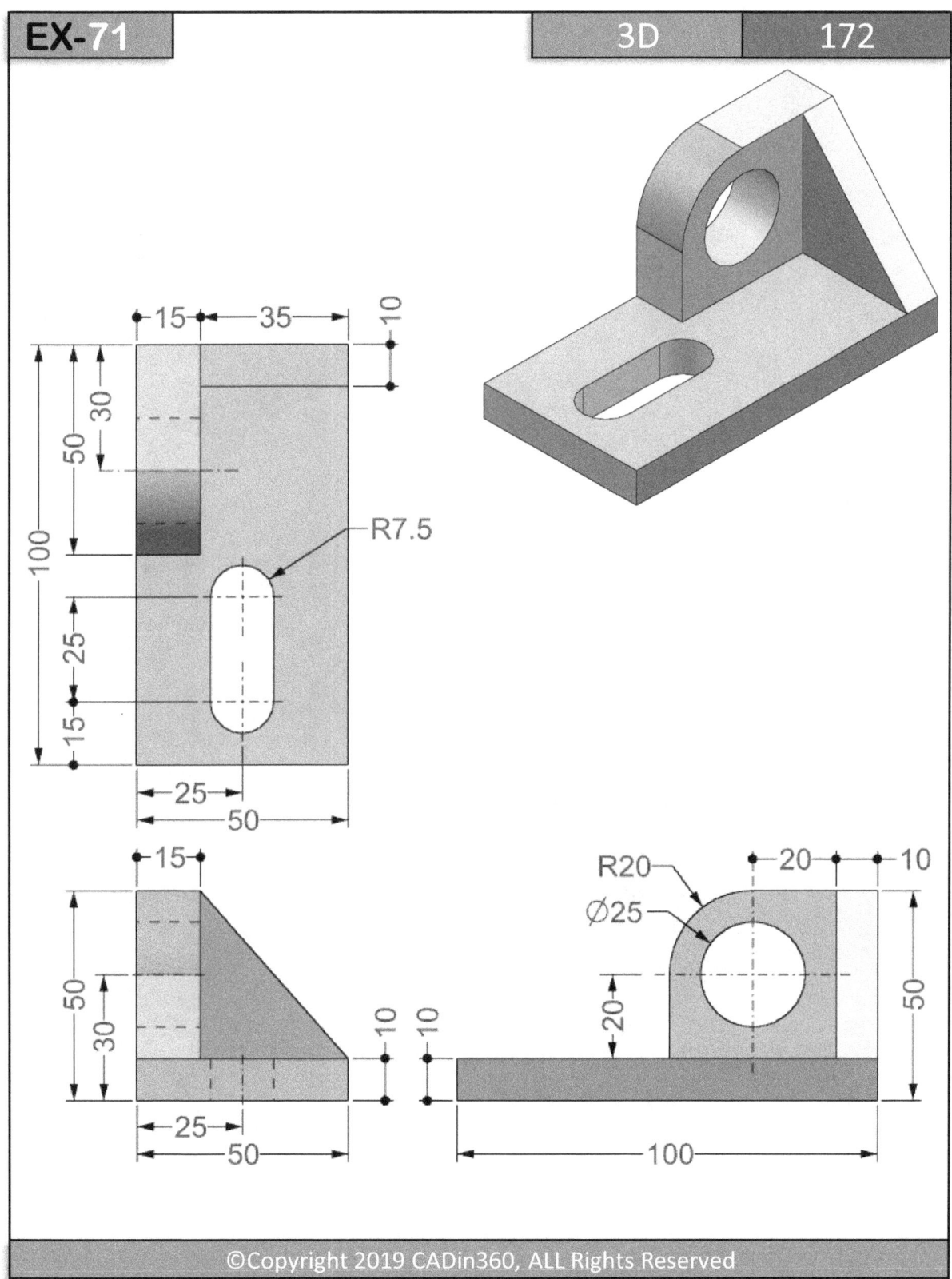

15 — 35 — 10

30

50

100

R7.5

25

15

25

50

15

50

30

10 10

25

50

R20

Ø25

20 — 20 — 10

50

100

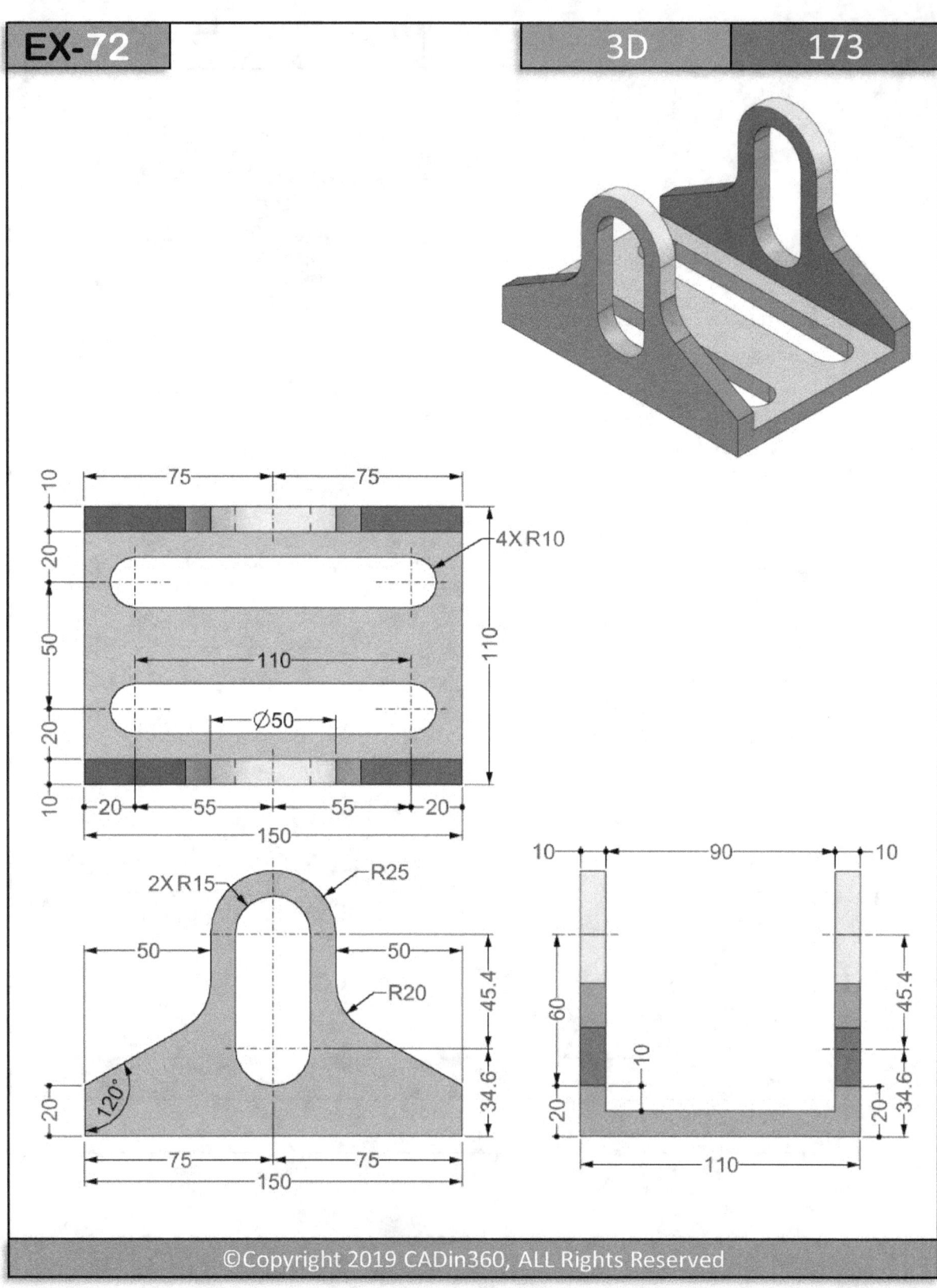

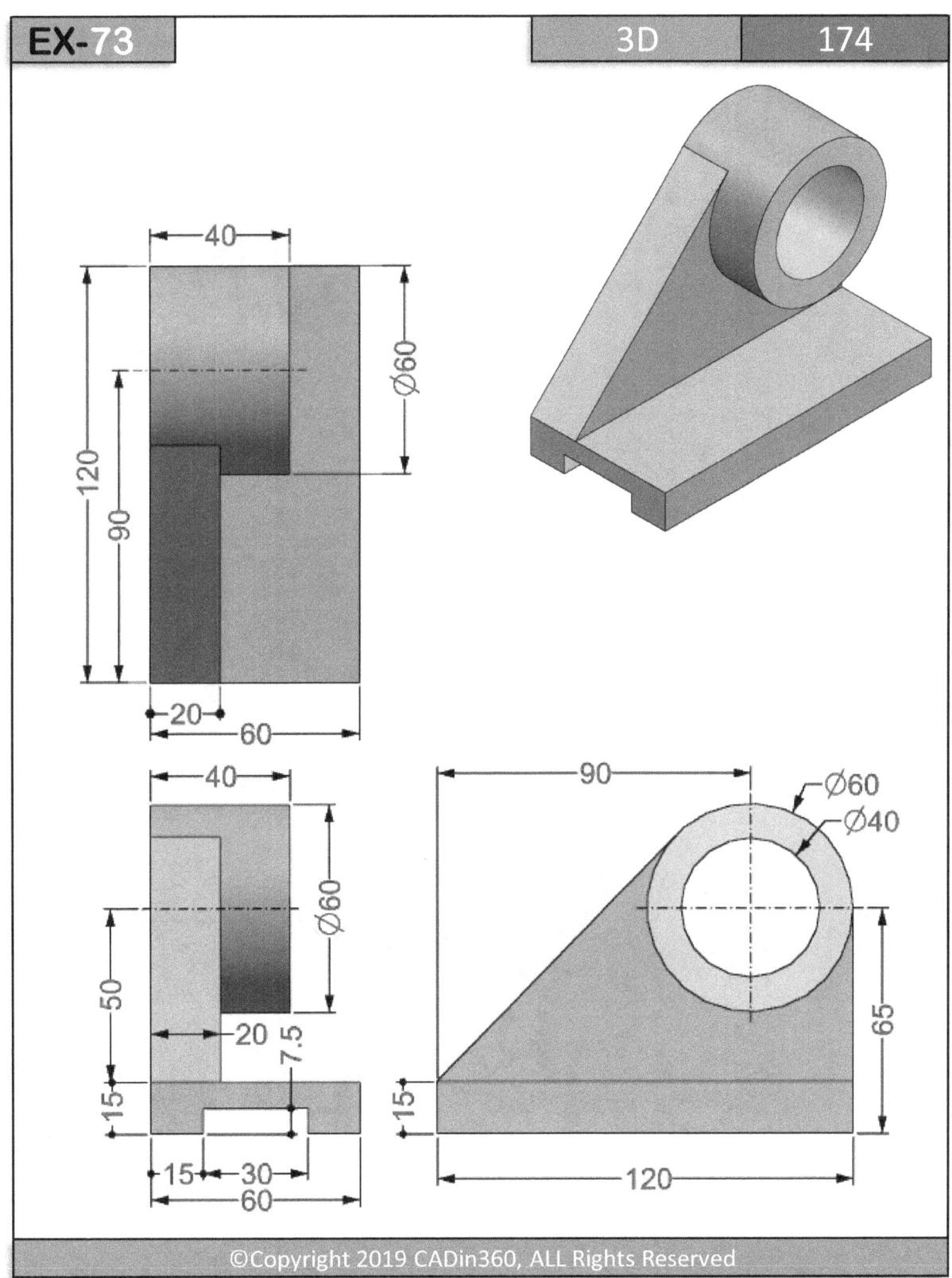

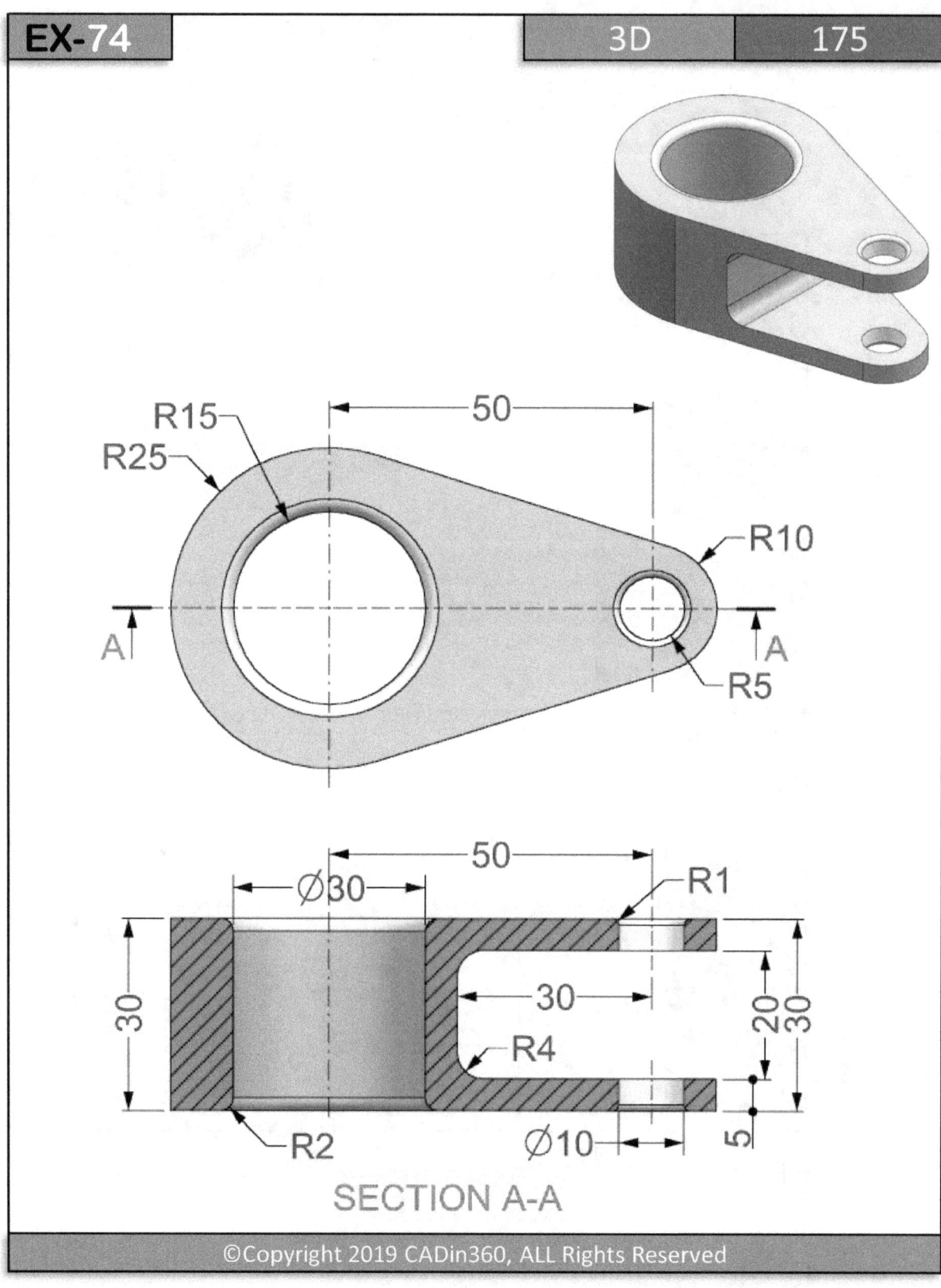

EX-74   3D   175

R15
R25
50
R10
A
R5
A

50
Ø30
R1
30
30
20
30
R4
R2
Ø10
5

SECTION A-A

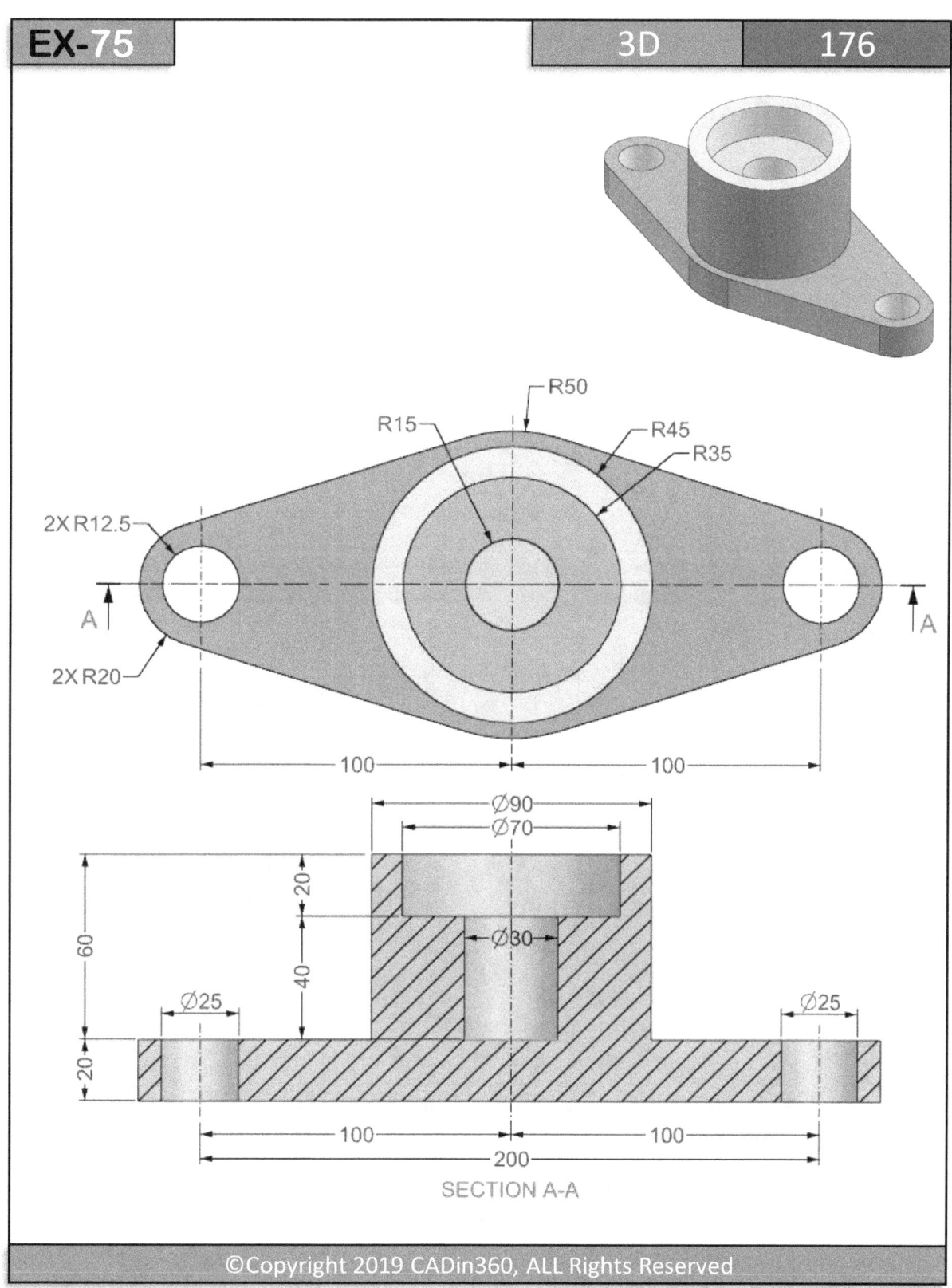

R50

R15

R45

R35

2X R12.5

2X R20

100

100

Ø90

Ø70

20

Ø30

60

40

Ø25

Ø25

20

100

100

200

SECTION A-A

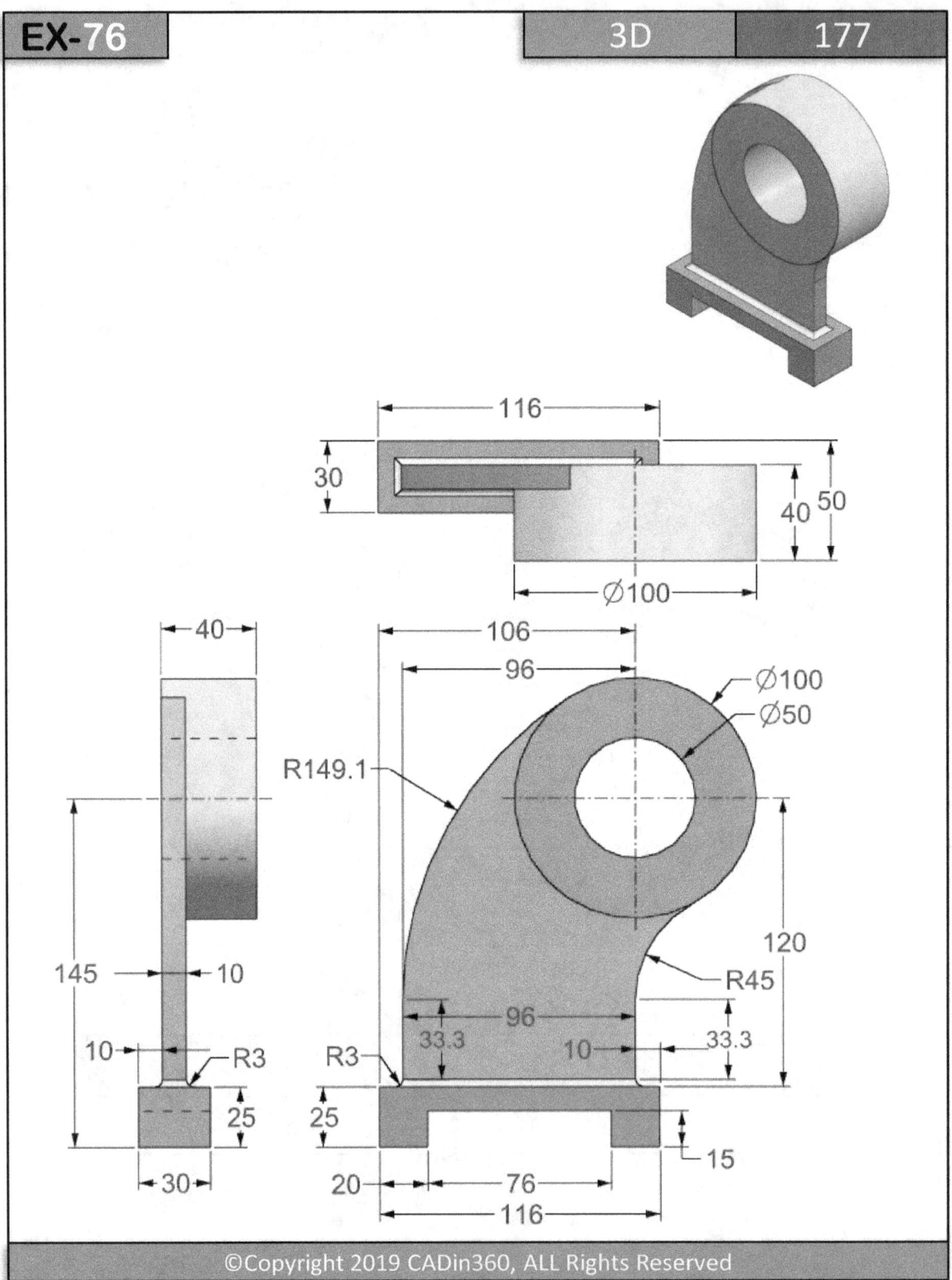

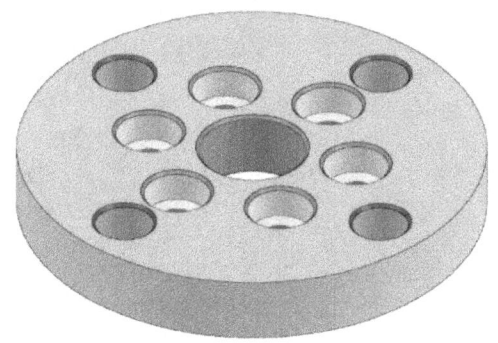

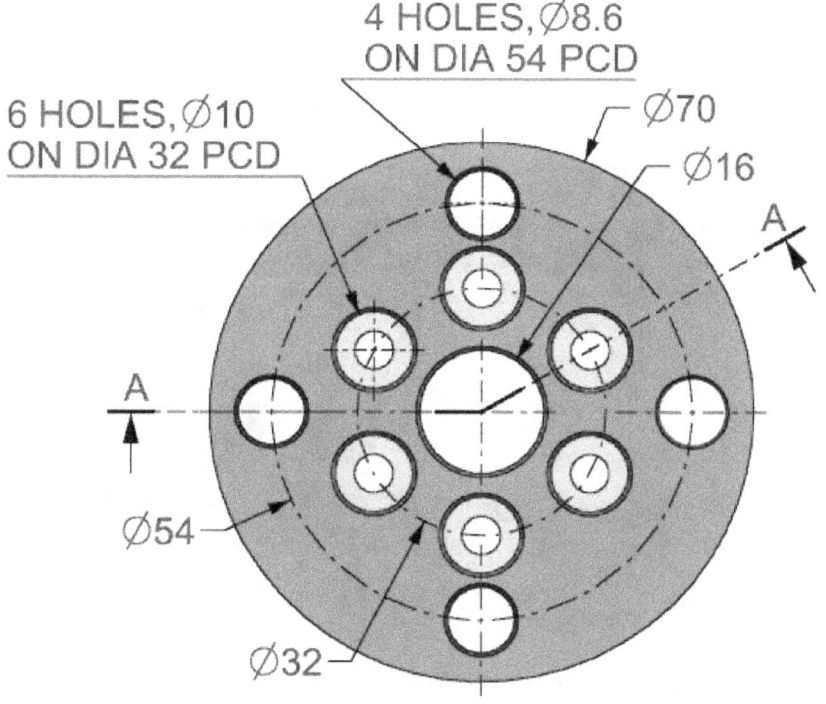

4 HOLES, ∅8.6
ON DIA 54 PCD

6 HOLES, ∅10
ON DIA 32 PCD

∅70

∅16

A

A

∅54

∅32

CHAMFER 0.5 X 45°

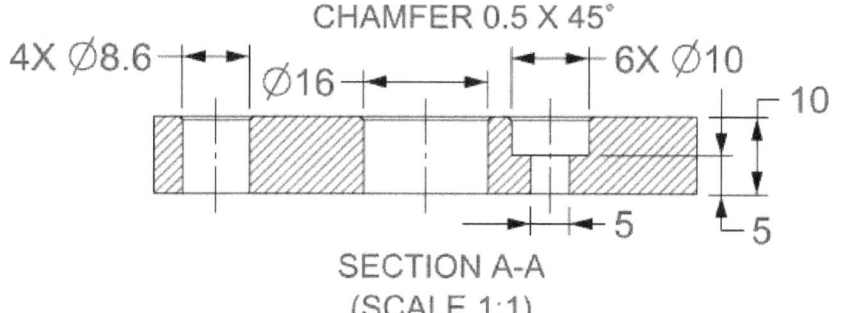

4X ∅8.6

∅16

6X ∅10

10

5

5

SECTION A-A
(SCALE 1:1)

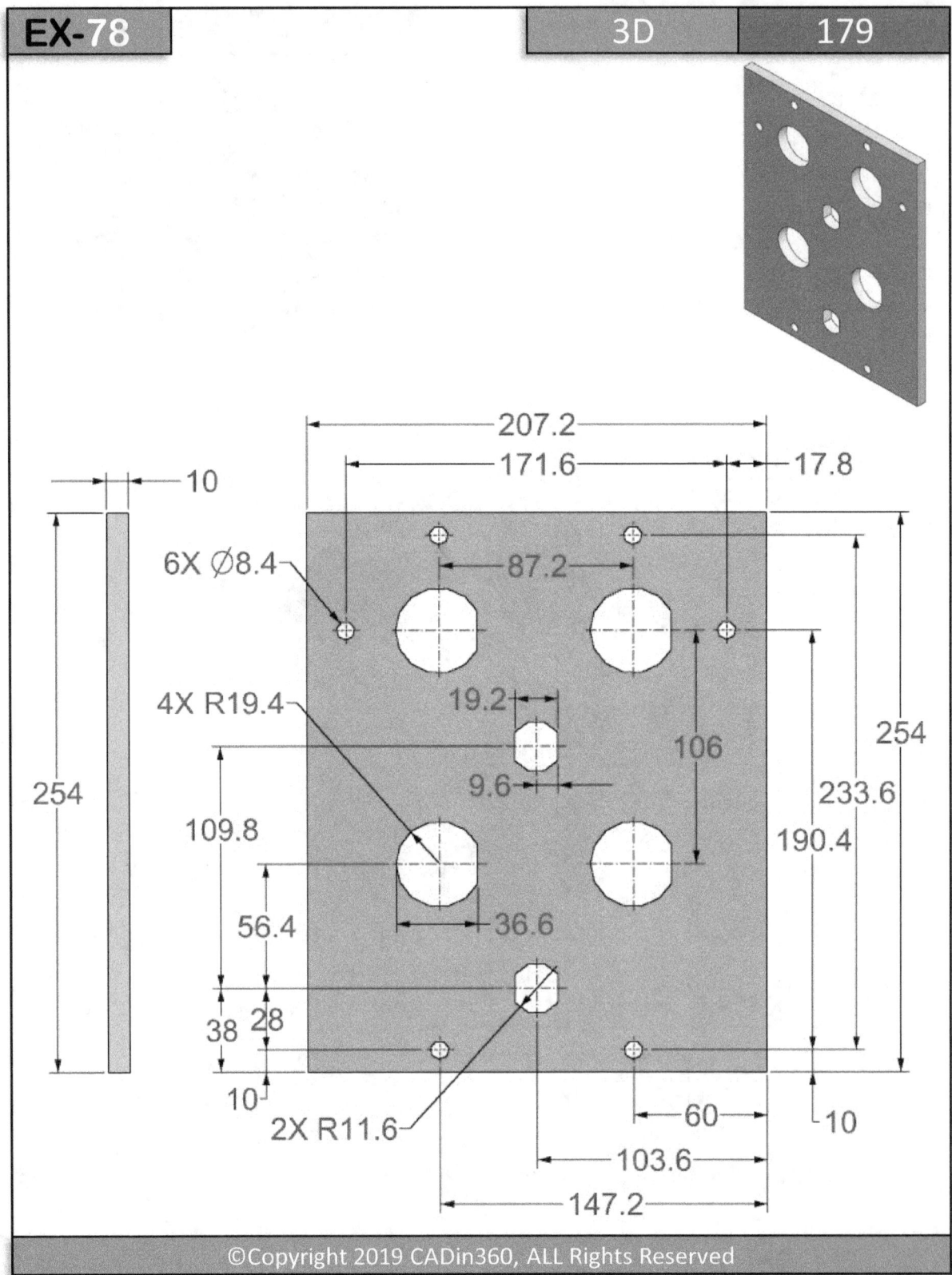

6X Ø8.4

207.2
171.6
17.8
10
87.2
4X R19.4
19.2
9.6
106
254
254
233.6
109.8
190.4
56.4
36.6
38 28
2X R11.6
10
60
10
103.6
147.2

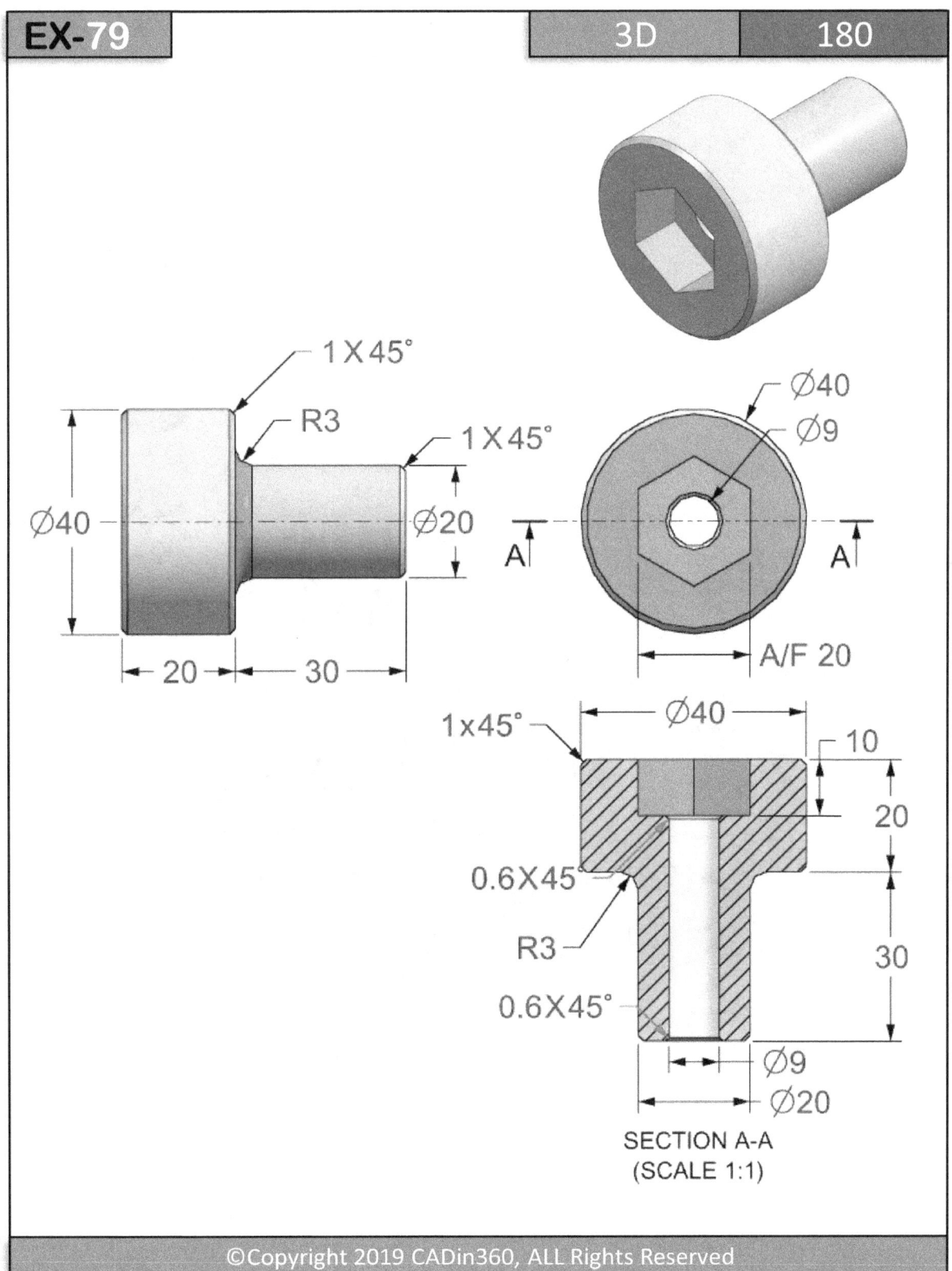

1 X 45°

R3

1 X 45°

Ø40

Ø20

20

30

Ø40
Ø9

A

A

A/F 20

1x45°

Ø40

10

20

0.6X45°

R3

30

0.6X45°

Ø9

Ø20

SECTION A-A
(SCALE 1:1)

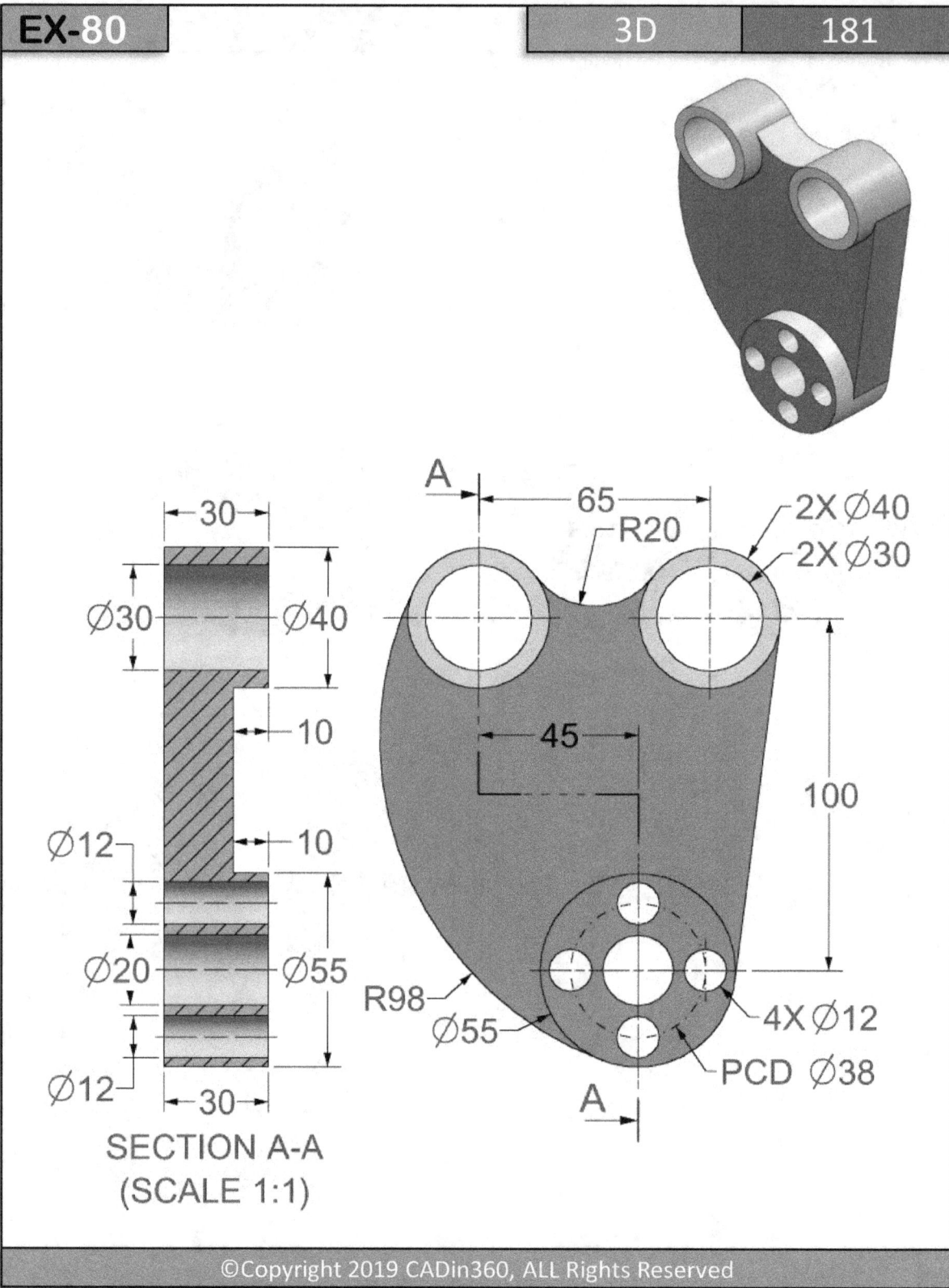

30

Ø30 Ø40

10

10

Ø12

Ø20 Ø55

Ø12

30

SECTION A-A
(SCALE 1:1)

A

65

R20

2X Ø40
2X Ø30

45

100

R98

Ø55

4X Ø12

PCD Ø38

A

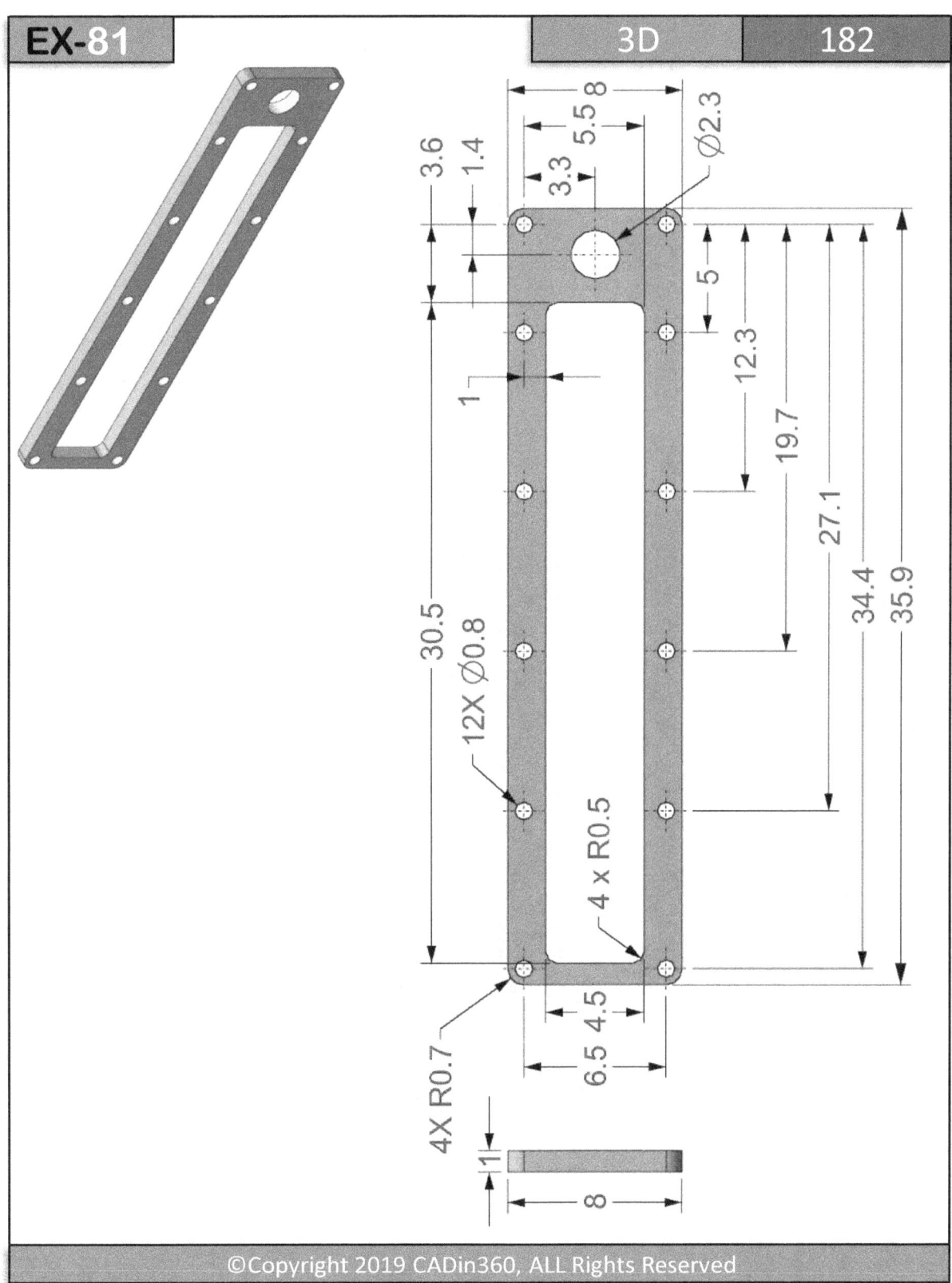

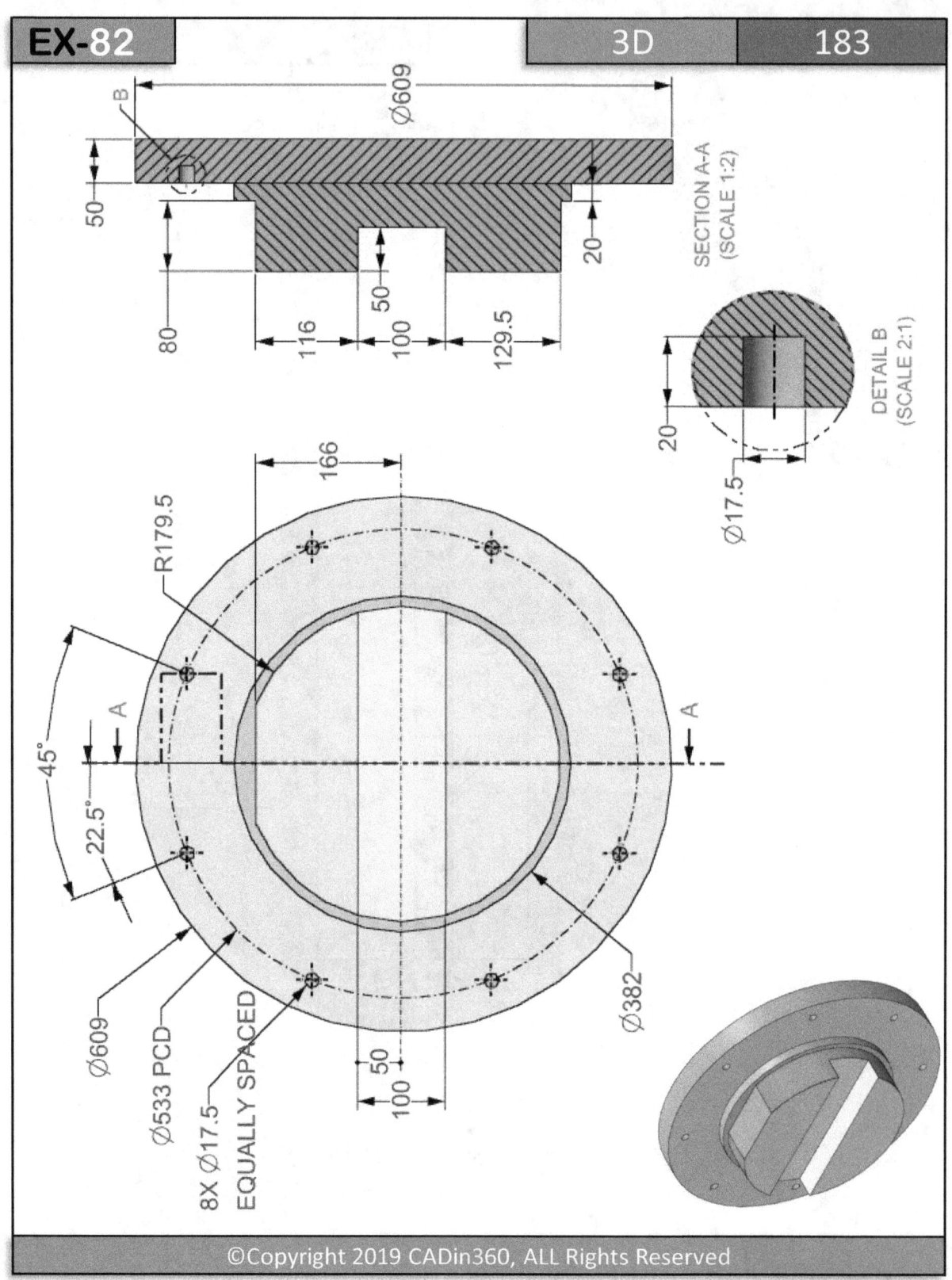

Ø609

50

80

116

50

100

129.5

20

B

SECTION A-A
(SCALE 1:2)

20

Ø17.5

DETAIL B
(SCALE 2:1)

166

R179.5

45°

22.5°

A

A

Ø609

Ø533 PCD

8X Ø17.5

EQUALLY SPACED

Ø382

50

100

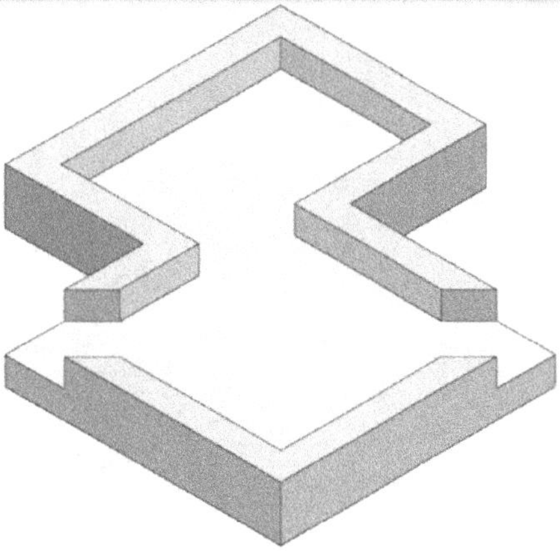

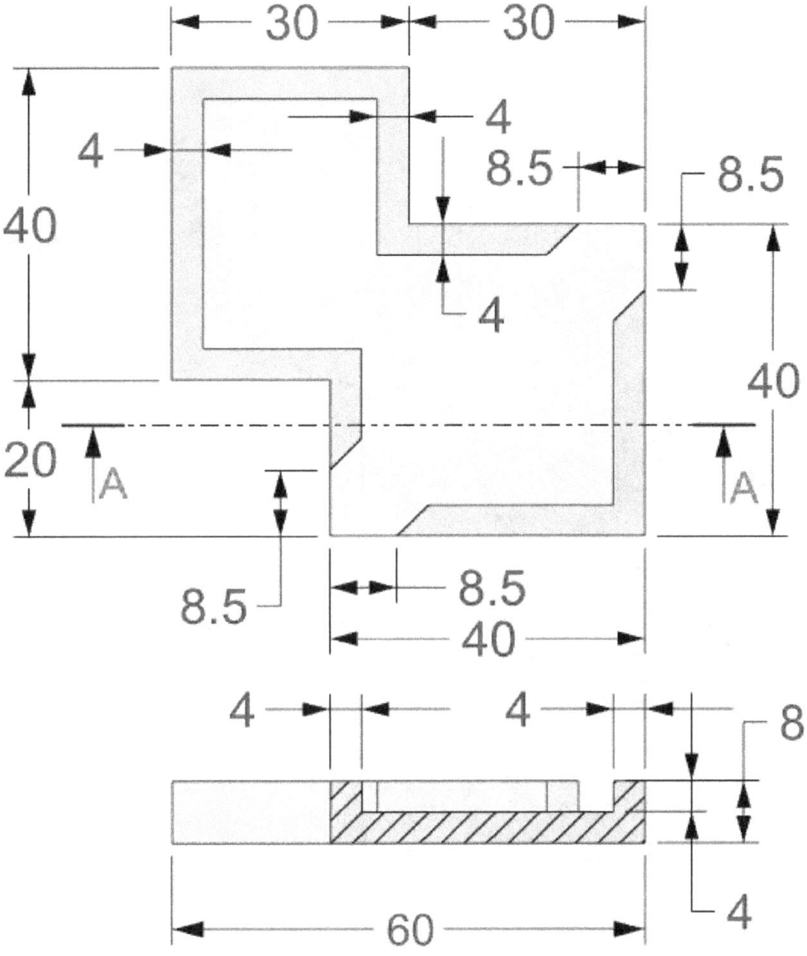

30 30

4

4

8.5 8.5

4

40

40

4

20 A A

8.5 8.5

8.5

40

4 4 8

60

SECTION A-A
(SCALE 1:1)

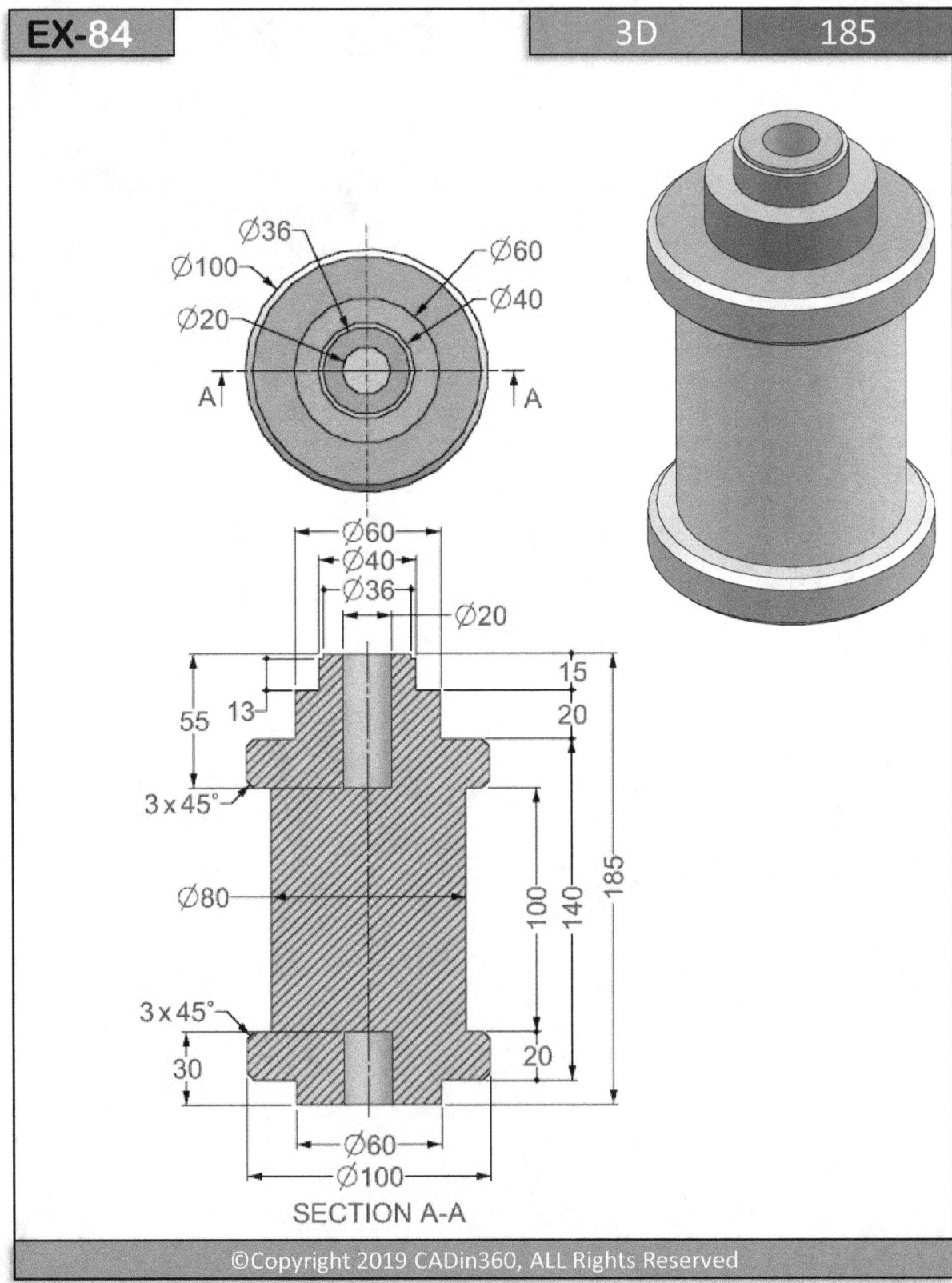

Ø36
Ø100
Ø20
Ø60
Ø40
A
A

Ø60
Ø40
Ø36
Ø20
15
13
55
20
3 x 45°
Ø80
100
140
185
3 x 45°
30
20
Ø60
Ø100

SECTION A-A

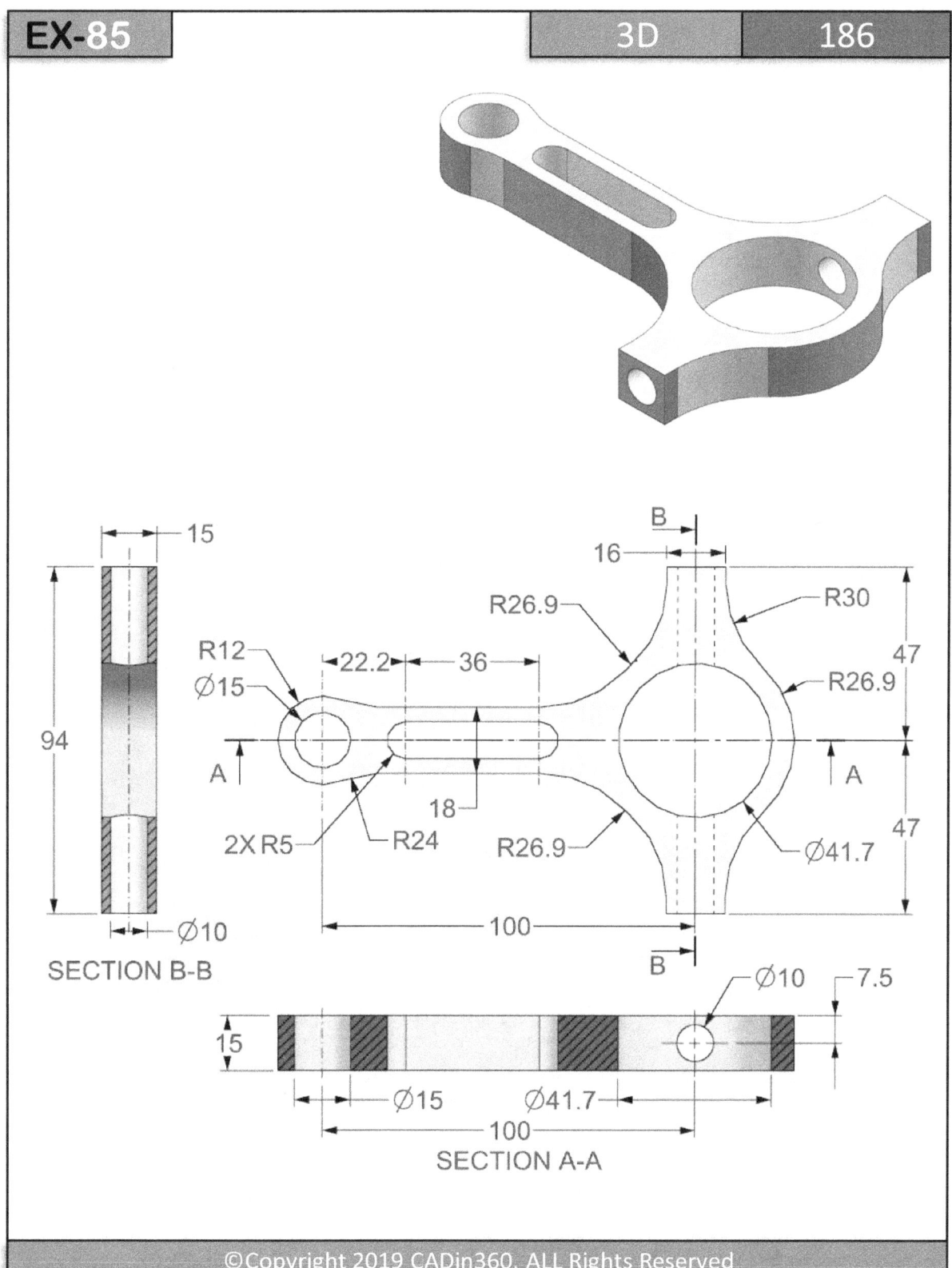

SECTION B-B

SECTION A-A

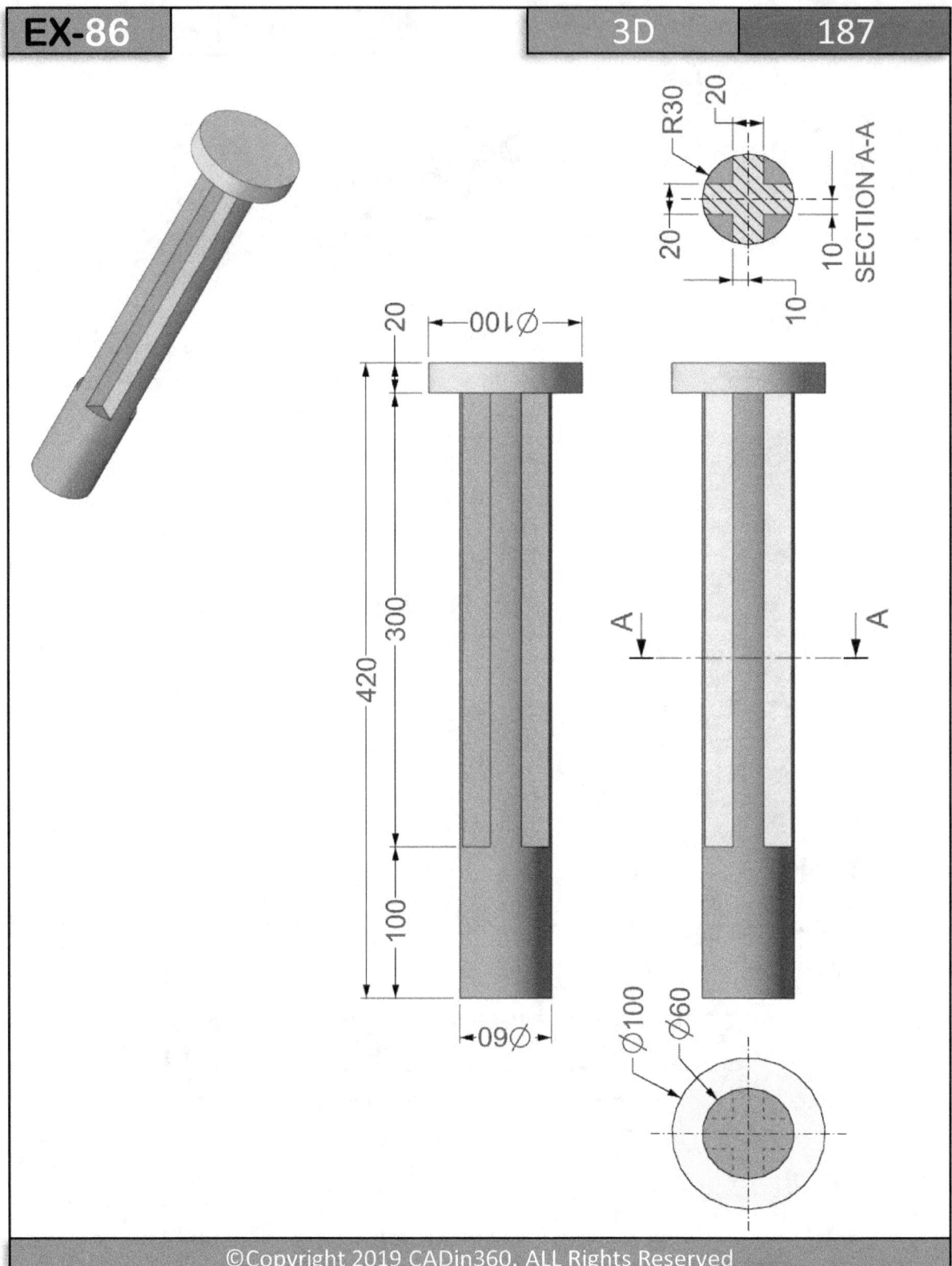

EX-86

SECTION A-A

R30
20
20
10
10

20
Ø100

420
300
100

Ø60

A
A

Ø100
Ø60

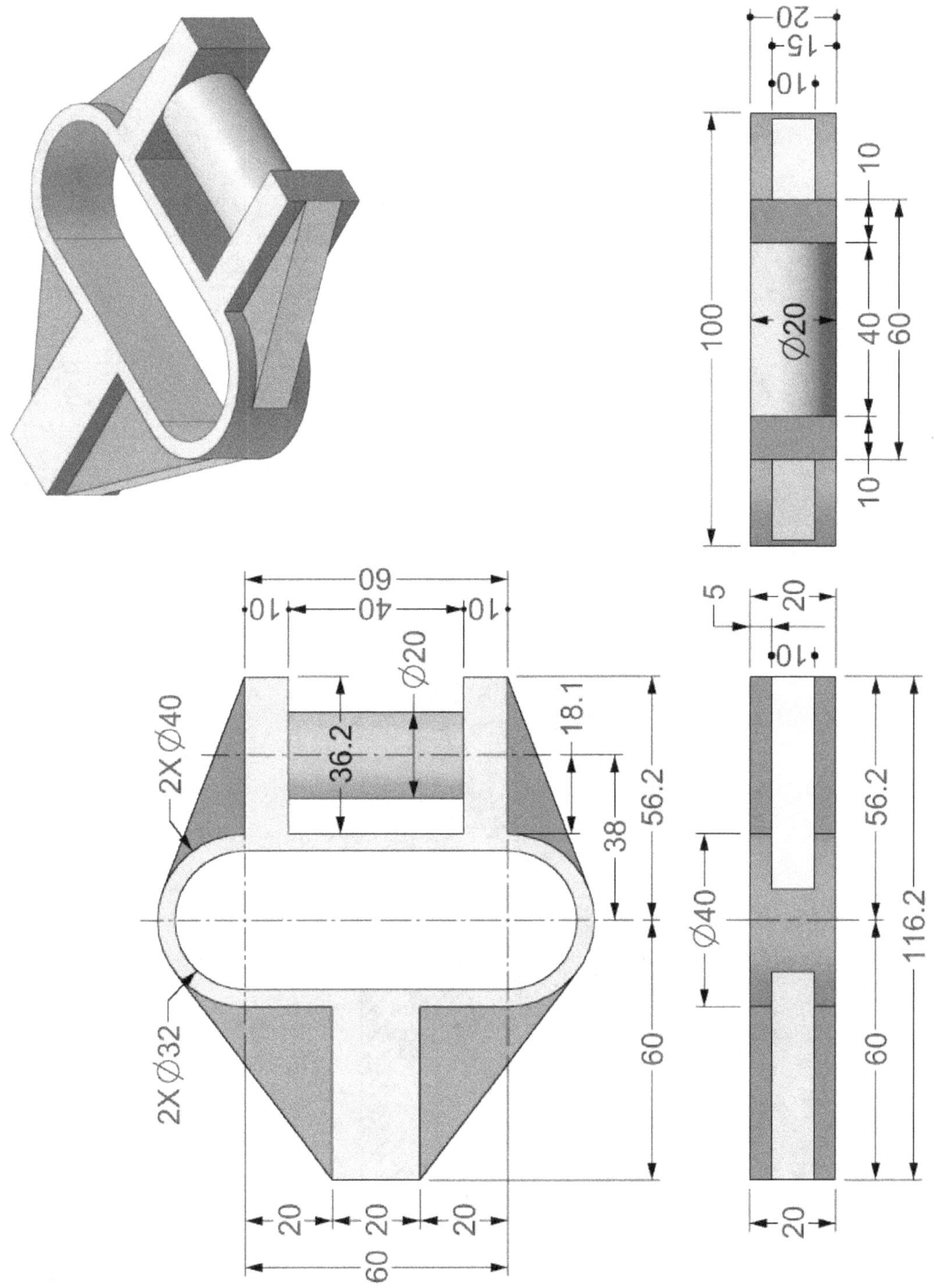
20
15
10
100
10
40
60
Ø20
10
60
10
40
10
Ø20
36.2
18.1
38
56.2
2X Ø40
2X Ø32
60
20
20
20
60
5
20
10
56.2
Ø40
116.2
60
20

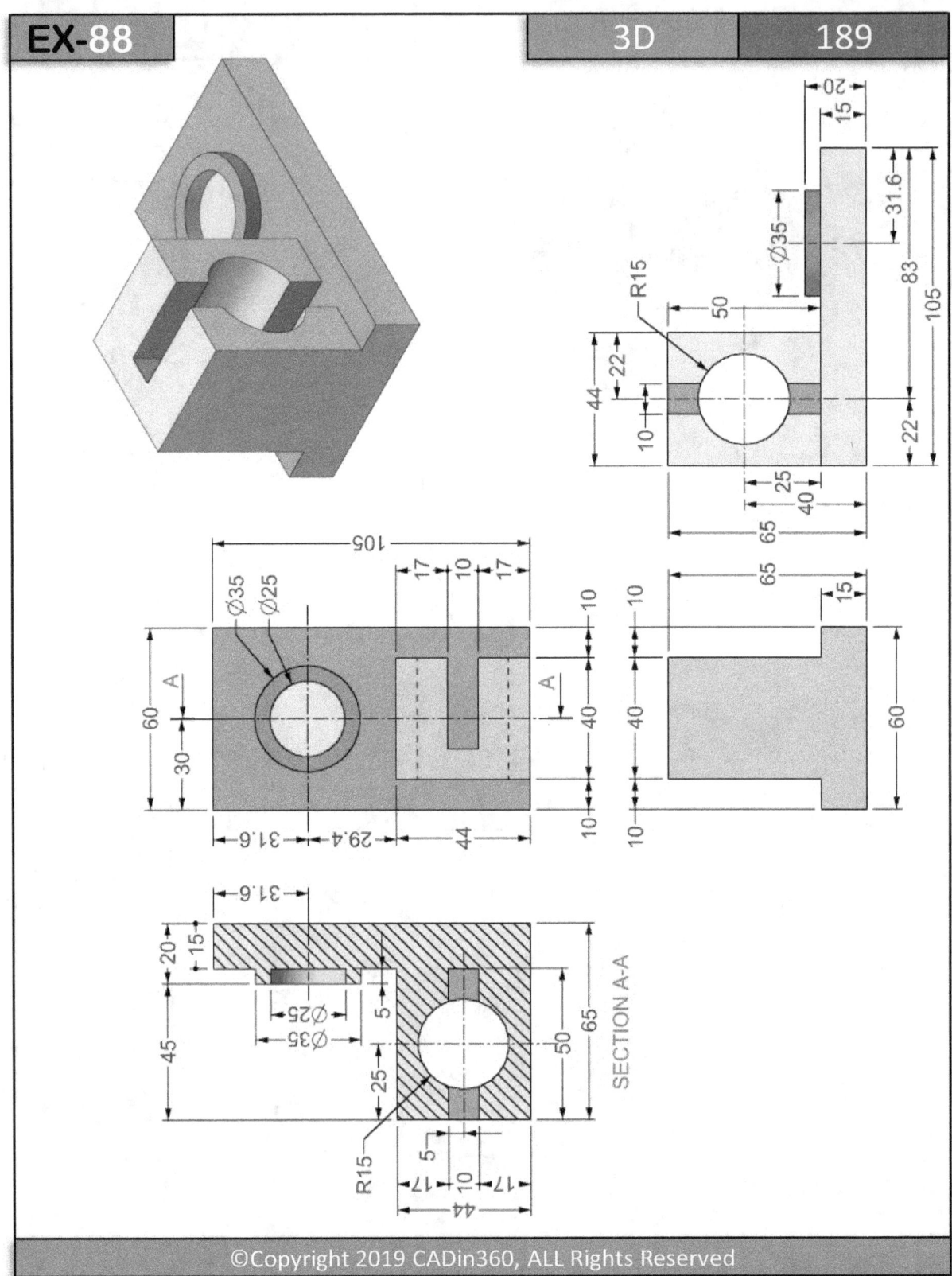

20
15
Ø35
31.6
83
105
R15
50
22
44
10
22
25
40
65

105
Ø35
Ø25
17
10
17
A
10
60
A
40
40
30
10
10
31.6
29.4
44
65
15
60

31.6
20
15
45
Ø25
Ø35
5
25
65
50
R15
5
17
10
17
44
SECTION A-A

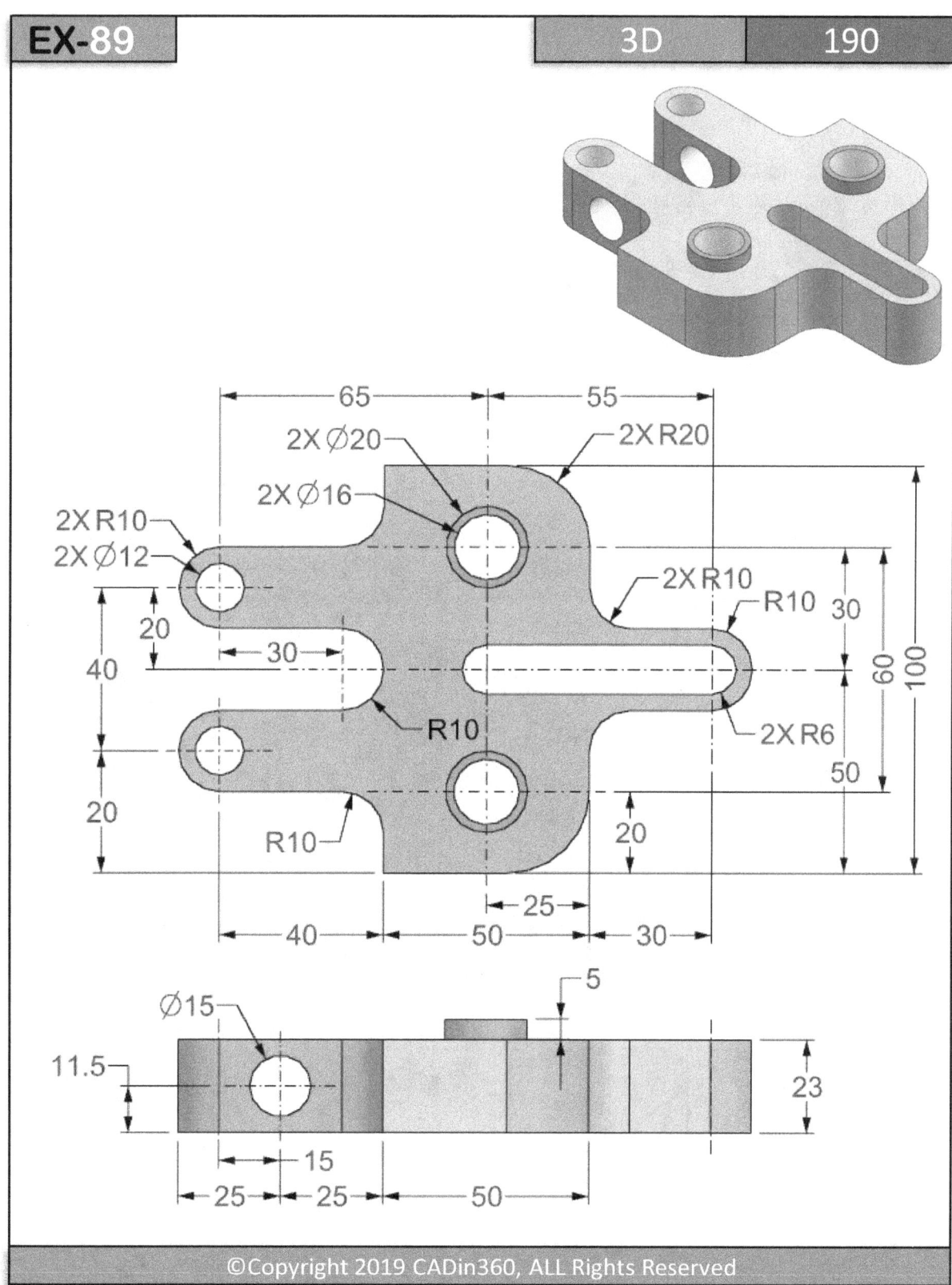

EX-89　　3D　　190

2X Ø20
2X R20
2X Ø16
2X R10
2X Ø12
65
55
2X R10
R10
30
2X R6
60
100
20
40
30
R10
40
50
20
R10
50
30
20
25

Ø15
5
11.5
23
15
25　25　50

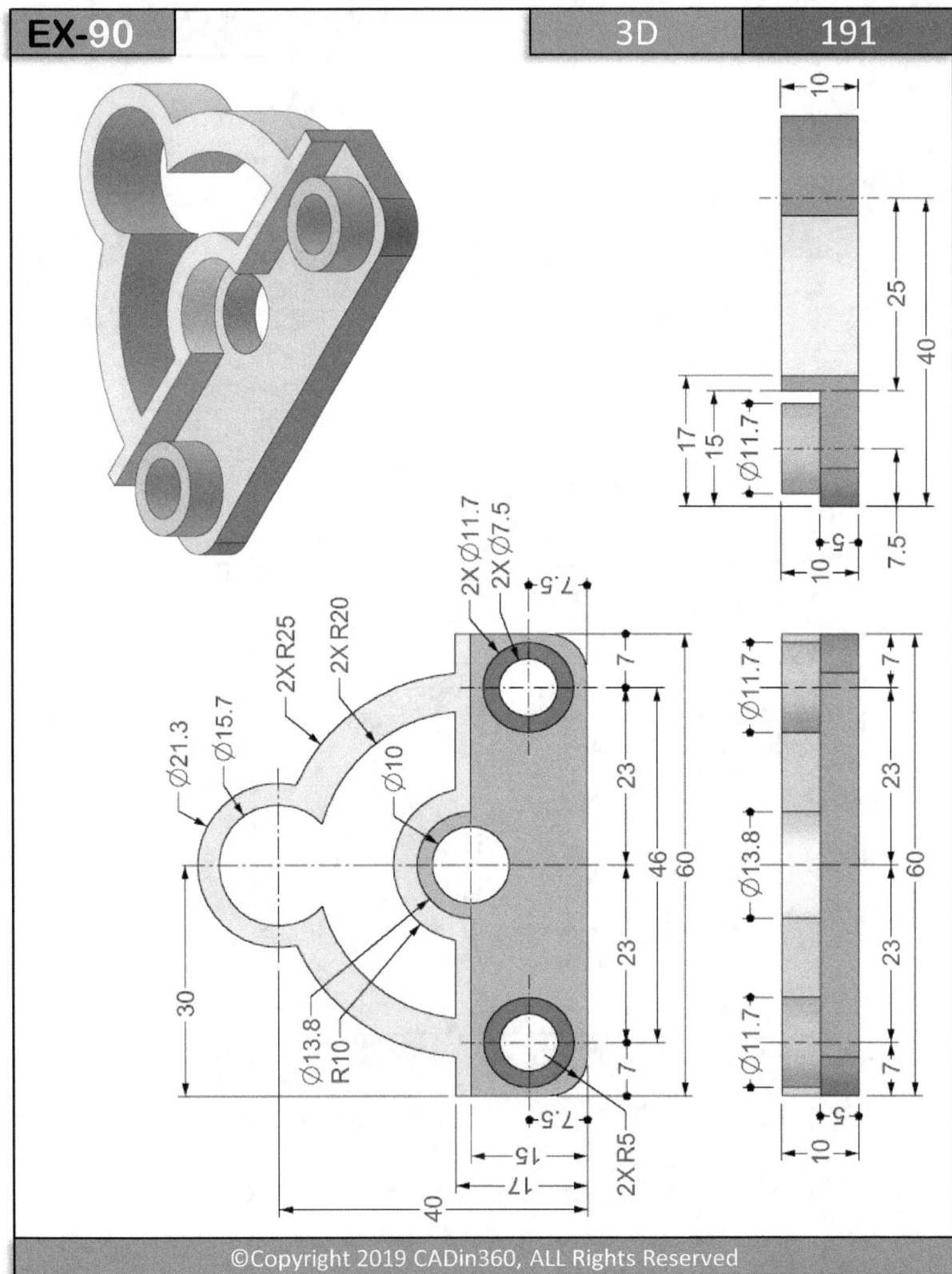

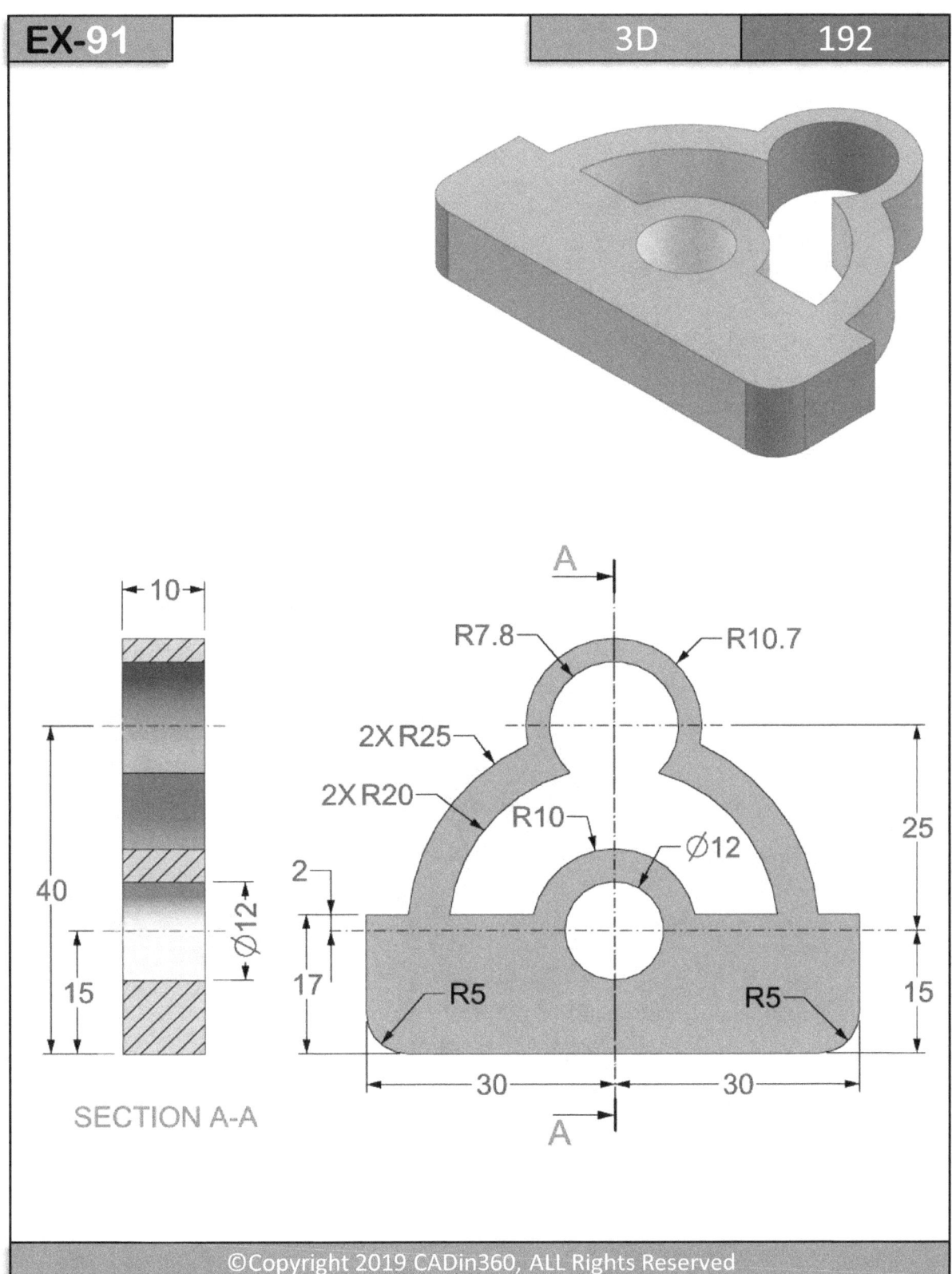

SECTION A-A

10

40

15

Ø12

A

R7.8 R10.7

2X R25

2X R20

R10

Ø12

2

17

R5 R5

30 30

25

15

A

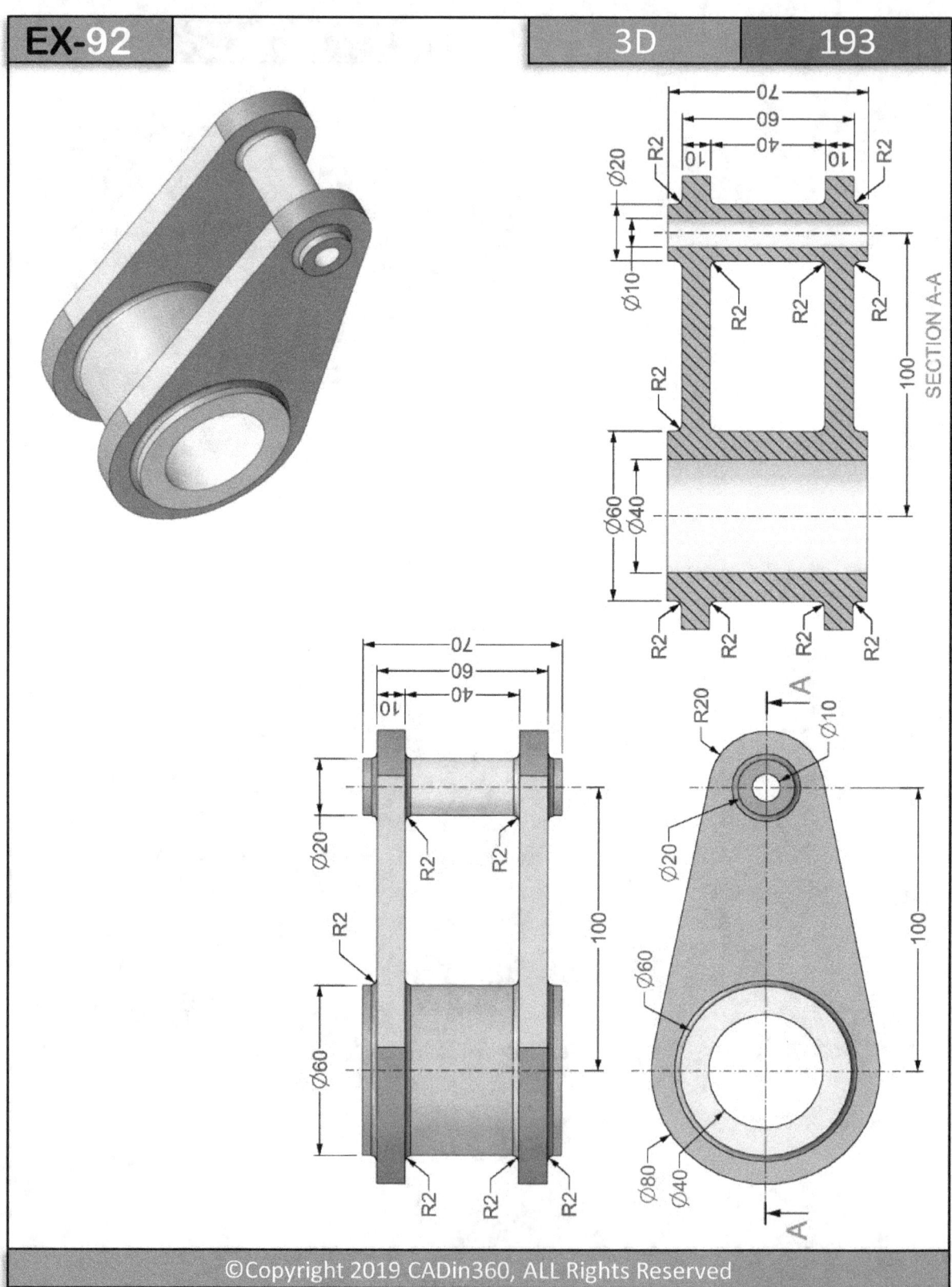

SECTION A-A

Ø20
R2
70
60
40
10
10
R2
Ø10
R2
R2
R2
R2
100
Ø60
Ø40
R2
R2
R2
R2
R2

Ø20
70
60
40
10
R2
R2
R2
100
Ø60
R2
R2
R2

R20
A
Ø10
Ø20
Ø60
100
Ø80
Ø40
A

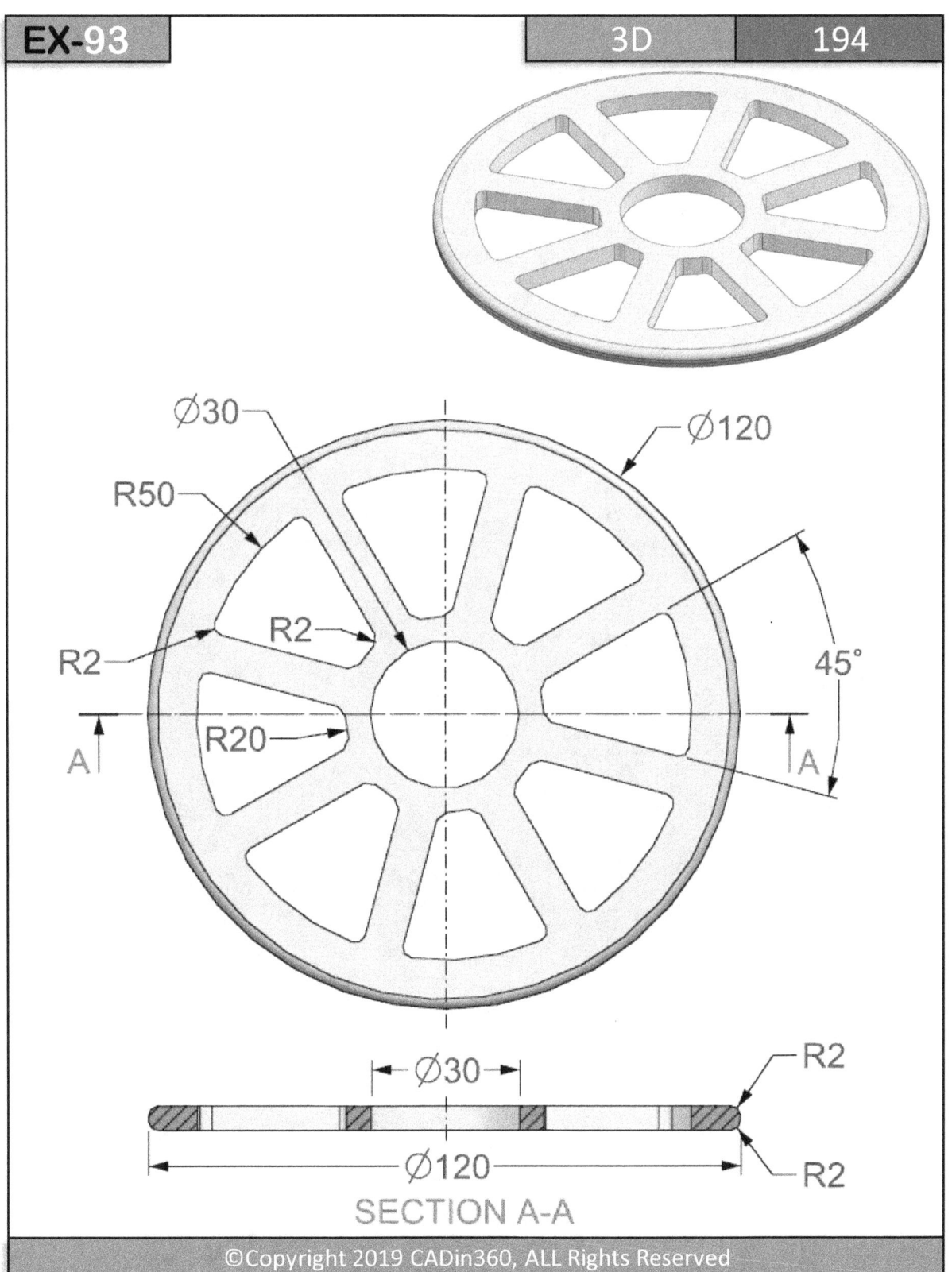

EX-93　3D　194

Ø30
Ø120
R50
R2
R2
45°
R20
A
A

Ø30
R2
Ø120
R2

SECTION A-A

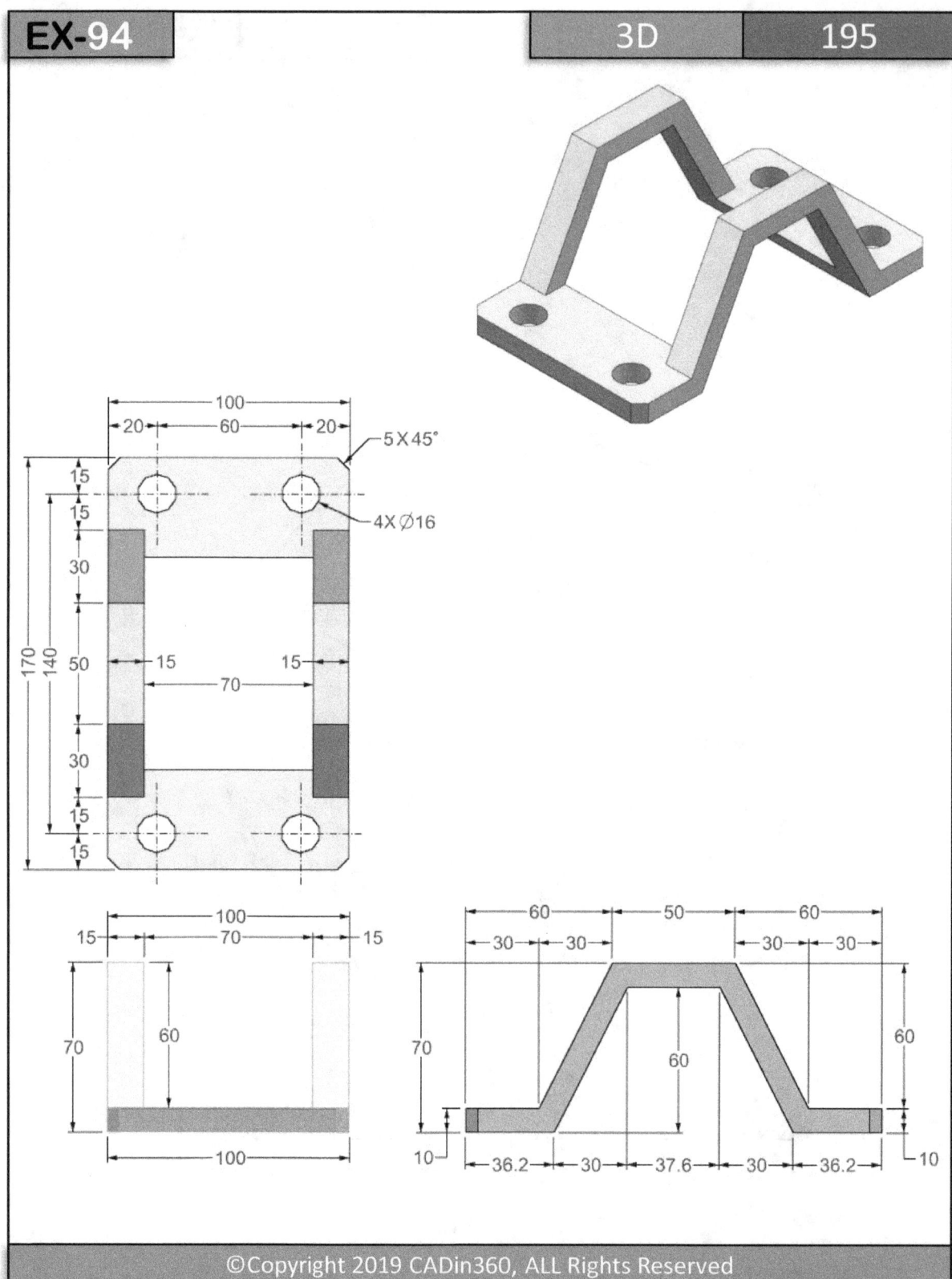

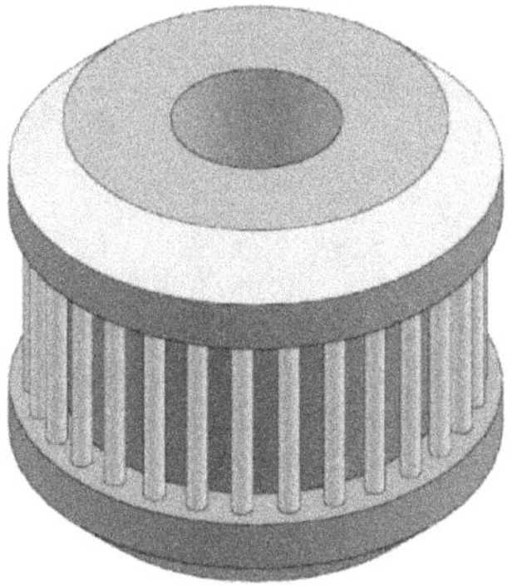

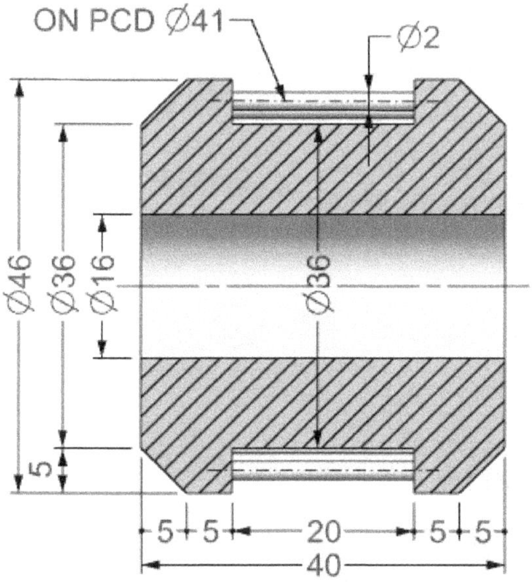

ON PCD ⌀41   ⌀2

⌀46
⌀36
⌀16

⌀36

5

5┤5├   20   ┤5├5
40

SECTION A-A

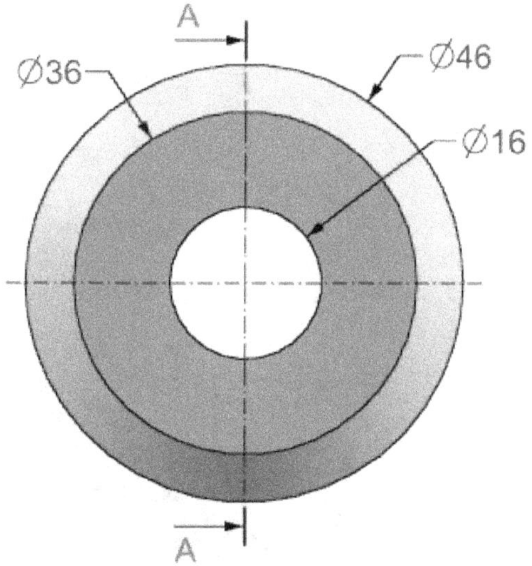

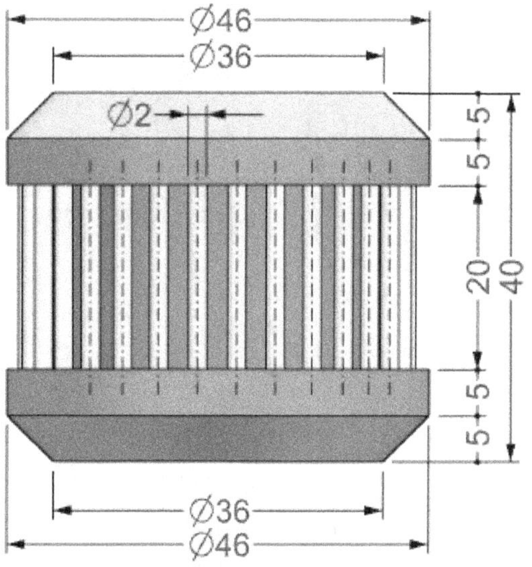

A

⌀36

⌀46

⌀16

A

⌀46
⌀36

⌀2

5┤5

5

20   40

5┤5

⌀36
⌀46

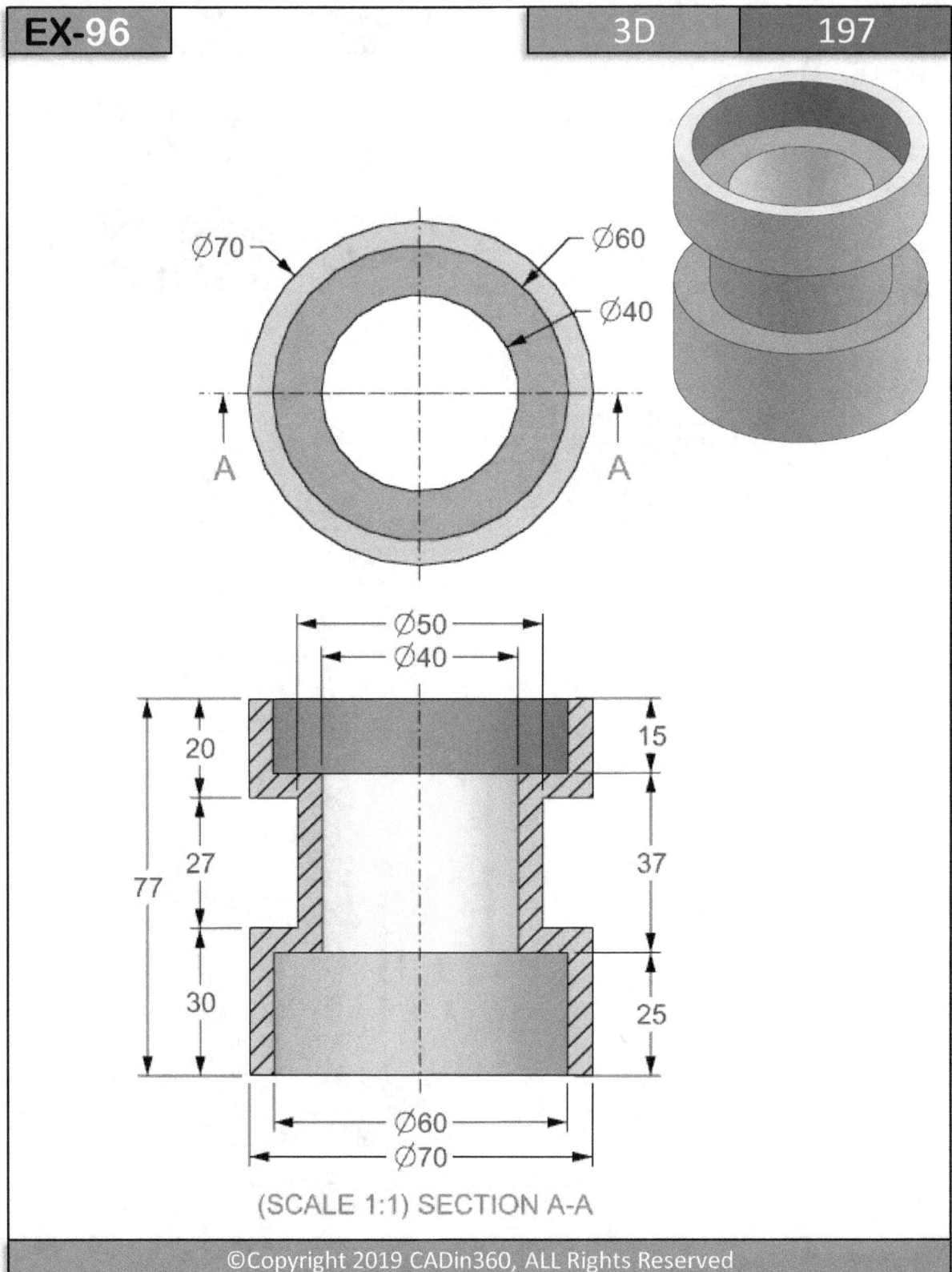

EX-96 3D 197

Ø70 Ø60
Ø40

A A

Ø50
Ø40

20 15

27 37

77

30 25

Ø60
Ø70

(SCALE 1:1) SECTION A-A

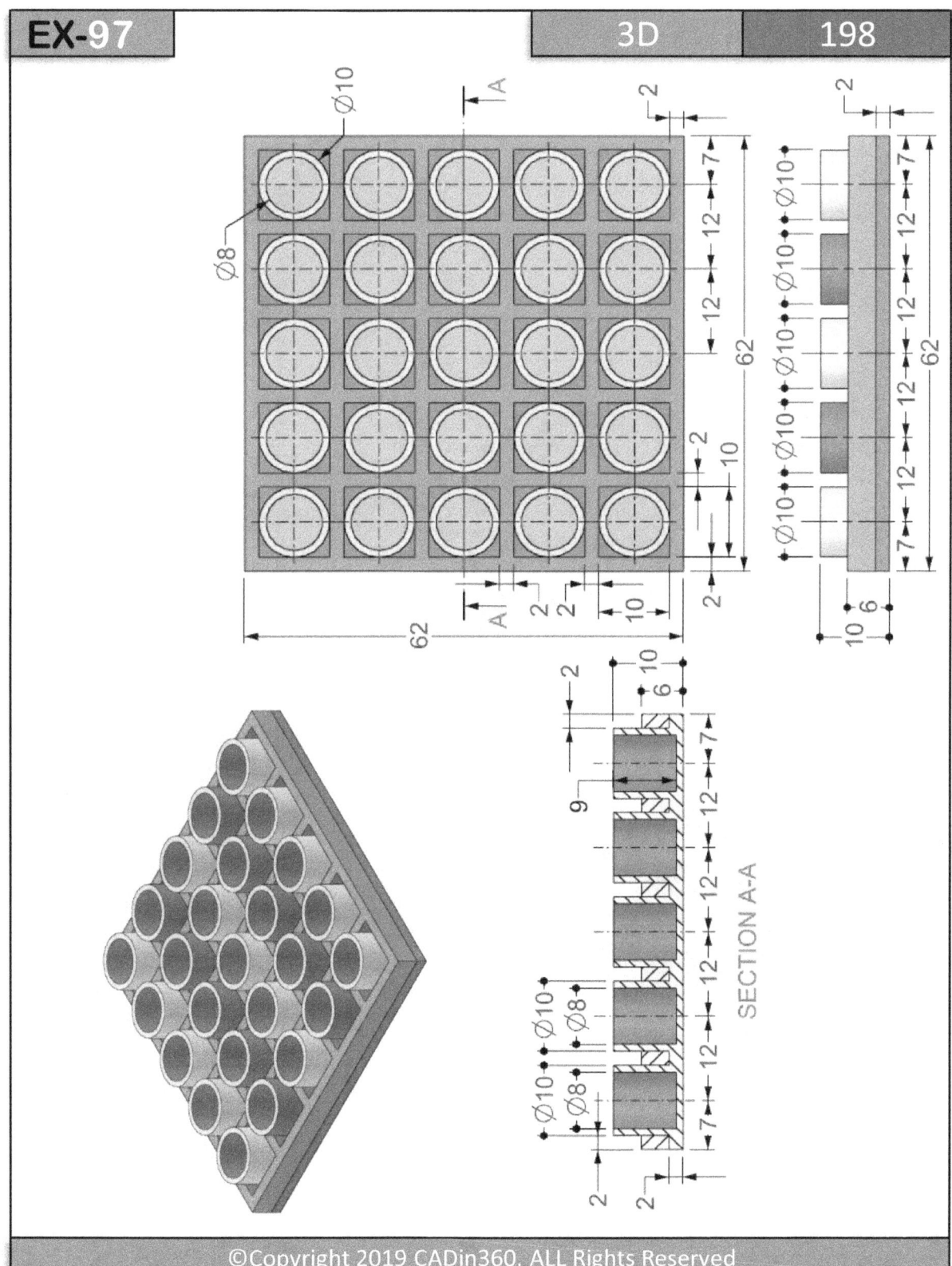

SECTION A-A

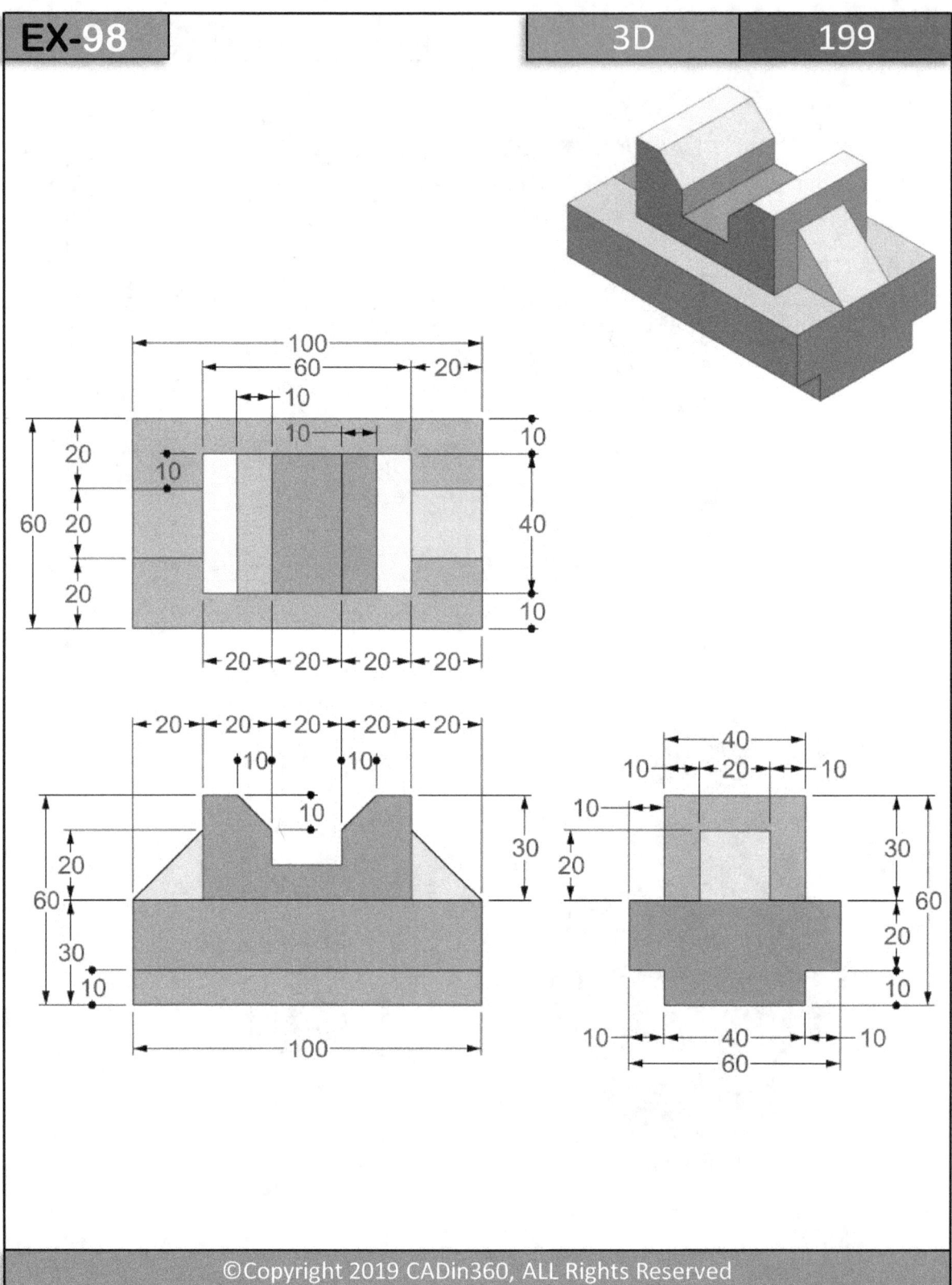

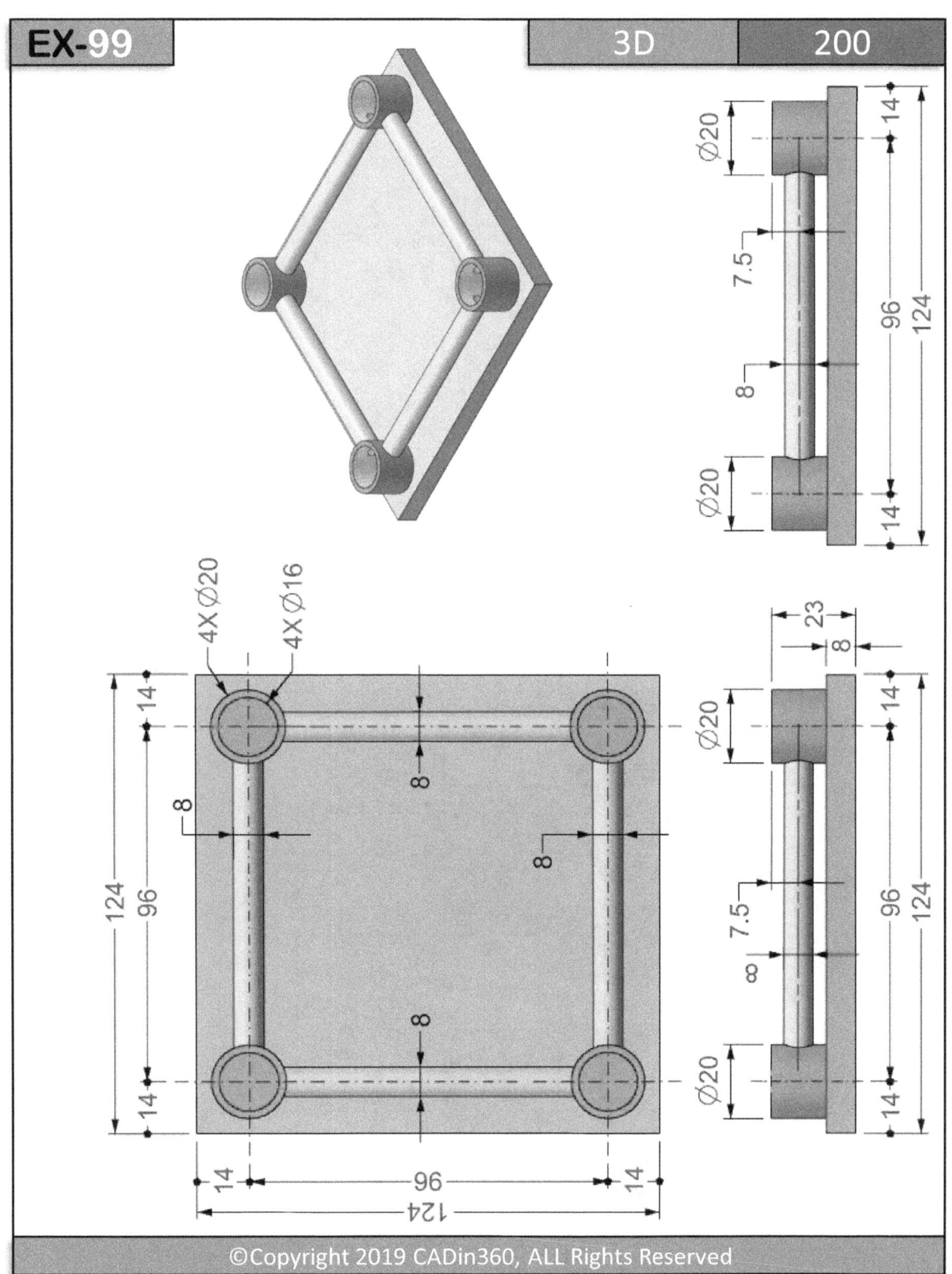

EX-99

3D

200

4X Ø20
4X Ø16

Ø20
Ø20
7.5
8
14
124
96

23
8
Ø20
14
124
96
7.5
8
Ø20
14

14
124
96
8
8
8
124
96
14
14
96
14

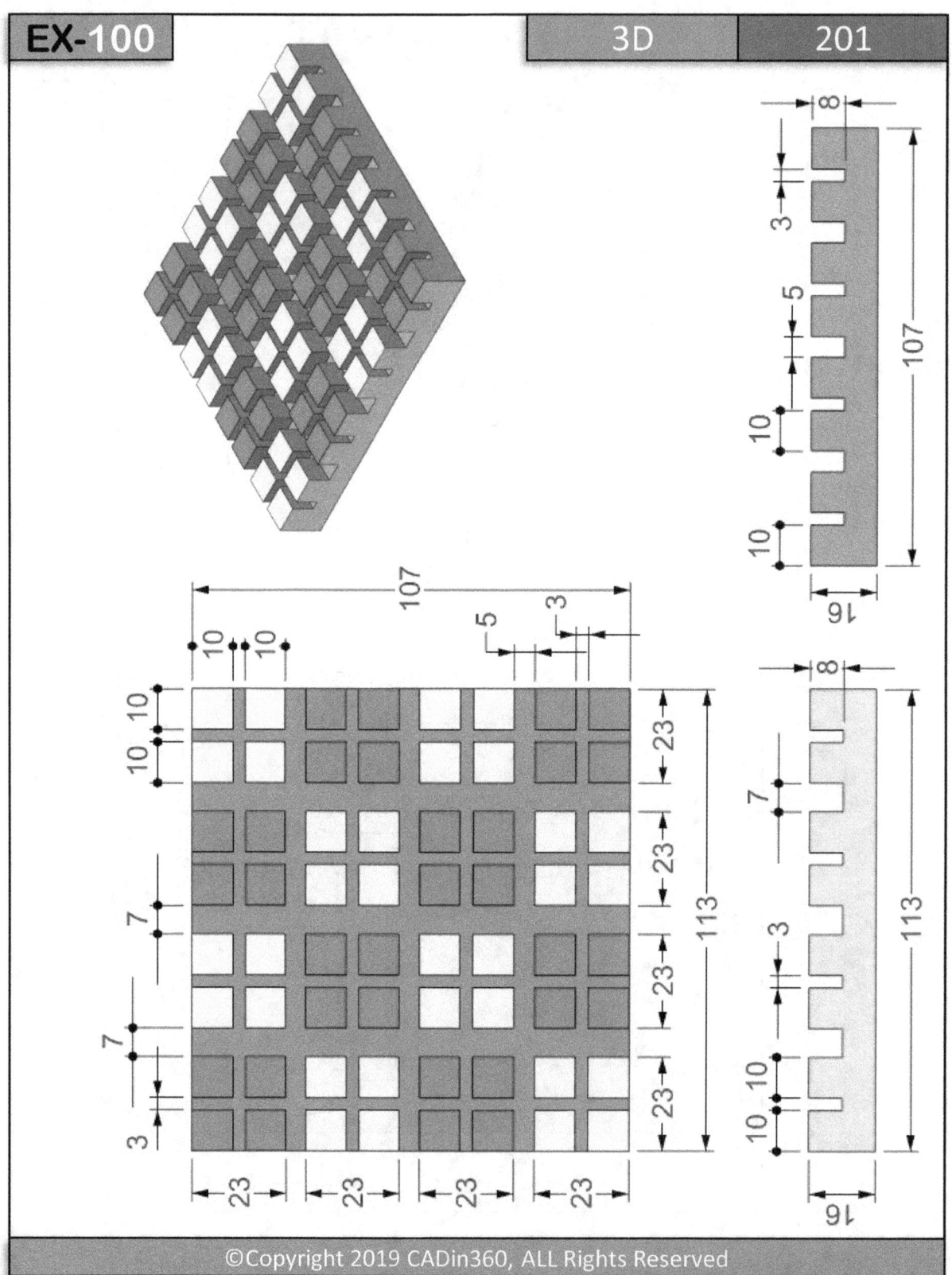

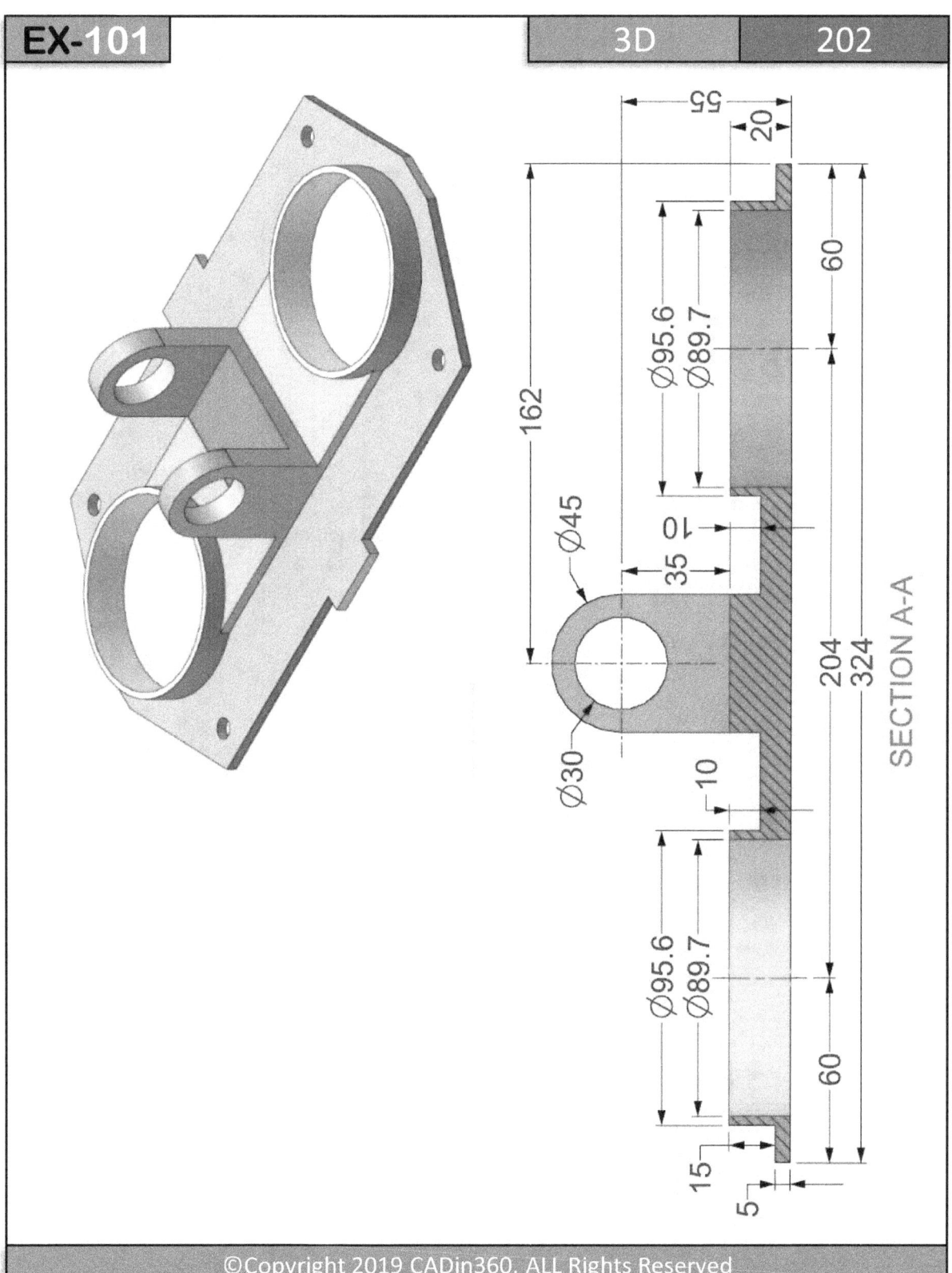

SECTION A-A

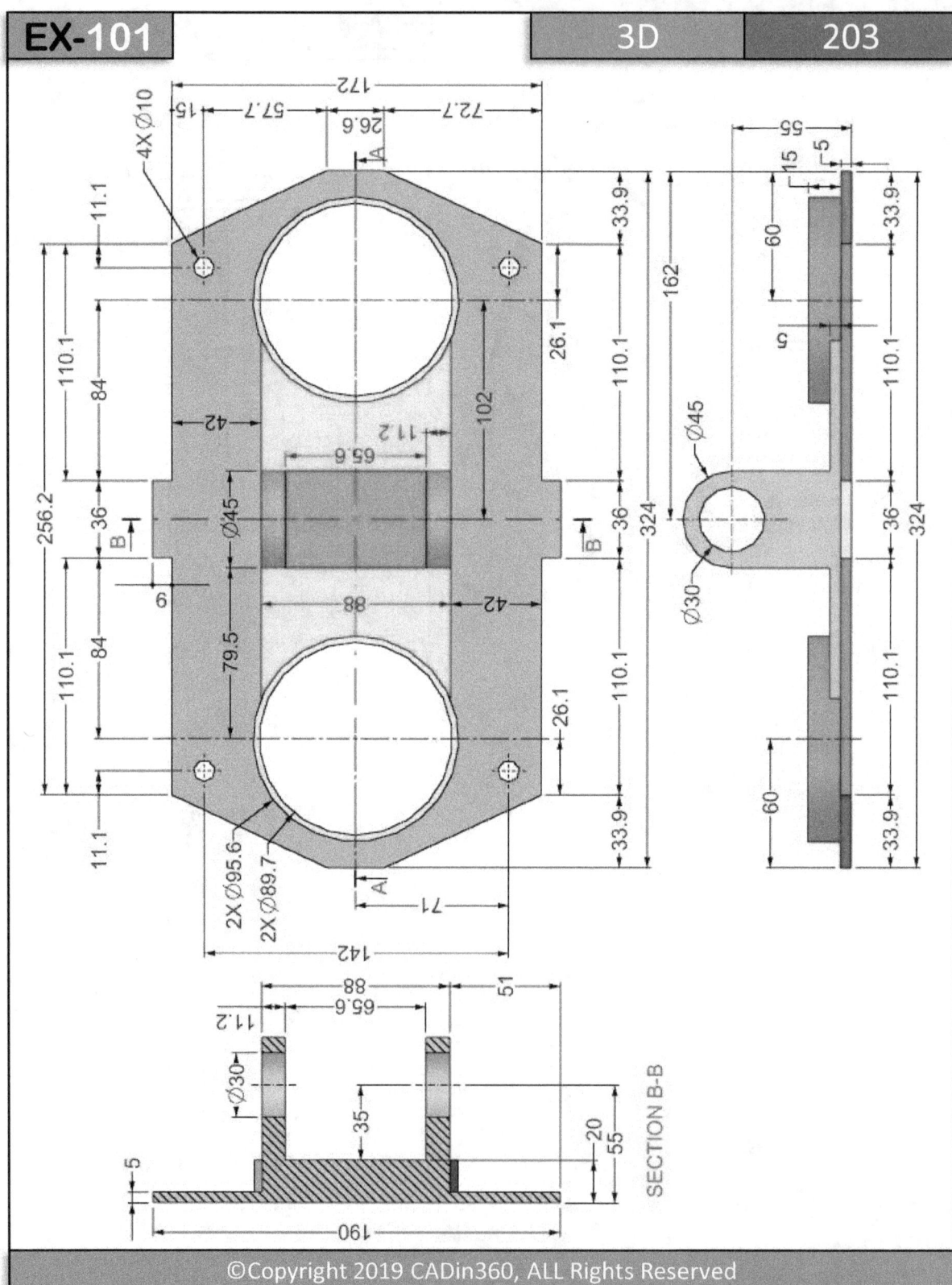

SECTION B-B

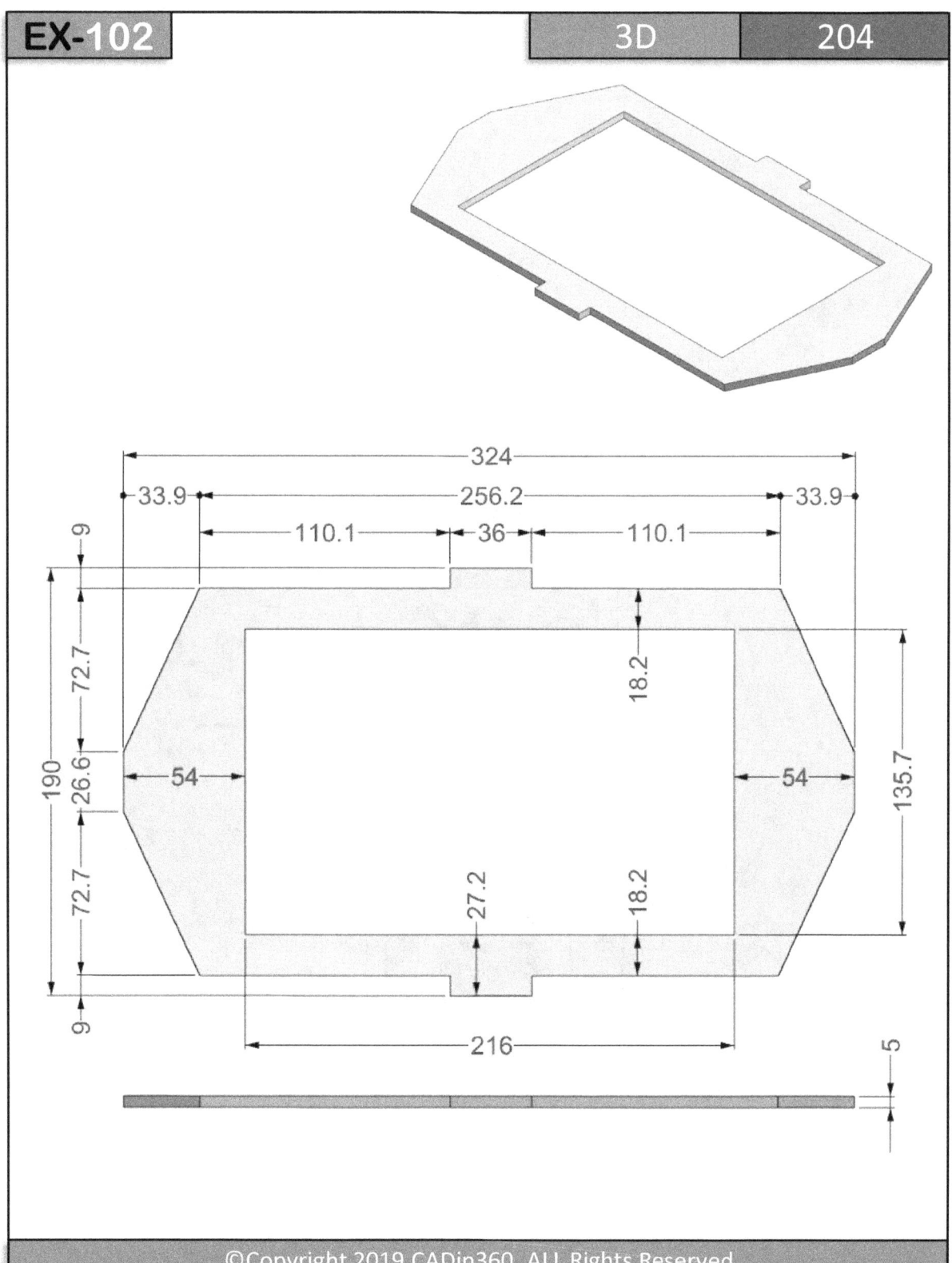

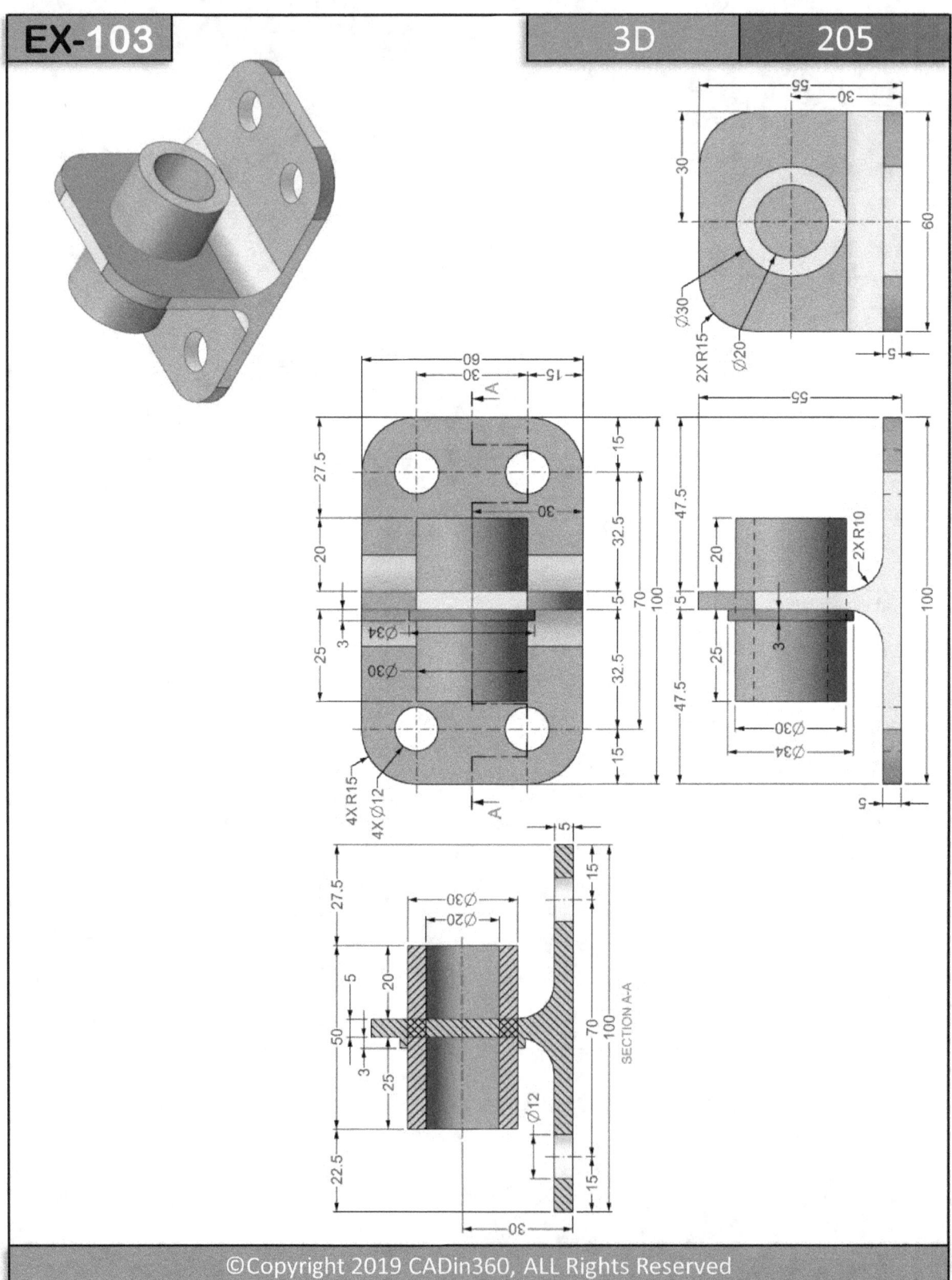

SECTION A-A

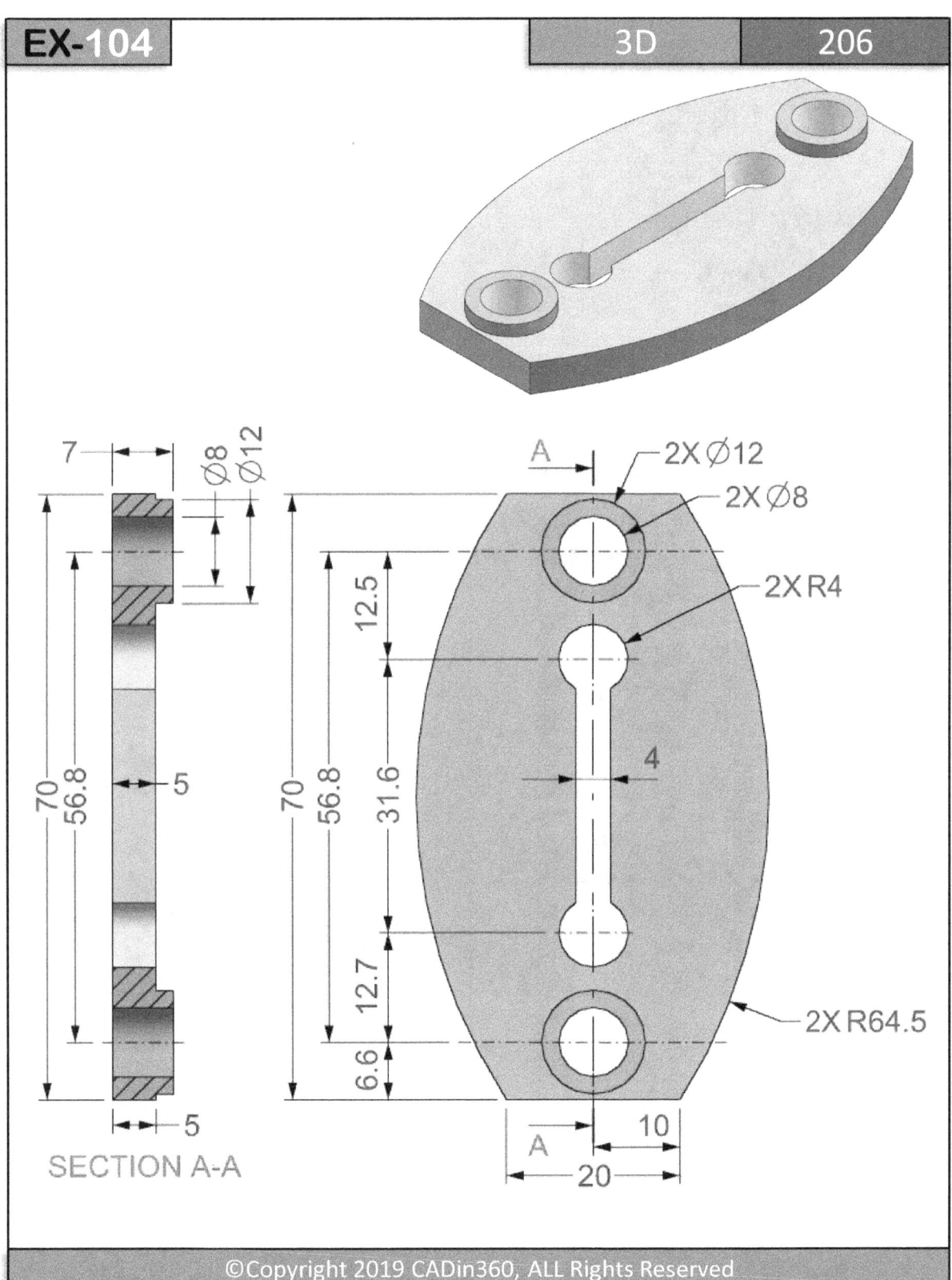

7

Ø8
Ø12

70
56.8

5

5

SECTION A-A

A

2X Ø12
2X Ø8

2X R4

12.5

70
56.8

31.6

4

12.7

6.6

2X R64.5

A

10

20

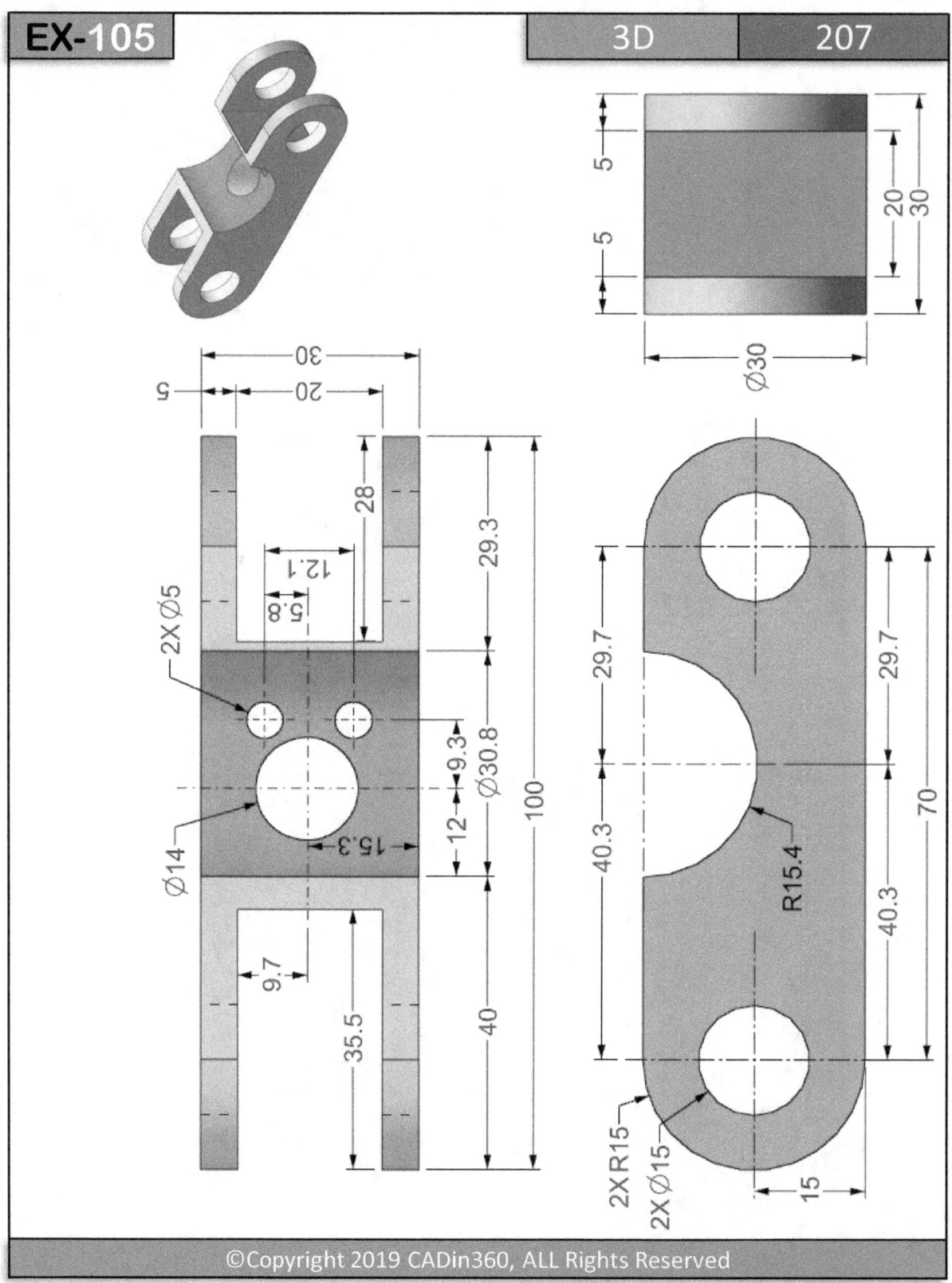

EX-105

3D 207

2X Ø5

Ø14

Ø30.8

R15.4

2X R15

2X Ø15

30
20
5
28
12.1
5.8
29.3
9.3
12
100
15.3
9.7
35.5
40

5
5
20
30
Ø30

29.7
40.3
29.7
70
40.3
15

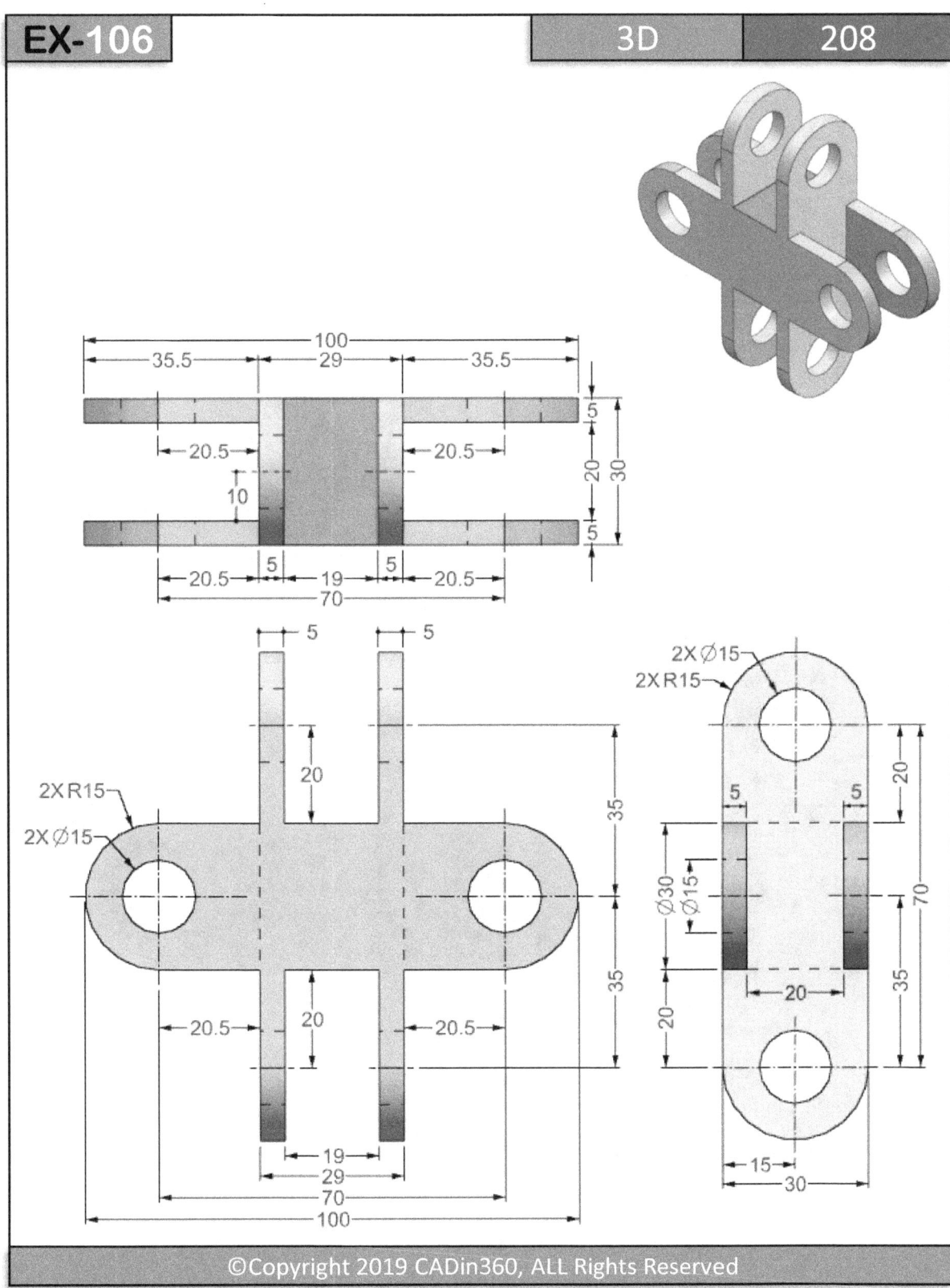

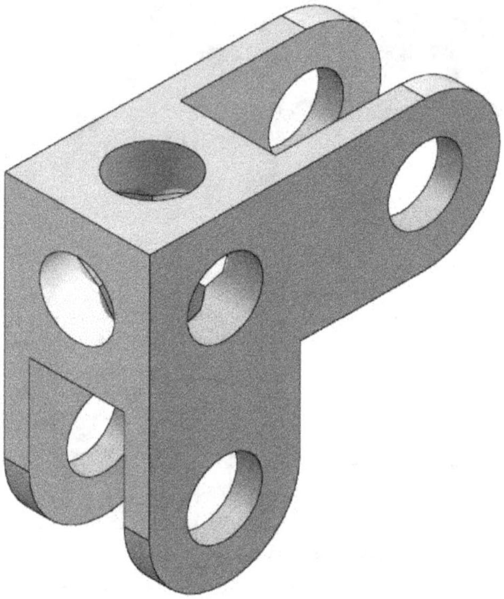

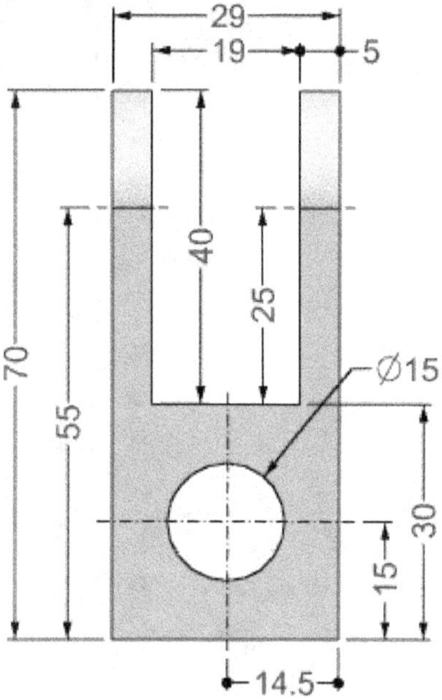

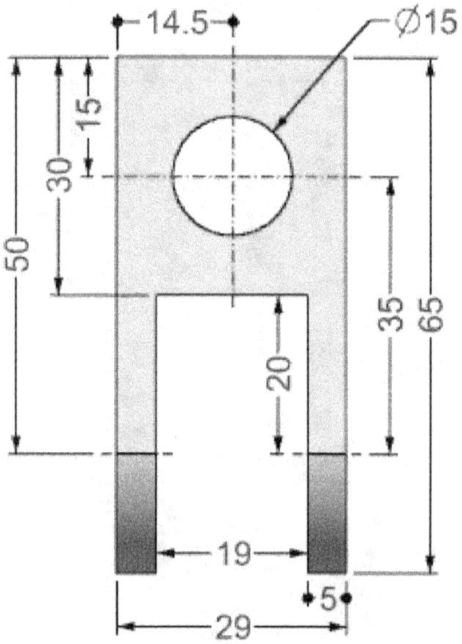

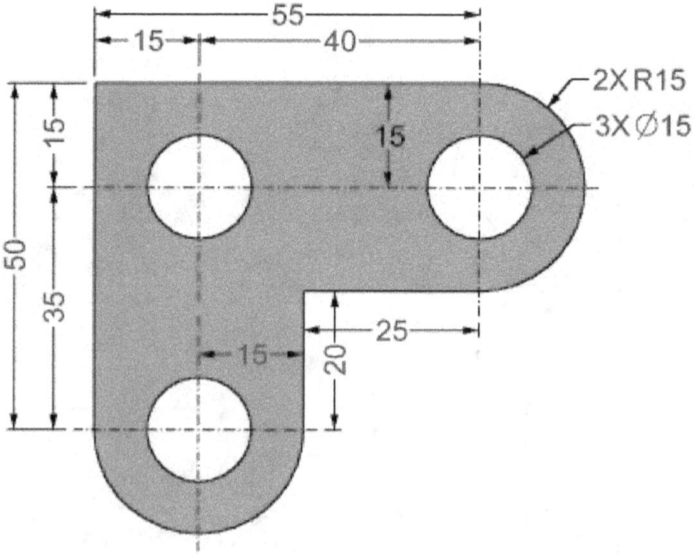

29
19
5
40
25
70
55
∅15
30
15
14.5

14.5
∅15
15
30
50
65
35
20
19
5
29

55
15
40
2X R15
3X ∅15
15
15
50
35
15
20
25
20

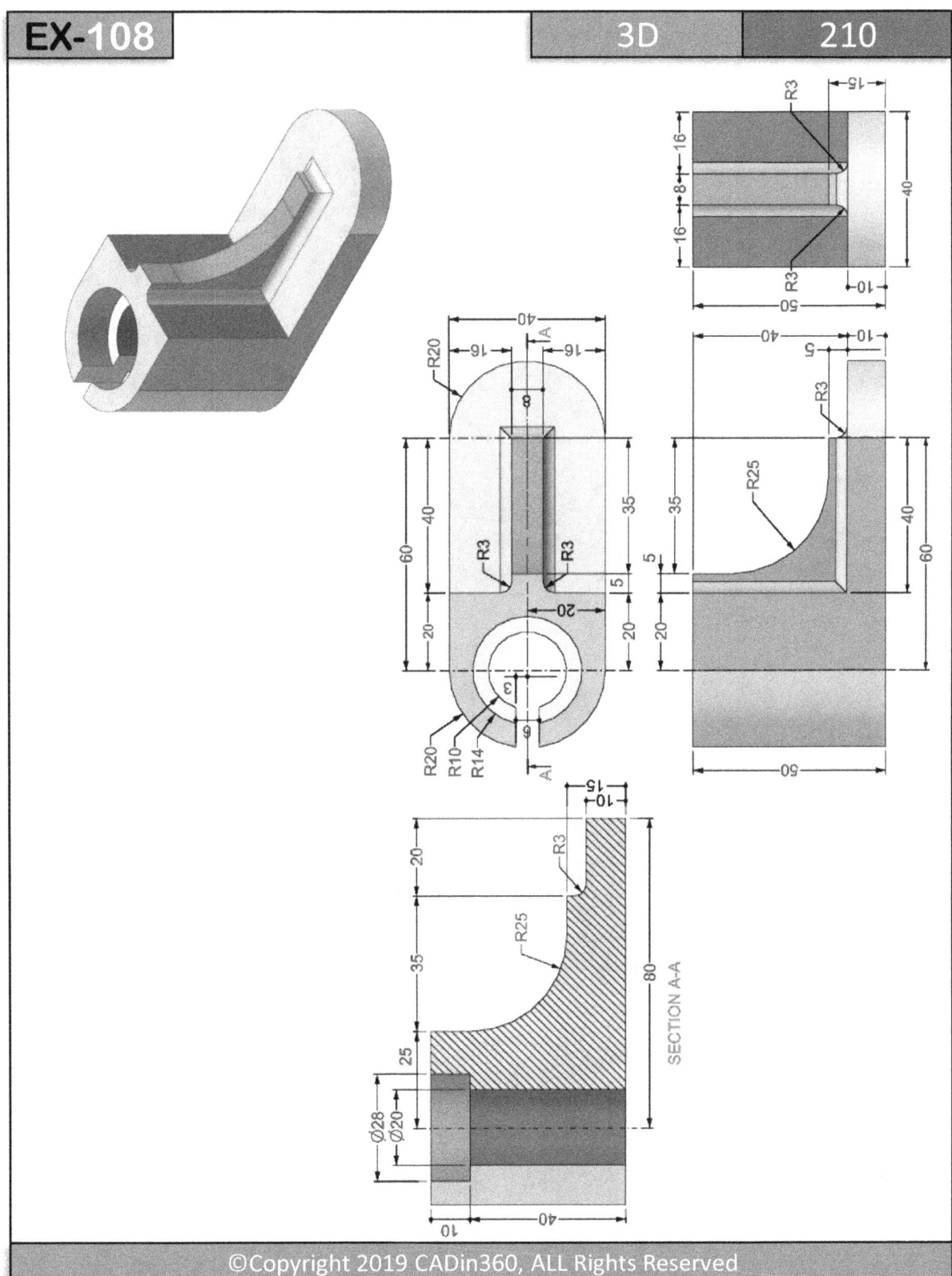

SECTION A-A

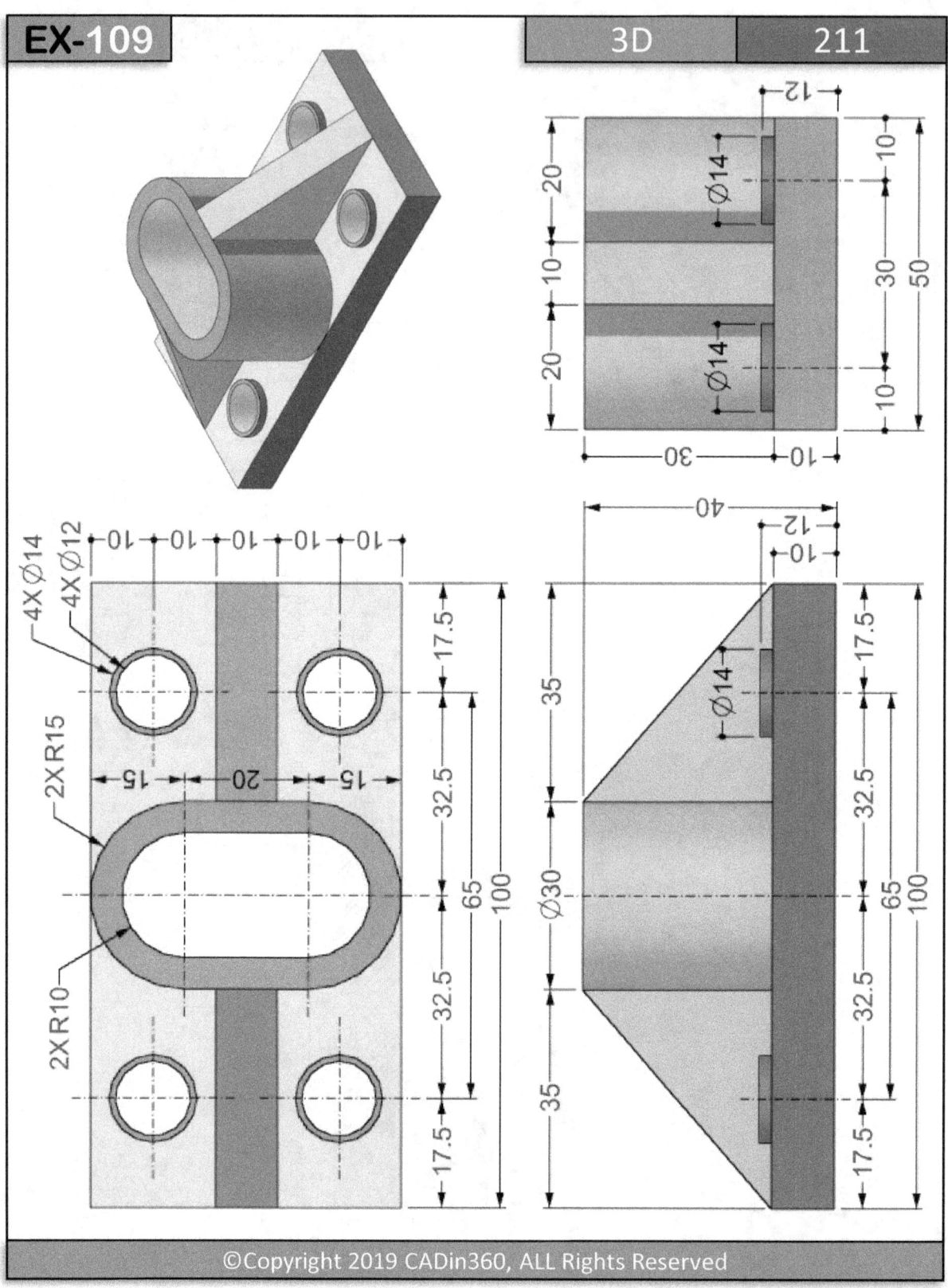

EX-109

3D 211

4X Ø14
4X Ø12
2XR15
2XR10

10 · 10 · 10 · 10 · 10

15 · 20 · 15

17.5
32.5
65
32.5
17.5
100

35
Ø30
35

40
12
10

17.5
32.5
65
32.5
17.5
100

Ø14

12
Ø14

20
10
20

Ø14

30
10

50

10
30

10

©Copyright 2019 CADin360, ALL Rights Reserved

70

35

10

40

110

30

30

R5

Ø30

R10

Ø40

Ø50

R10

4X R10

Ø22

A

4X Ø15

Ø30

Ø50

50

70

25

10

A

25

10

50

70

R10

R10

R5

110

Ø22

70

30

30

40

10

SECTION A-A

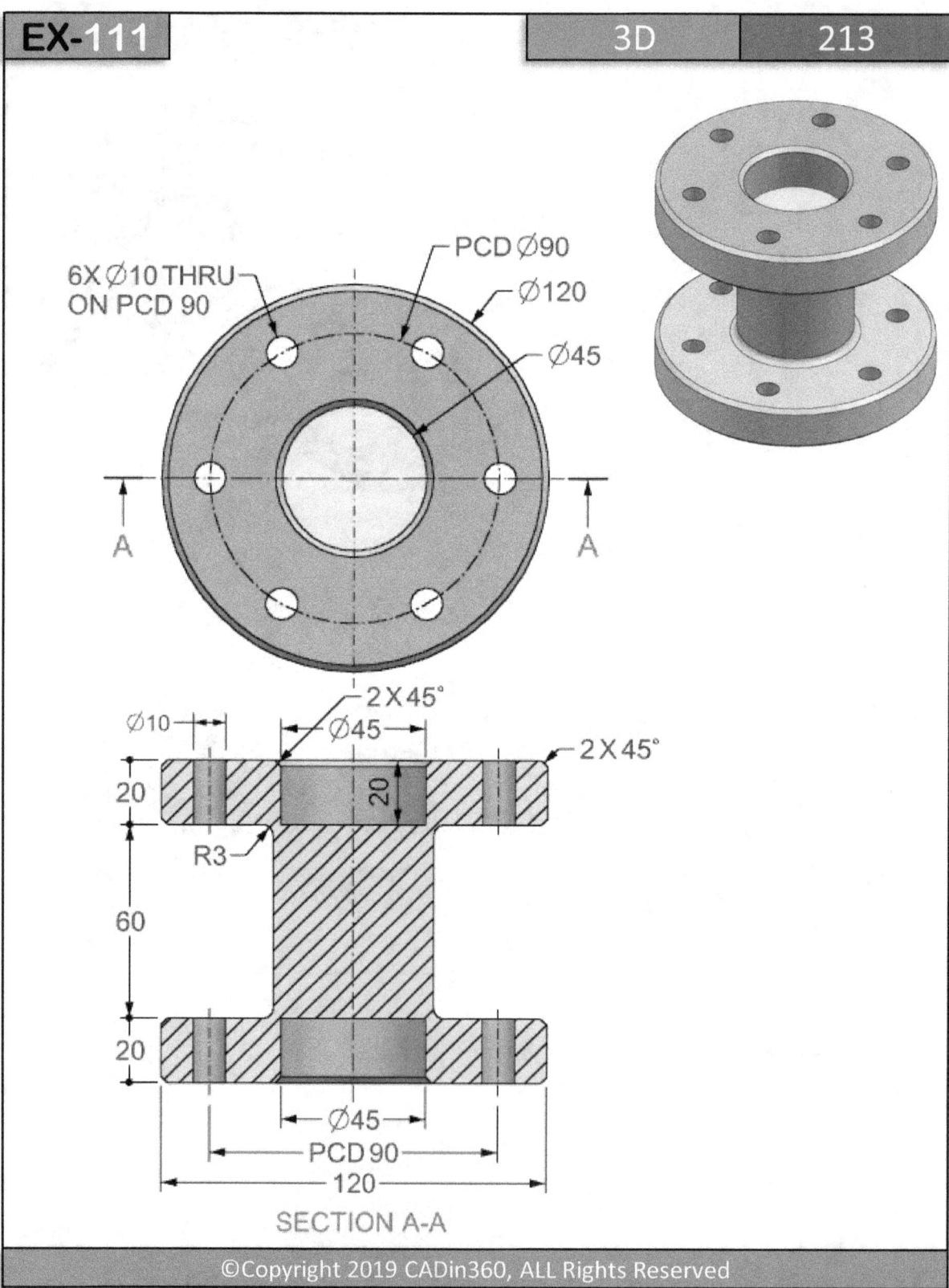

EX-111 3D 213

6X Ø10 THRU
ON PCD 90

PCD Ø90

Ø120

Ø45

2 X 45°

Ø45

Ø10

20

20

2 X 45°

R3

60

20

Ø45

PCD 90

120

SECTION A-A

A   A

Ø120
6X Ø10
6X Ø8
PCD Ø90
Ø68
Ø45

A | A

Ø120
Ø68
Ø10
R2
10
10
20
120
60
20
10
PCD 90

Ø120
PCD 90
Ø68
Ø45
Ø10
40
2 X 45°
20
R3
60
R3
Ø8
Ø55
Ø50
20
Ø45
Ø8

SECTION A-A

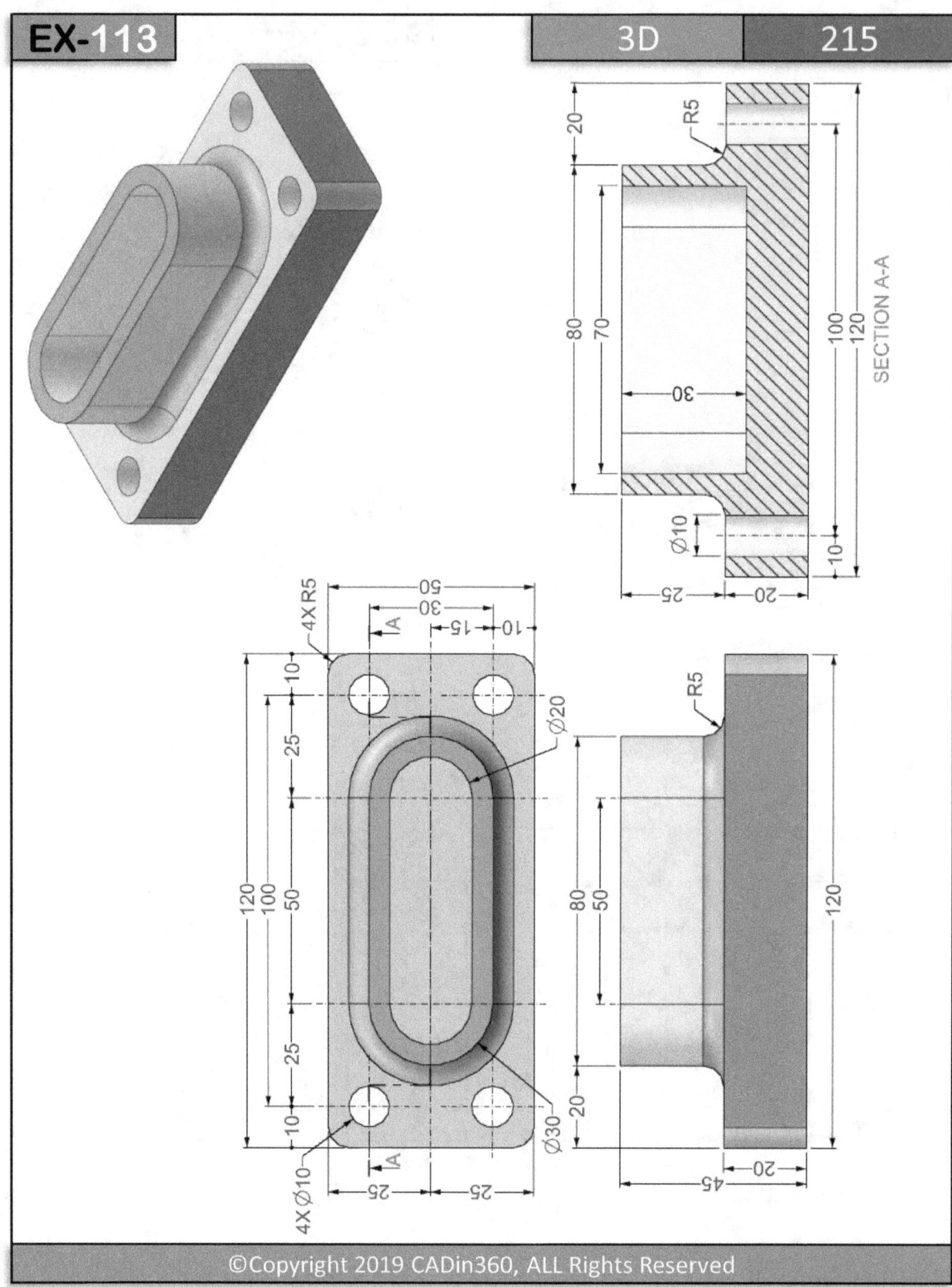

SECTION A-A

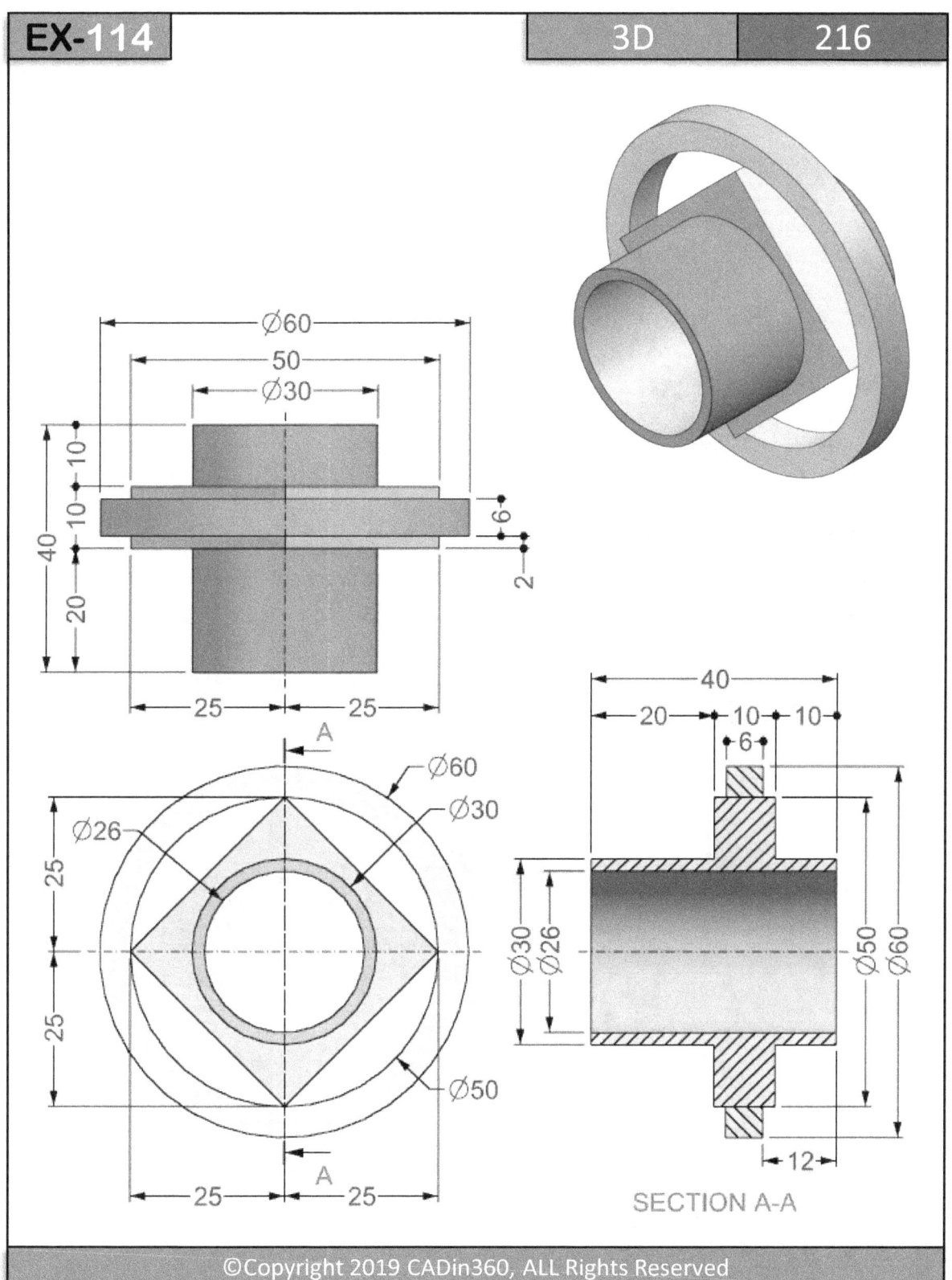

Ø60
50
Ø30
10
10
40
10
20
6
2
25   25

A
Ø26
Ø60
Ø30
25
25
25
Ø50
25   25
A

40
20   10   10
6
Ø30
Ø26
Ø50
Ø60
12
SECTION A-A

SECTION A-A

50
10 10 10 20
Ø20
R2
Ø40
45
20
10

R2
Ø30
15
10
Ø30
Ø50
20 10 20

R15
R10
15
A
45
7.5
2X Ø50
Ø40
A

Ø30
10 10 10 20
45
R2
R2

Ø190
Ø55

Ø140
70
2X R20
2X R25
25
50
Ø55
Ø75
Ø100
Ø180
Ø190

SECTION A-A

A
R25
25
50
A
Ø75
Ø190

Ø75
Ø190
Ø55
Ø180
Ø100

25 30
15   15
15
30   30   60
40
15   30
55

75
Ø50
25
75 100
Ø50   75
25
15 30 15
60

Ø50
Ø45
25
30
50   10
80 100
20
25   10
Ø50   10
Ø45   30
R15

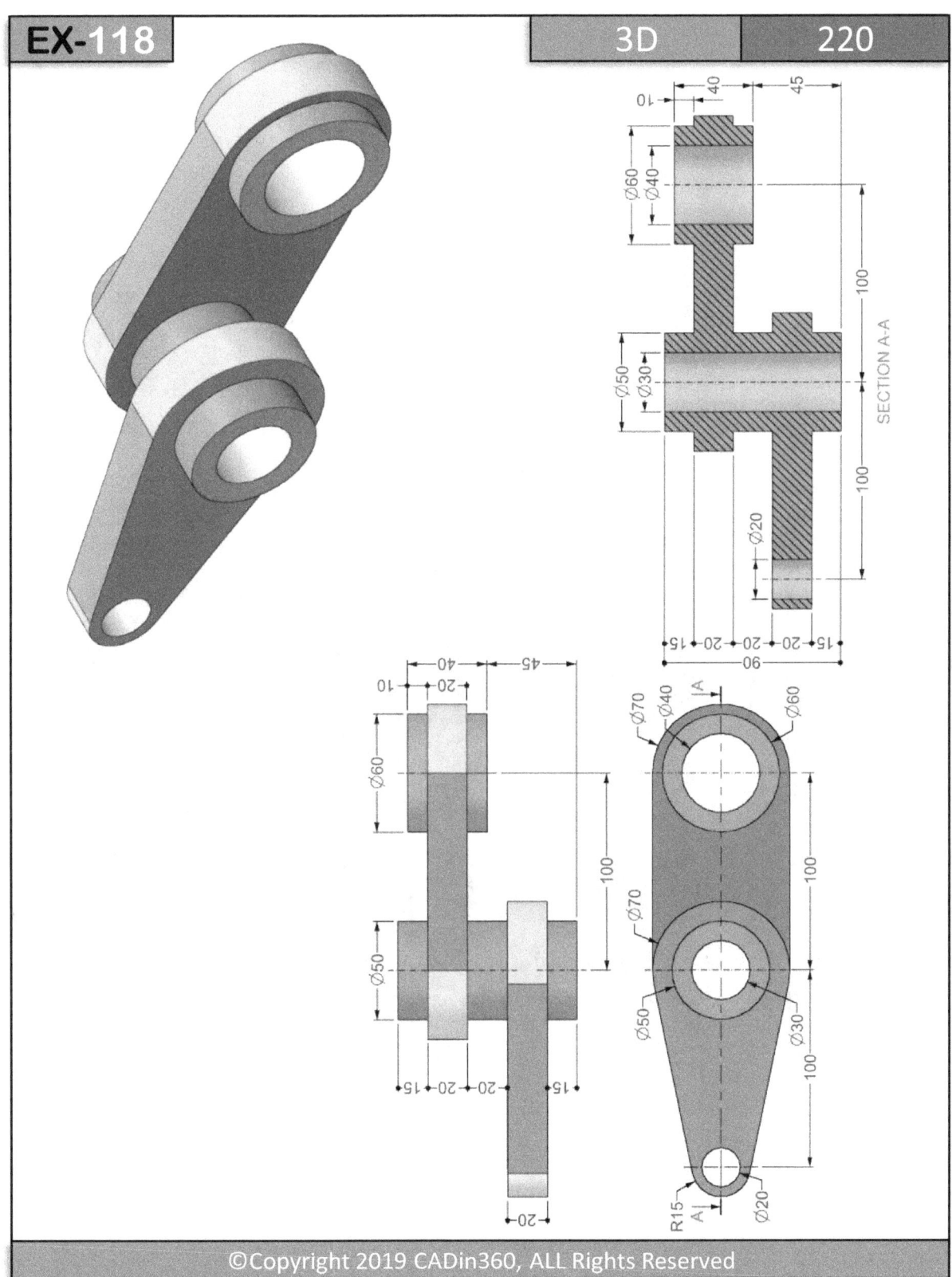

EX-118

3D

220

SECTION A-A

©Copyright 2019 CADin360, ALL Rights Reserved

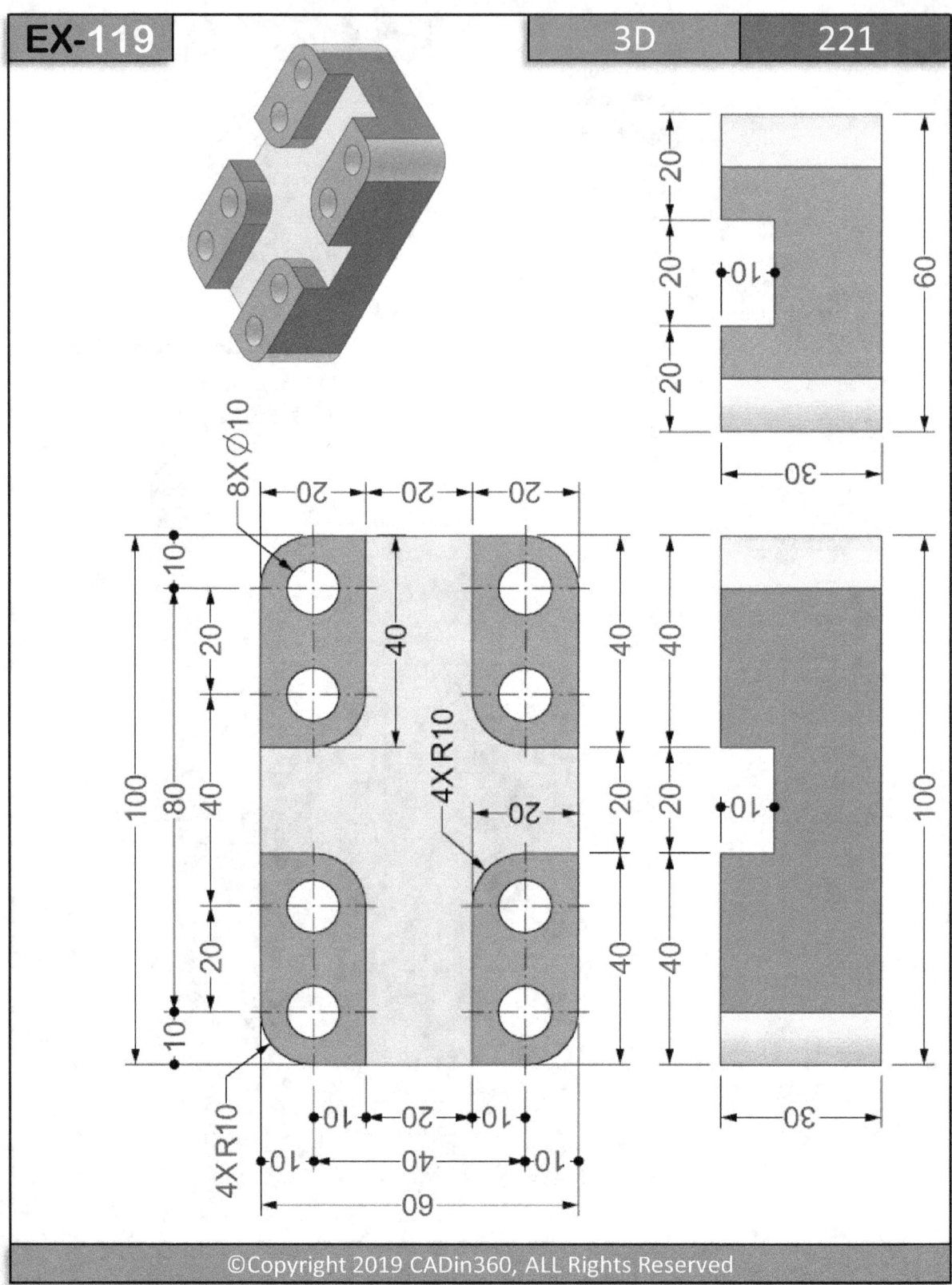

8X ⌀10

20 20 20

10

20

40

40

100

80

40

40

20

20

20

40

40

40

4X R10

20

4X R10

10 20 10

10 40 10

60

20

20

20

10

60

30

20

20

20

10

100

30

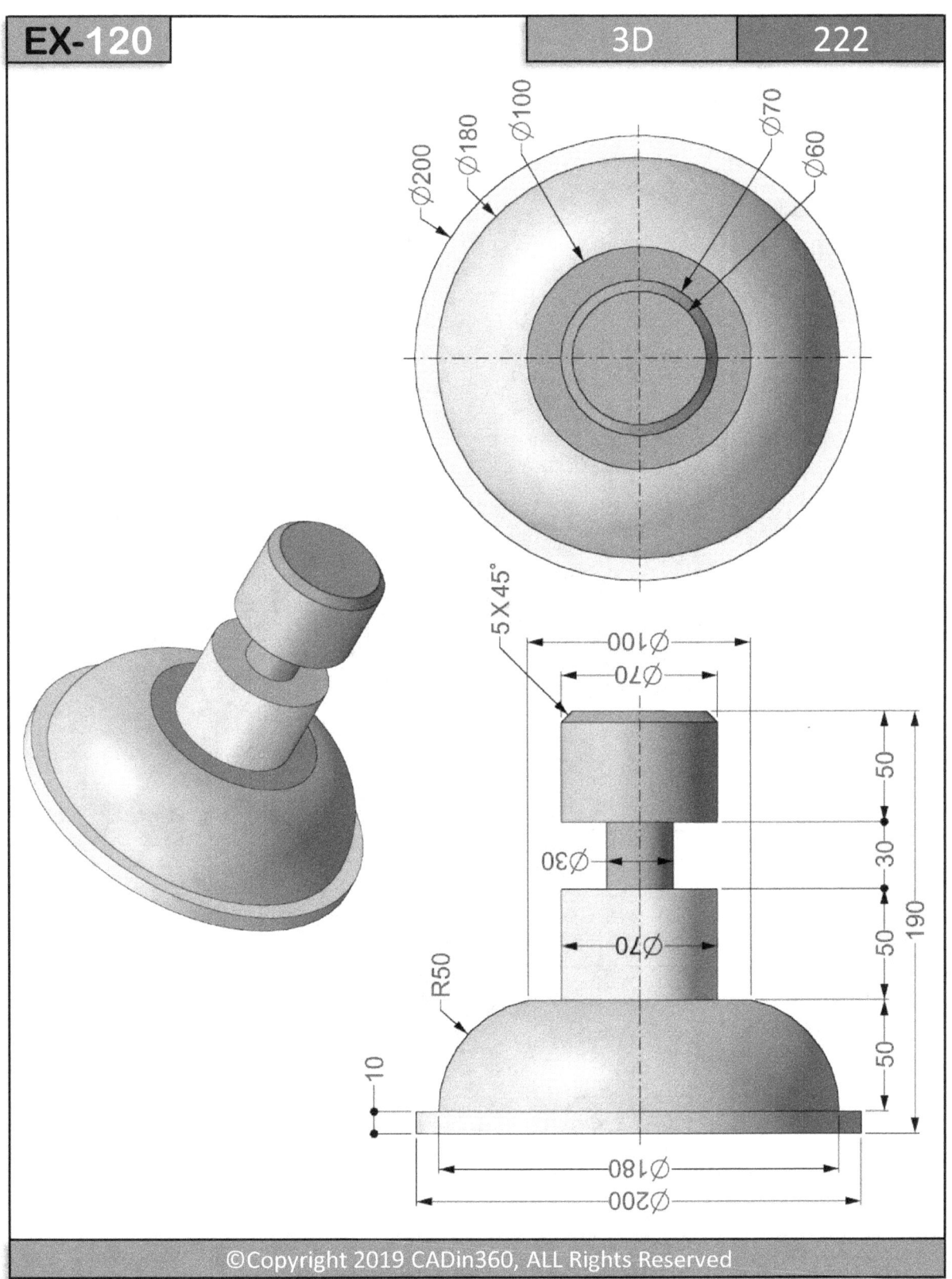

Ø200
Ø180
Ø100
Ø70
Ø60

5 X 45°
Ø100
Ø70
Ø30
Ø70
R50
10
Ø180
Ø200

50
30
50
50
190

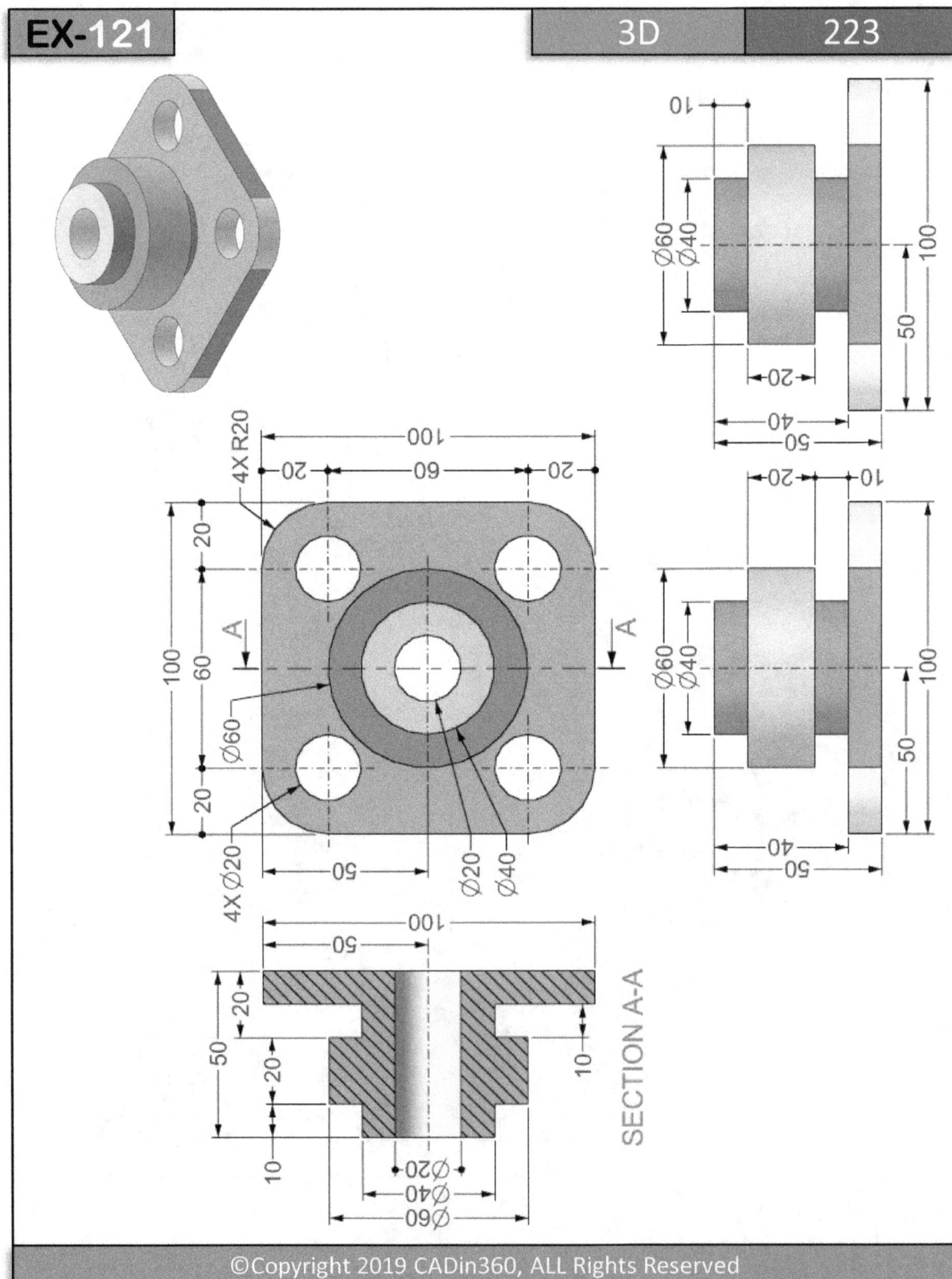

EX-121

3D 223

4X R20

100
20 60 20

A A

100
60

20 20

Ø60

4X Ø20

50

Ø20
Ø40

Ø60
Ø40

10

100

50

20

40
50

20

Ø60
Ø40

20 10

100

50

40
50

100
50

20

50

20

10

Ø20
Ø40
Ø60

SECTION A-A

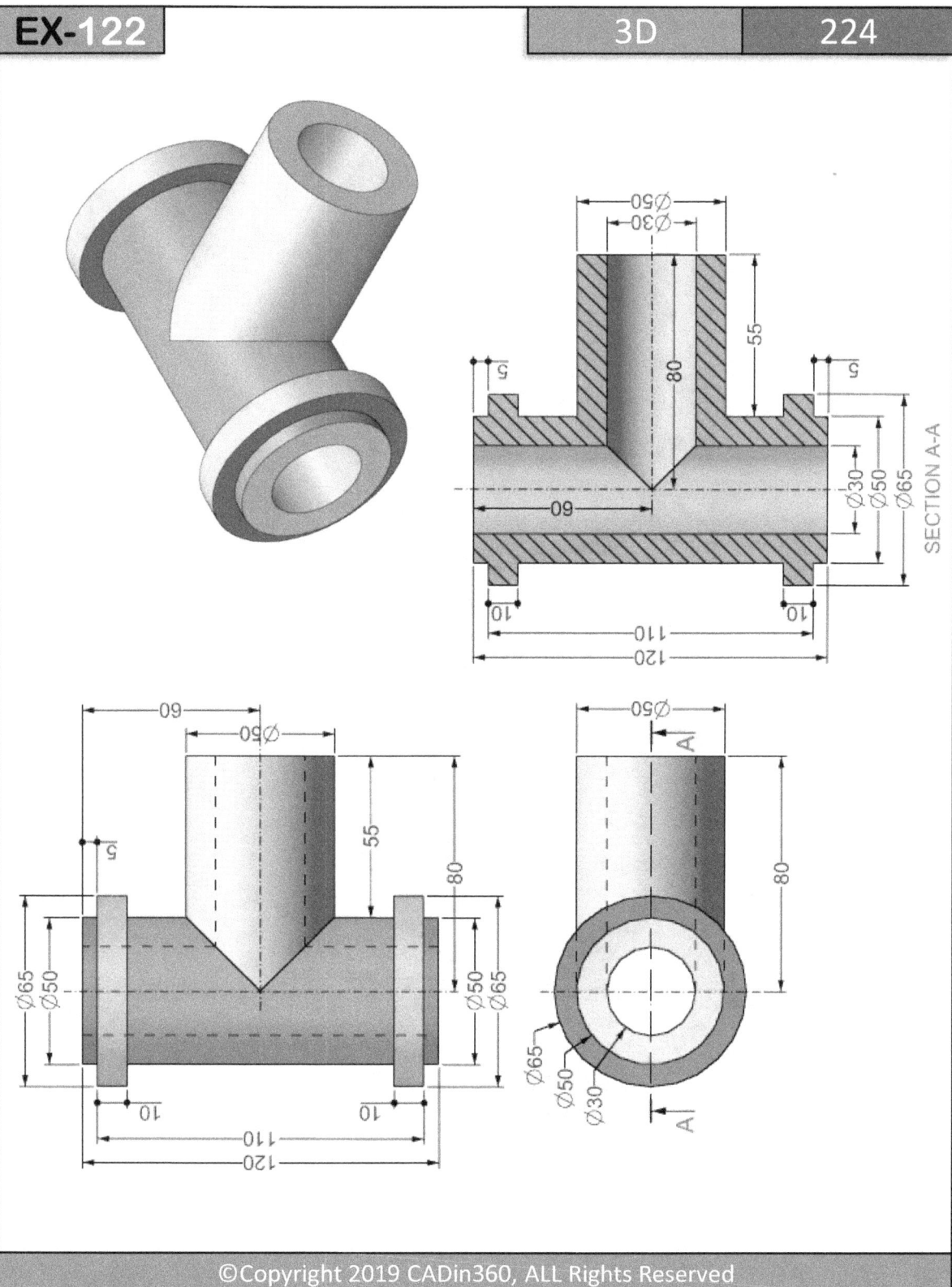

SECTION A-A

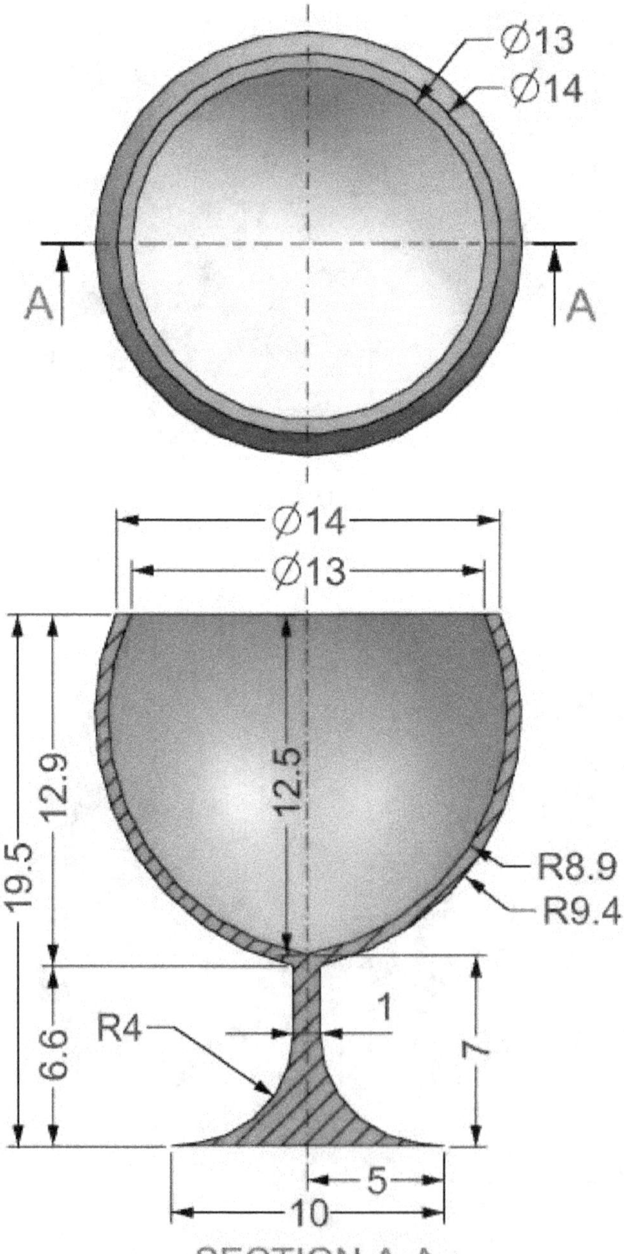

Ø13
Ø14

A | A

Ø14
Ø13

12.9
19.5
12.5

R8.9
R9.4

6.6
7
1

R4
5
10

SECTION A-A

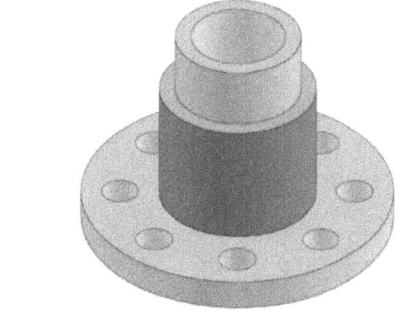

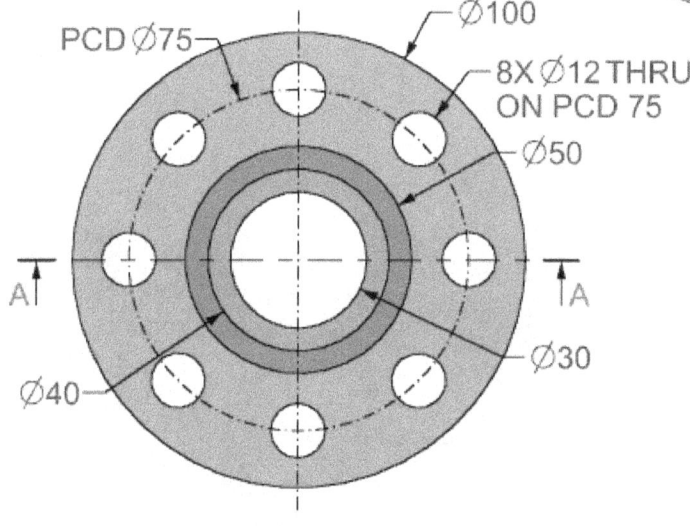

PCD Ø75

Ø100

8X Ø12 THRU
ON PCD 75

Ø50

A

A

Ø30

Ø40

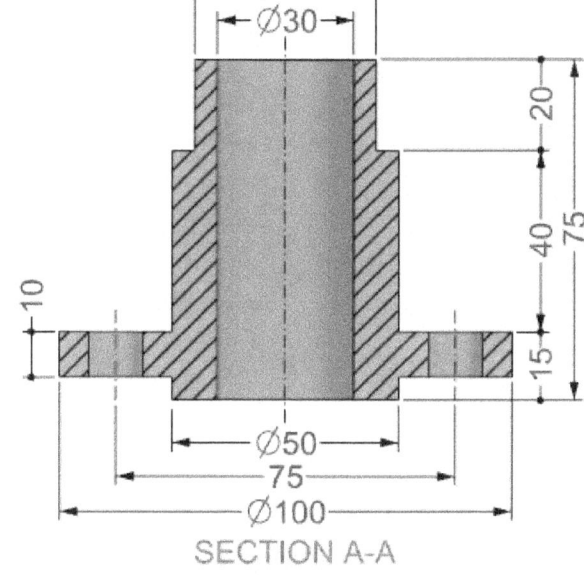

Ø40

Ø30

20

75

40

10

15

Ø50

75

Ø100

SECTION A-A

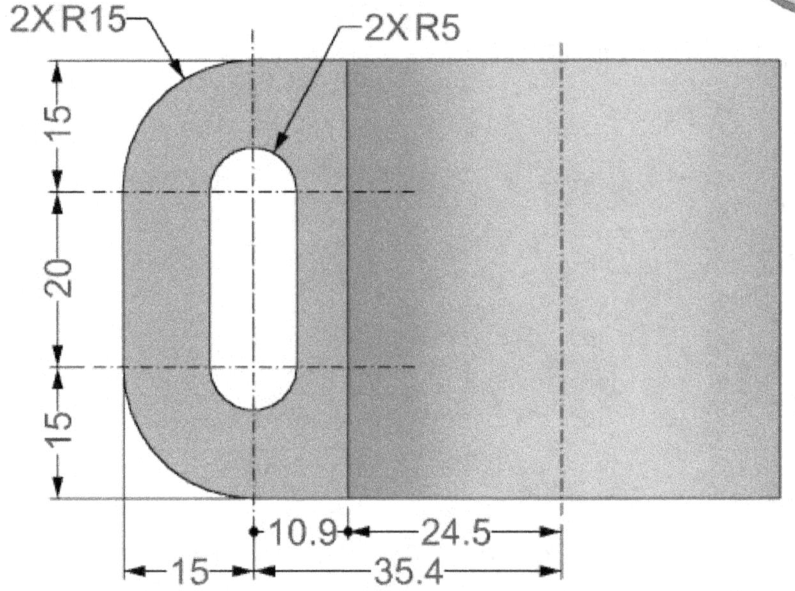

2X R15

2X R5

15

20

15

10.9

24.5

15

35.4

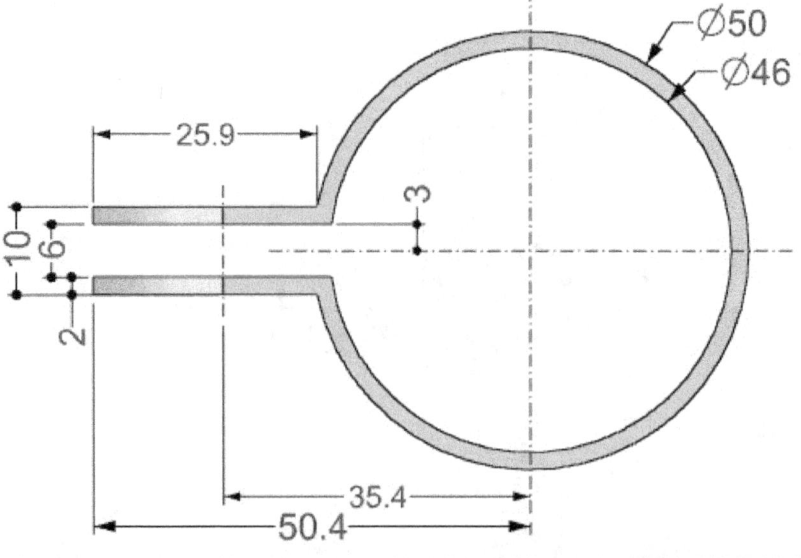

Ø50

Ø46

25.9

3

10

6

2

35.4

50.4

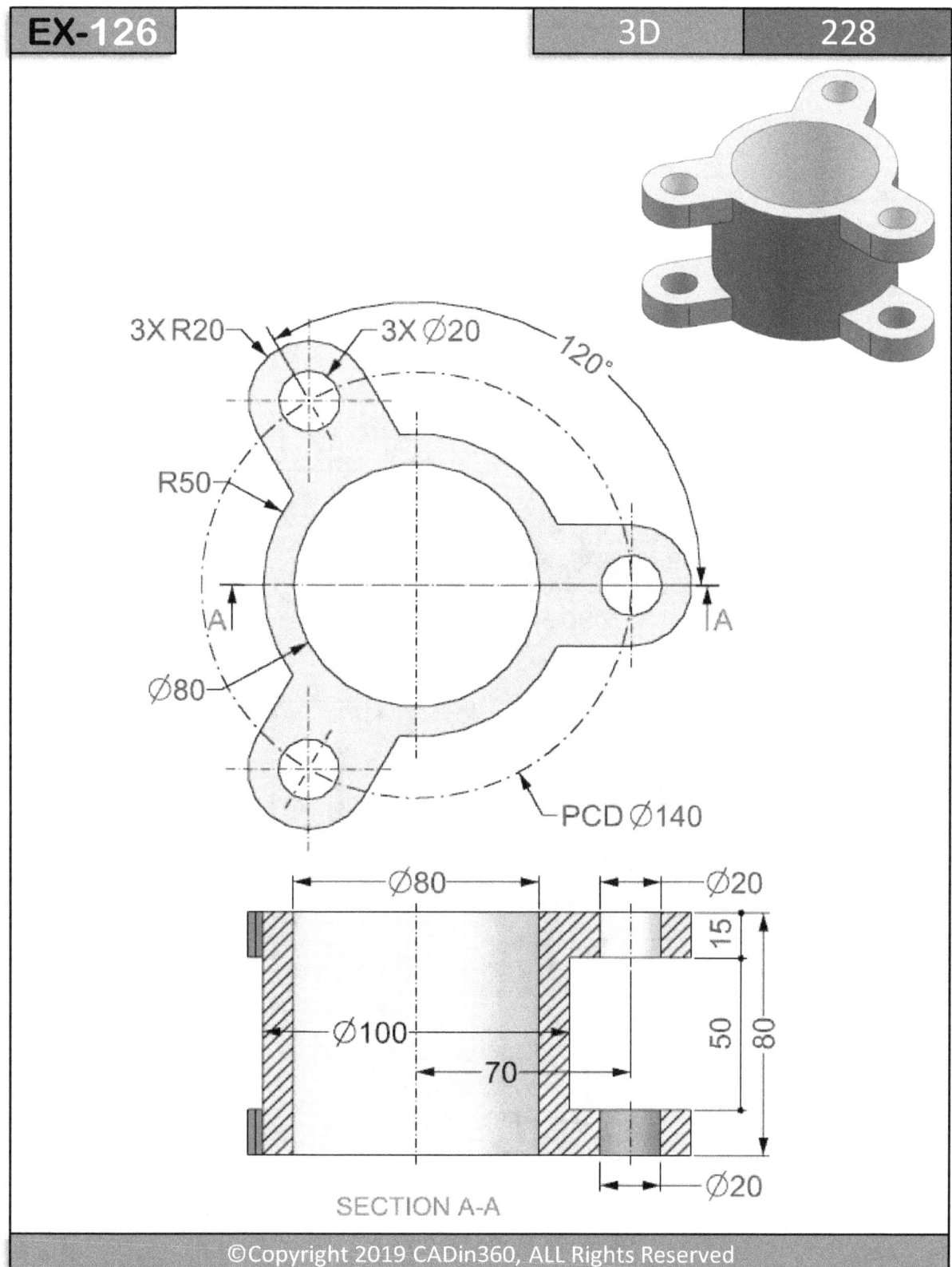

3X R20   3X Ø20   120°

R50

A

Ø80

PCD Ø140

Ø80   Ø20

15

Ø100

50   80

70

Ø20

SECTION A-A

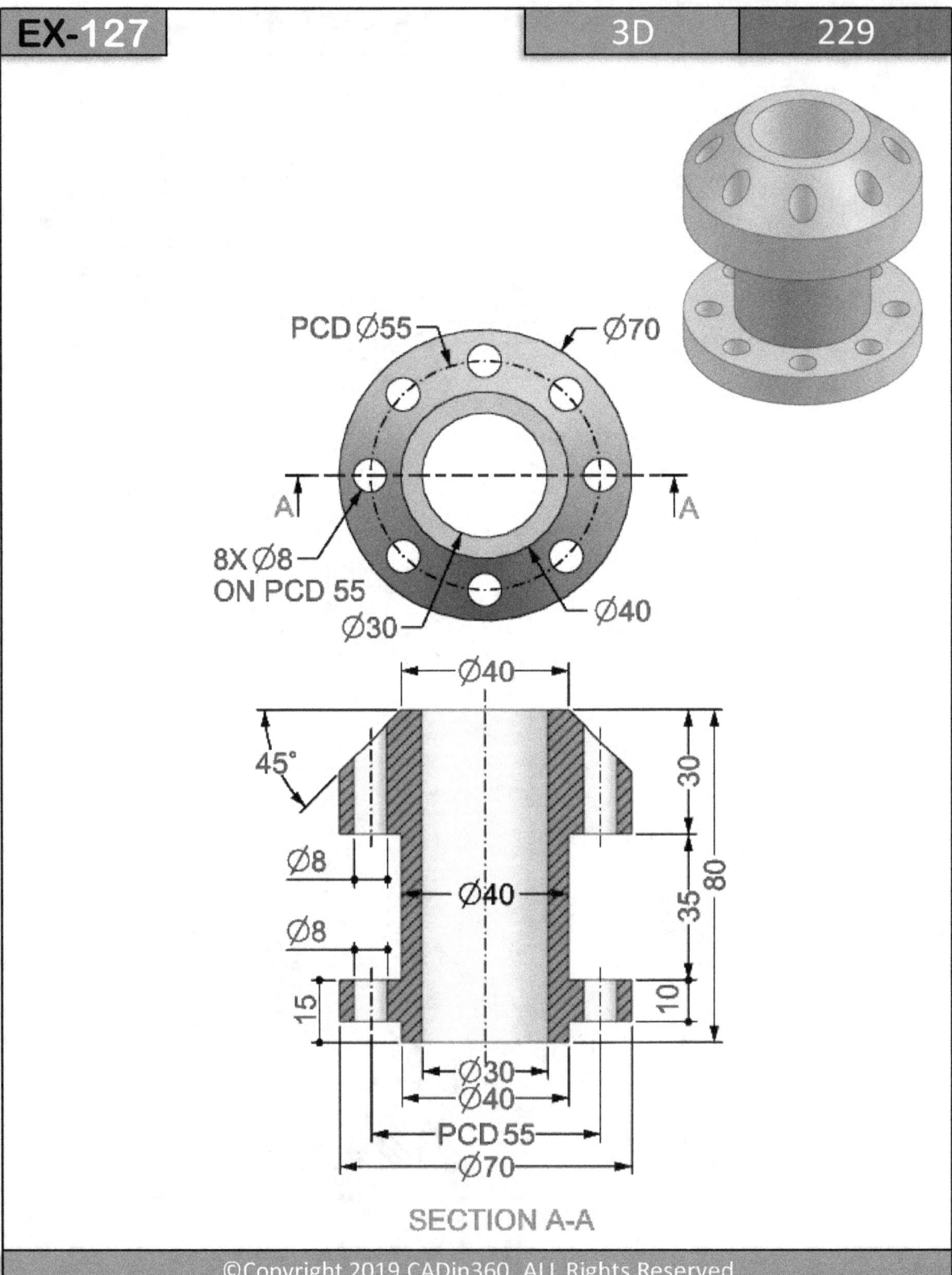

PCD Ø55   Ø70

8X Ø8
ON PCD 55

Ø30   Ø40

A   A

Ø40

45°

Ø8

Ø8

Ø40

15   Ø30
Ø40
PCD 55
Ø70

30

80

35

10

SECTION A-A

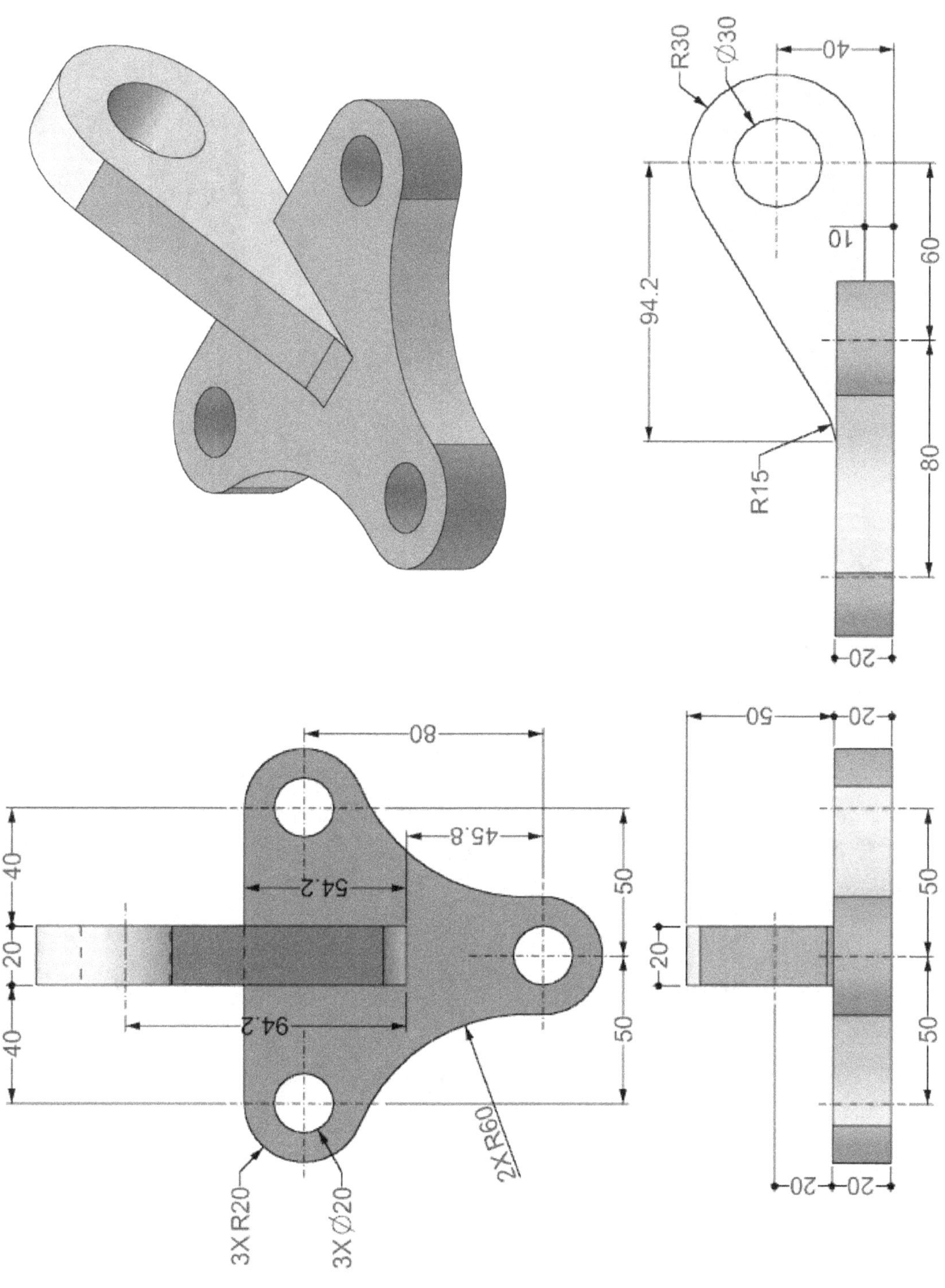

R30
Ø30
40
10
60
94.2
R15
80
20

80
45.8
54.2
94.2
40
20
40
50
50
2X R60
3X R20
3X Ø20

50
20
20
50
50
20
20

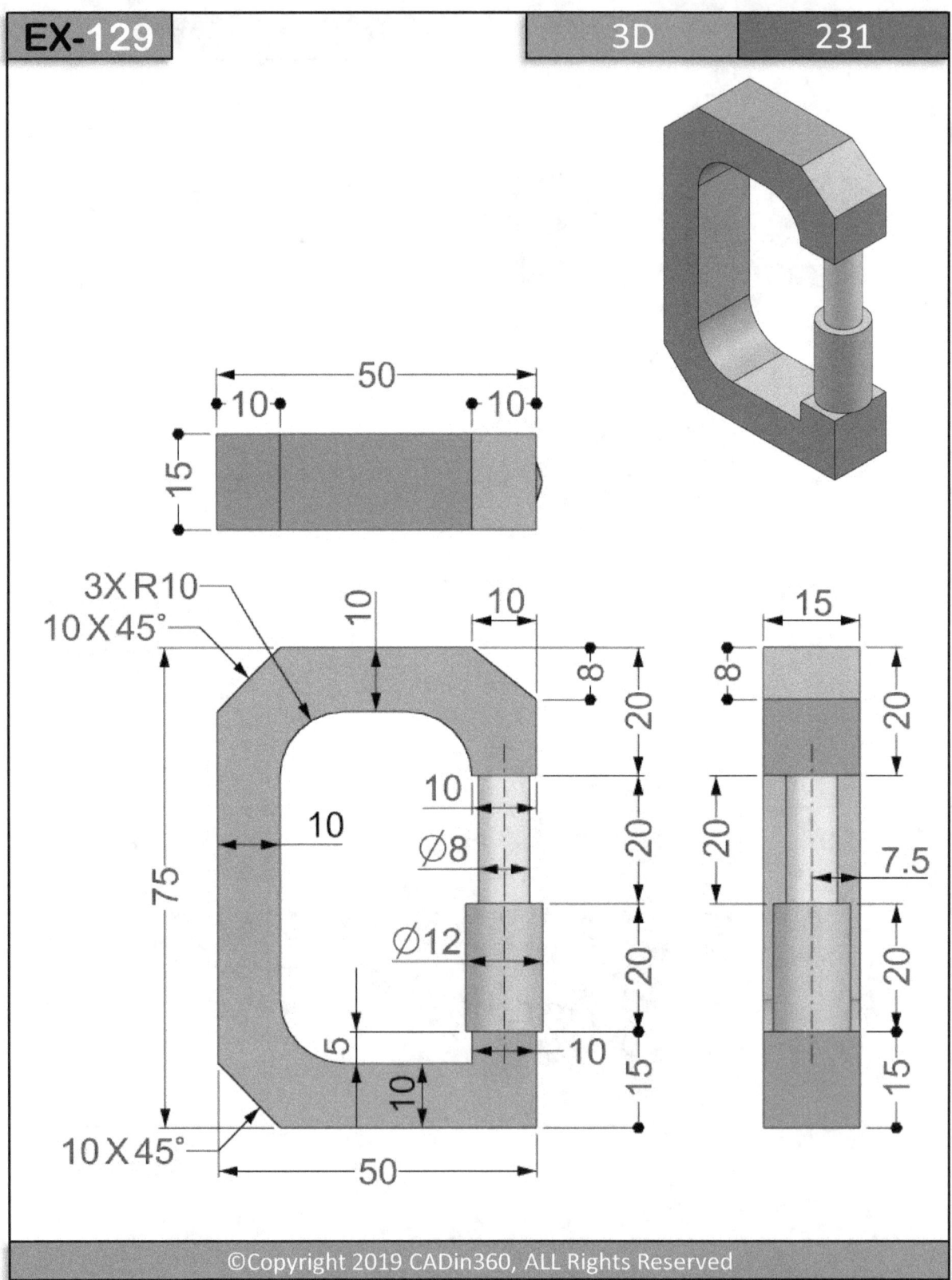

50

10   10

15

3X R10

10 X 45°

10

75

10 X 45°

50

10

10

8

20

Ø8

20

Ø12

20

10

15

5

10

15

8

8

20

20

20

20

7.5

15

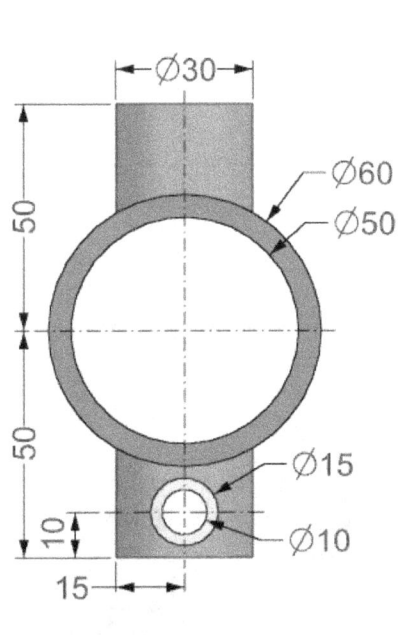

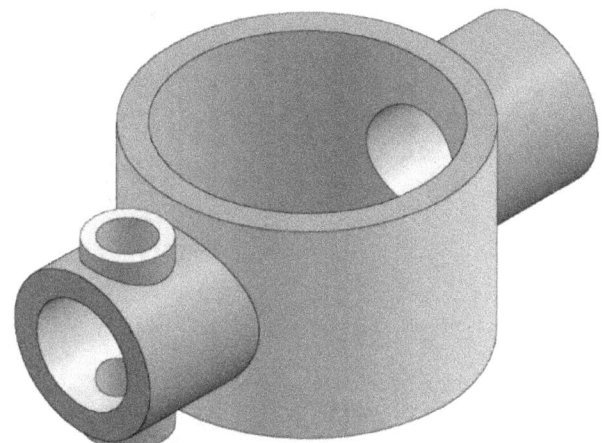

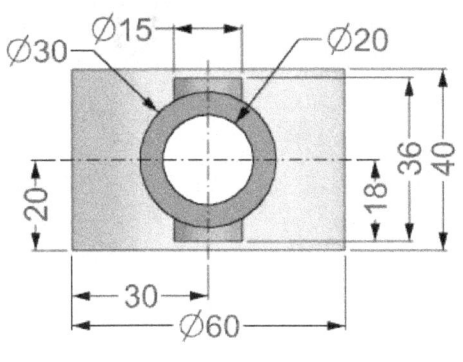

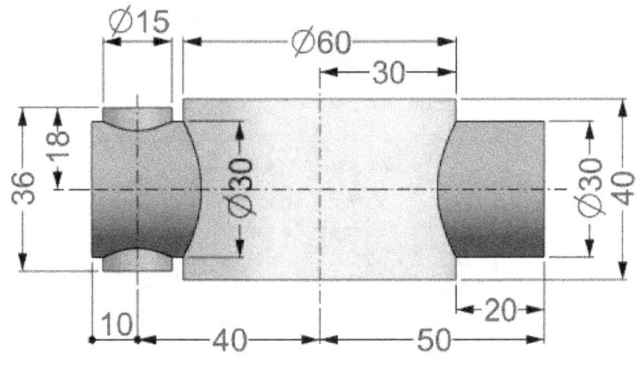

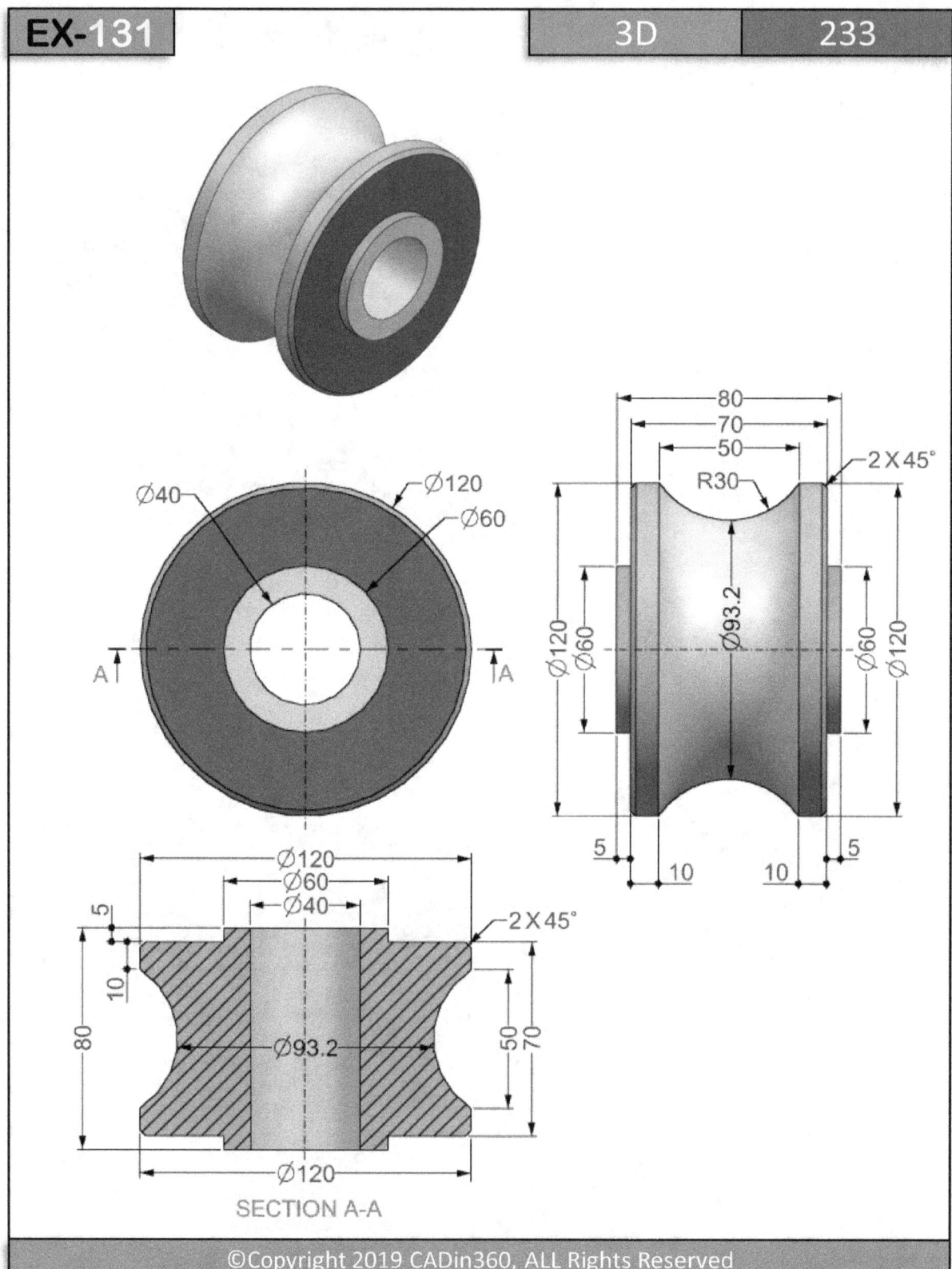

EX-131   3D   233

Ø40   Ø120
Ø60

A   A

Ø120
Ø60
Ø40
2 X 45°

5
10

Ø93.2
80
50
70

Ø120

SECTION A-A

80
70
50
R30   2 X 45°

Ø120
Ø60
Ø93.2
Ø60
Ø120

5   5
10   10

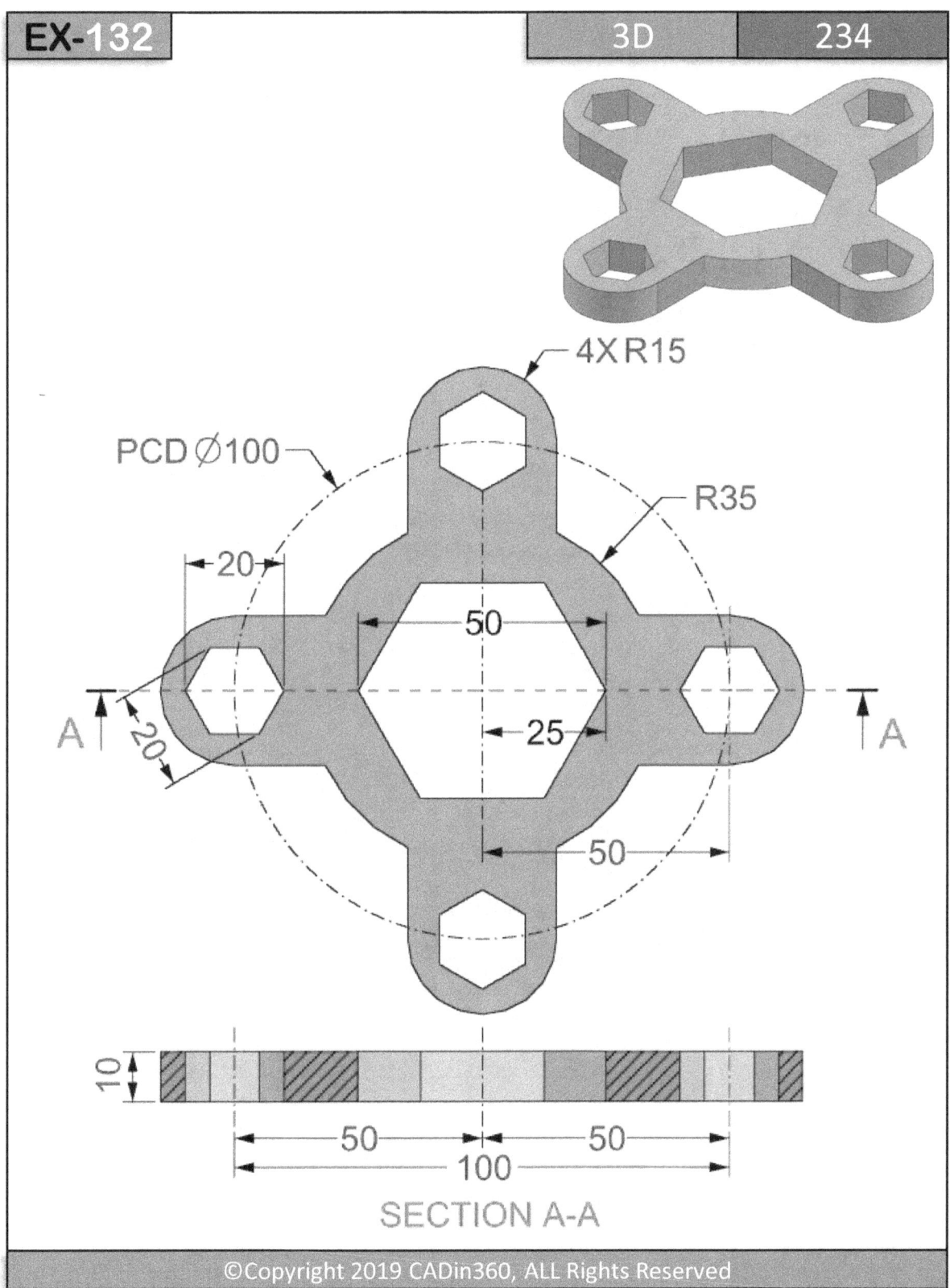

EX-132   3D   234

4X R15

PCD Ø100

R35

20

50

25

50

20

50

A   A

10

50   50

100

SECTION A-A

R60

4X R20

Ø20

Ø120

34.6

20

10

30

39.6

39.6

R10

25

15

40

30

10

5

5

5

5

39.6

10

39.6

Ø20

30

10

39.6

5

30

10

Ø120

15

39.6

R10

25

40

2X R6

2X R5

A

20

10

10

15

5

10

10

34.6

10.2

10   20   10

A

6

B

55

3

SECTION A-A
(SCALE 1:1)

R1

1

1

1

DETAIL B
(SCALE 2:1)

SHELL THICKNESS = 1MM
ALL INSIDE WALL THICKNESS

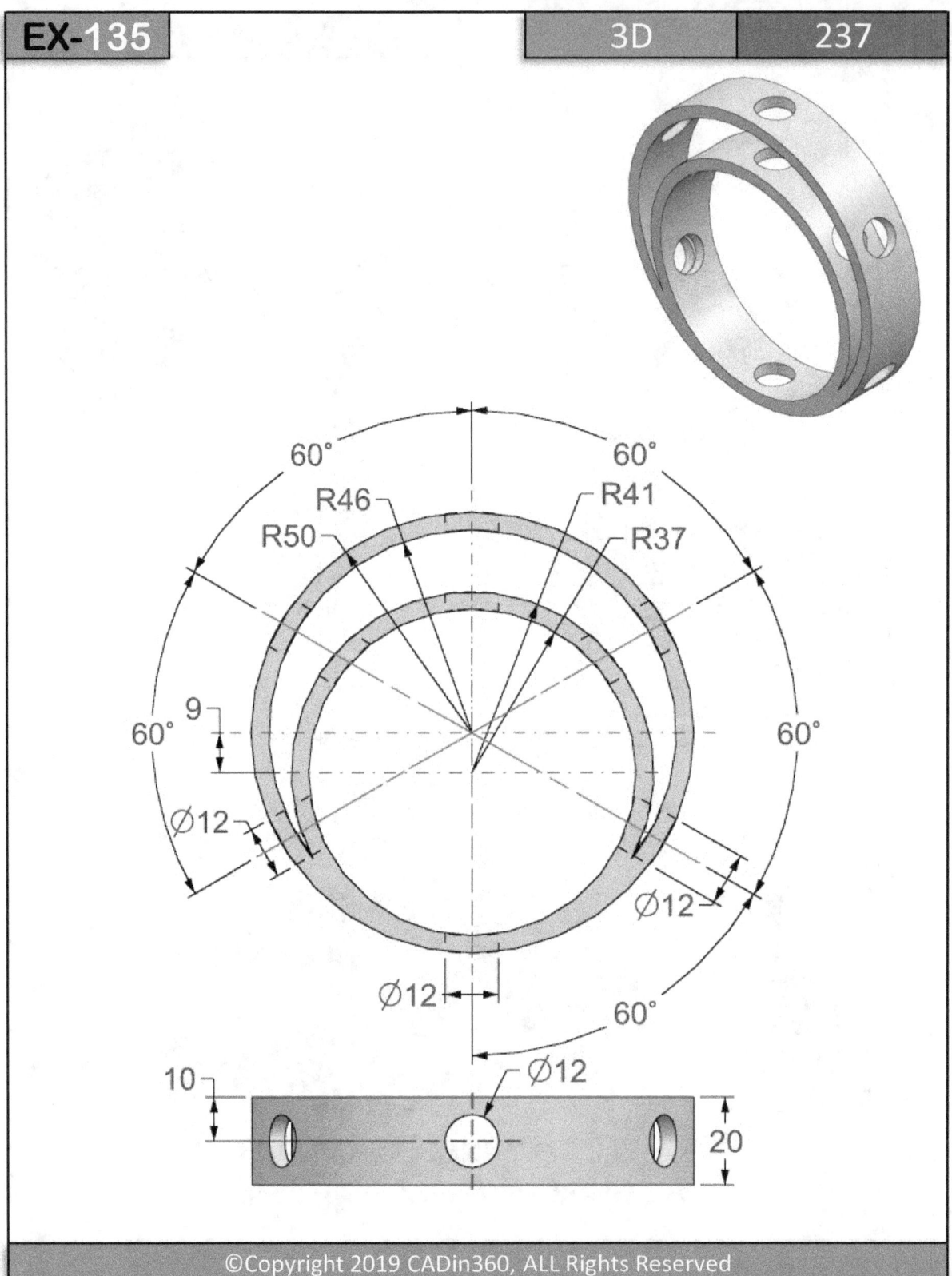

60°   60°

R46   R41

R50   R37

60°   60°

9

∅12

∅12

∅12

60°

10

∅12

20

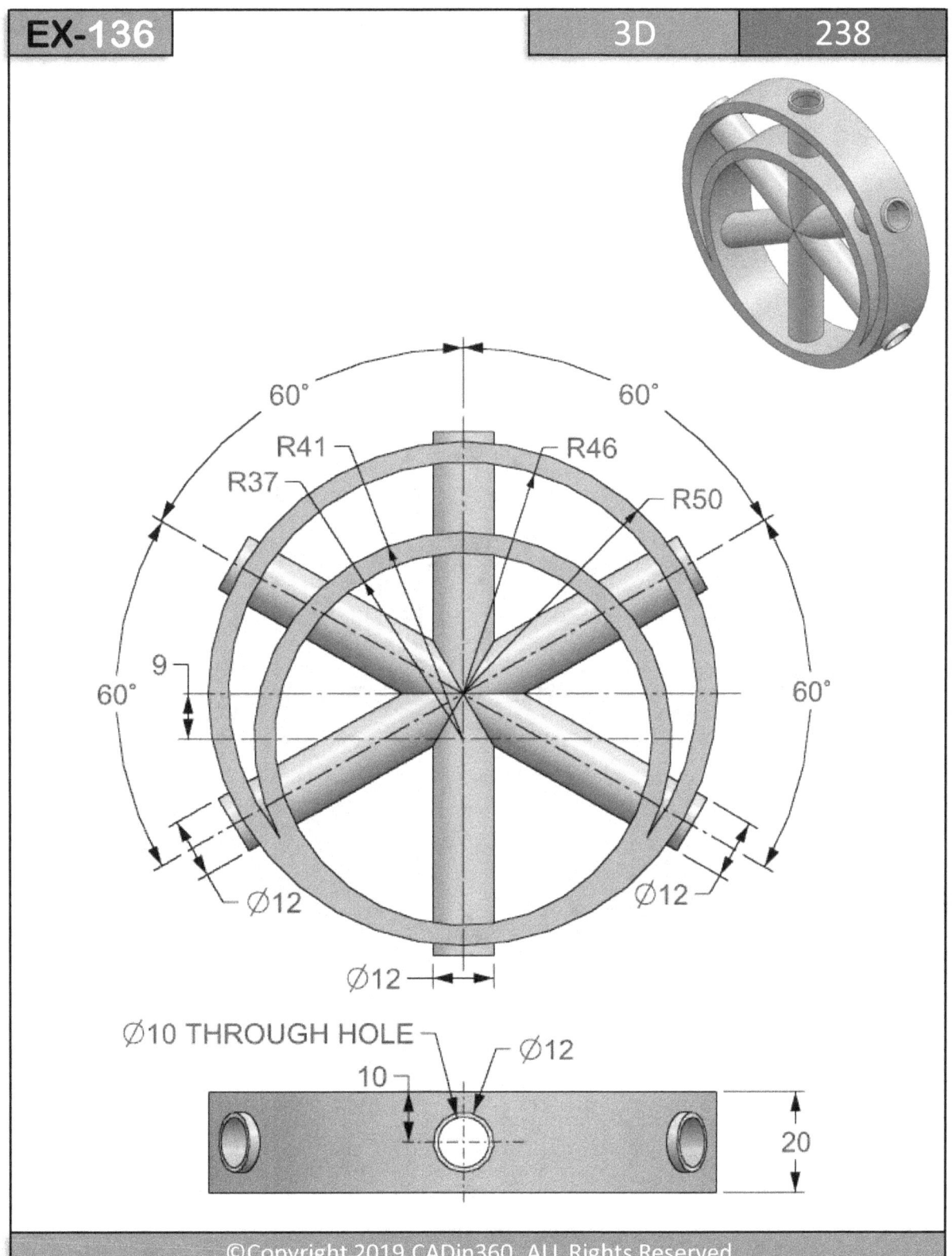

60°   60°

R41   R46

R37   R50

60°   9   60°

Ø12   Ø12

Ø12

Ø10 THROUGH HOLE   Ø12

10

20

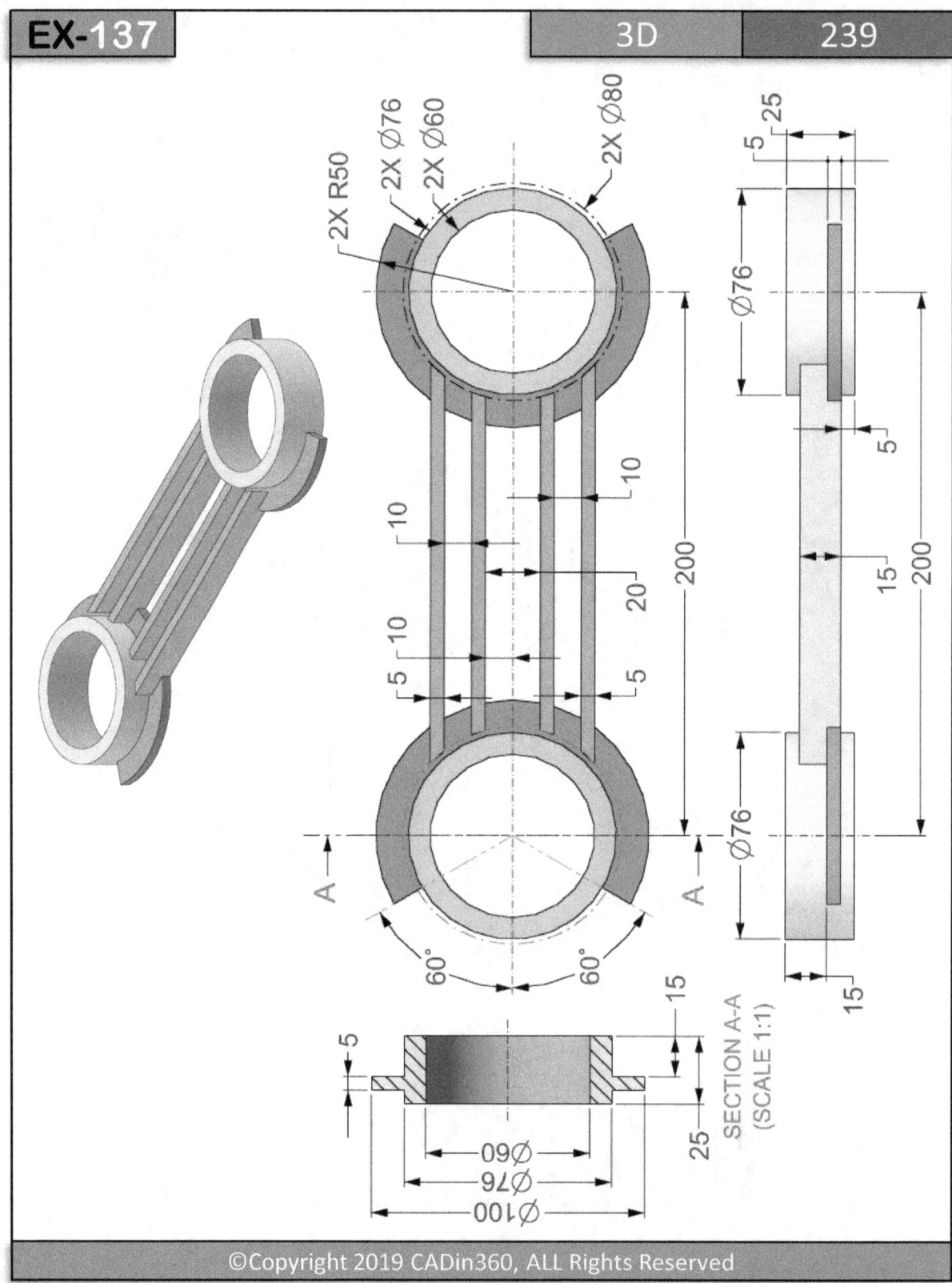

2X R50
2X Ø76
2X Ø60
2X Ø80

10
10
10
20
200
5
5

A
A
60°
60°

Ø76
5
5
15
200
15
25

SECTION A-A
(SCALE 1:1)

15

5
Ø60
Ø76
Ø100
25

Ø76

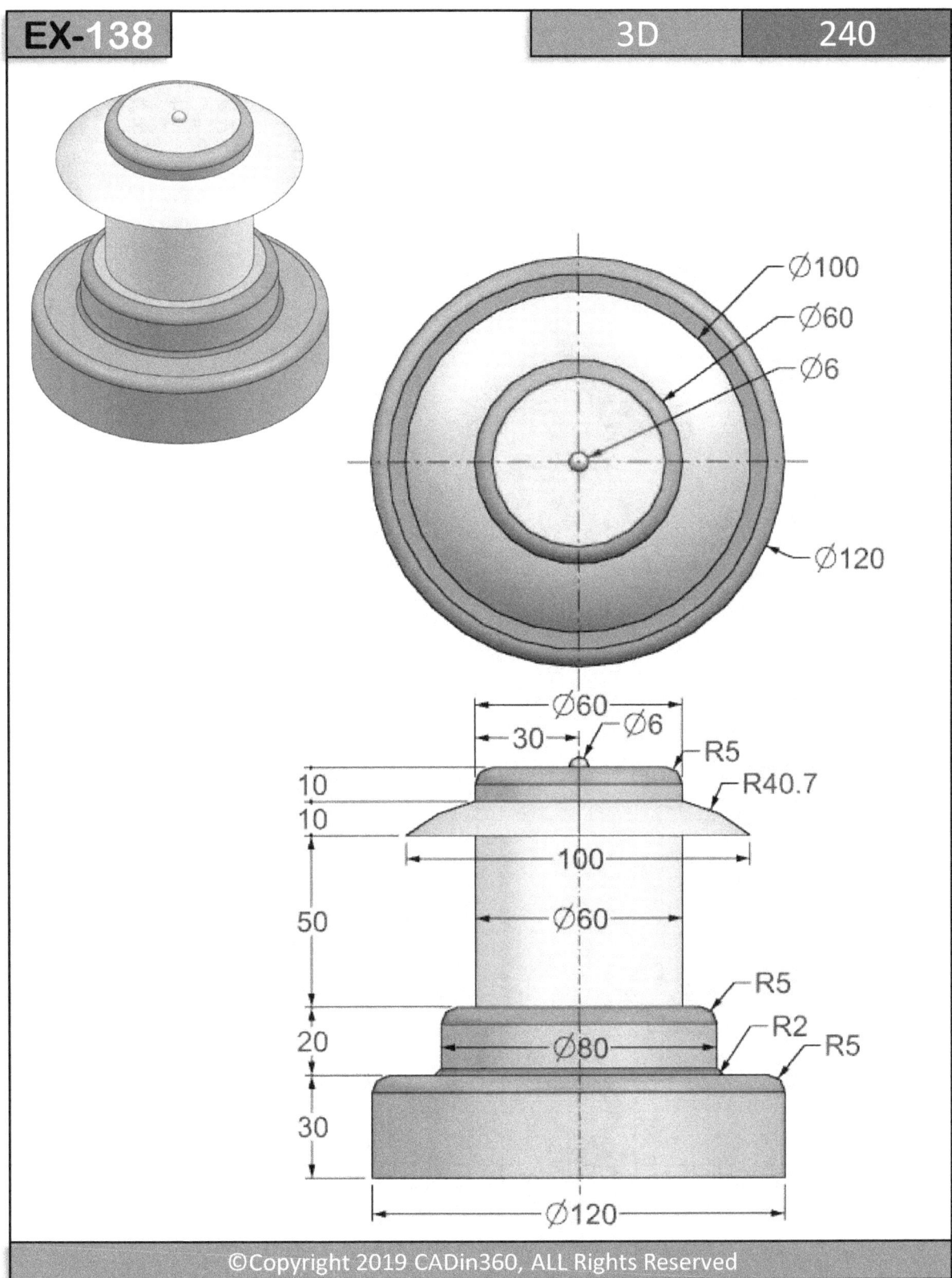

Ø100
Ø60
Ø6
Ø120

Ø60
30
Ø6
R5
R40.7
10
10
100
50
Ø60
R5
20
Ø80
R2
R5
30
Ø120

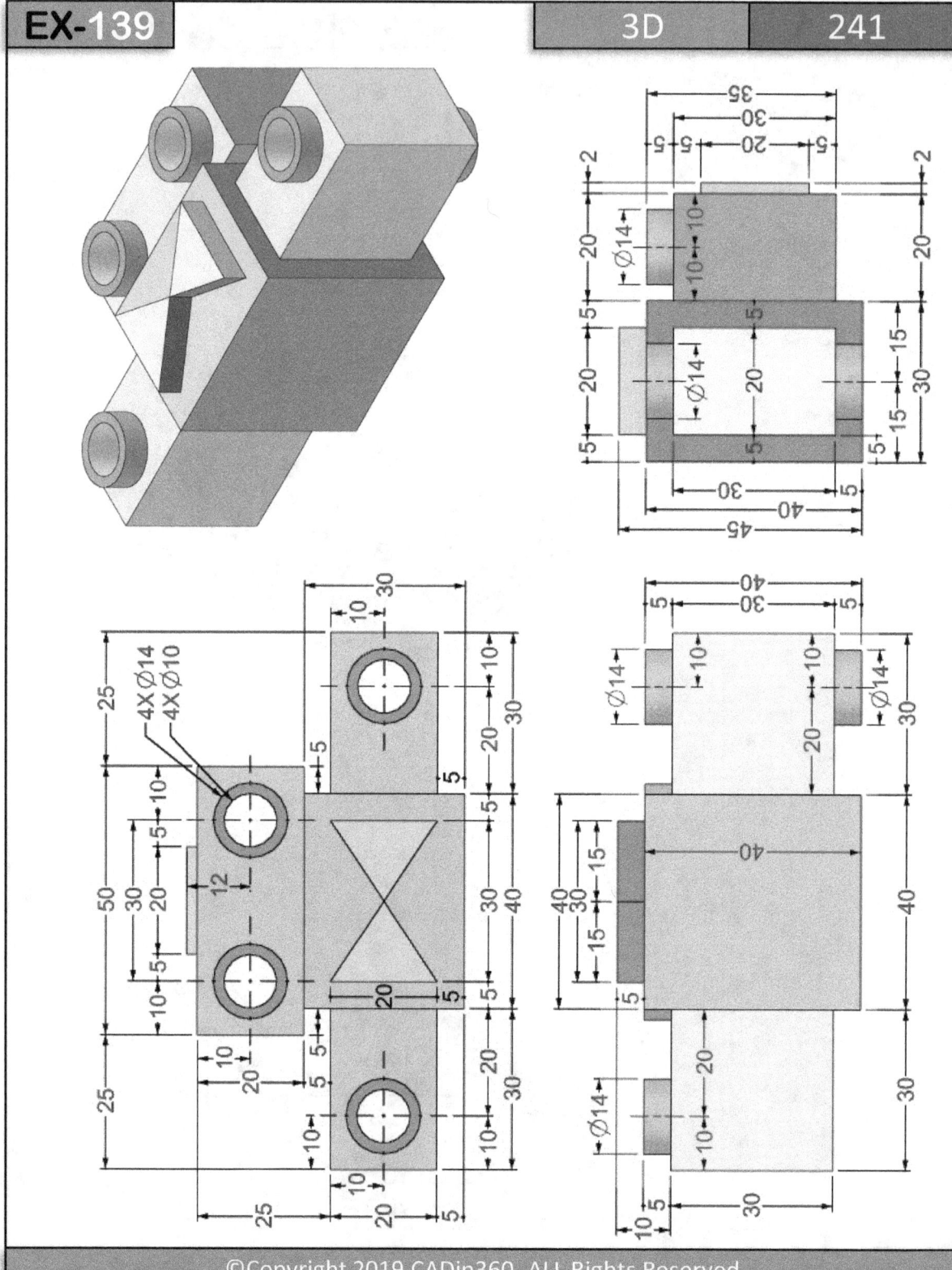

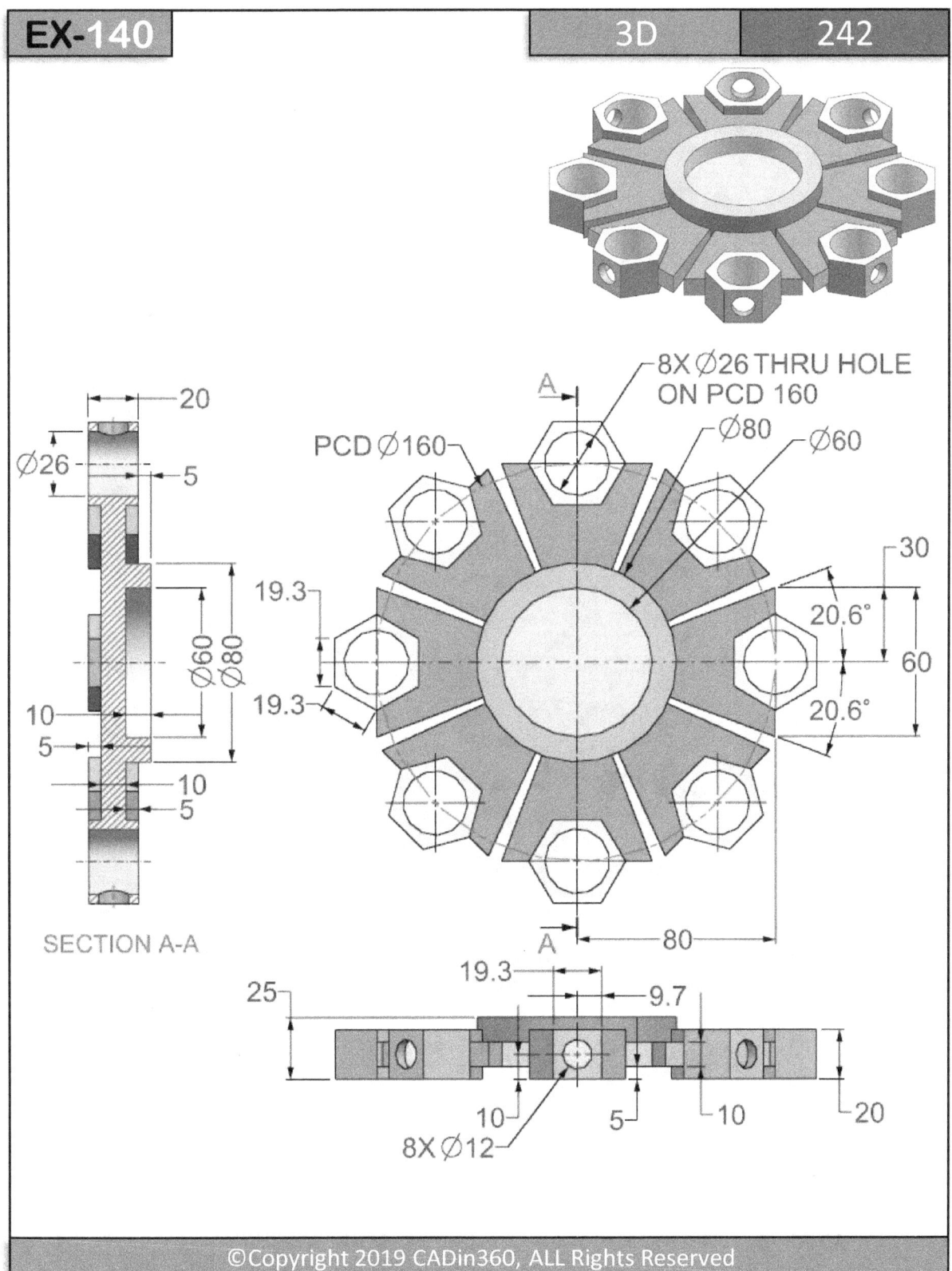

EX-140

8X Ø26 THRU HOLE
ON PCD 160

PCD Ø160

Ø80

Ø60

20

Ø26

5

Ø60
Ø80

10

5

10

5

SECTION A-A

A

19.3

19.3

30

20.6°

60

20.6°

A

80

25

19.3

9.7

10

5

10

20

8X Ø12

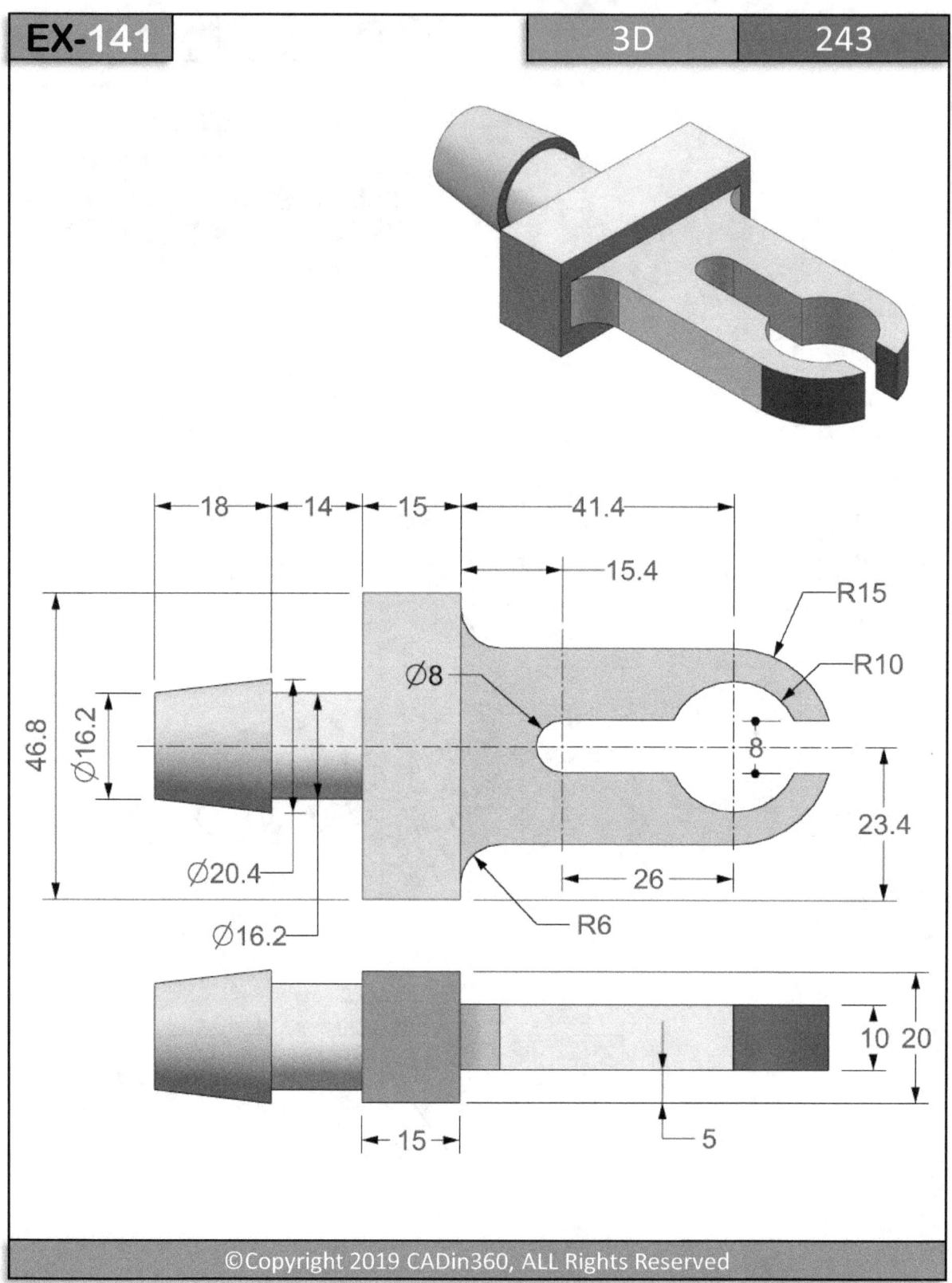

⌀27

5
10

R22.8
R21.8

DETAIL A-A
SCALE 5:1

13.2
10
8
A-A
R5
⌀21.5
⌀8
4
30°
11.1
R12.8
15
22.5
6
9
6
9
24
R2
⌀15
⌀21
⌀27

10
8.5
8.5
6
9
6 24
9
R2
⌀15
⌀21
⌀27

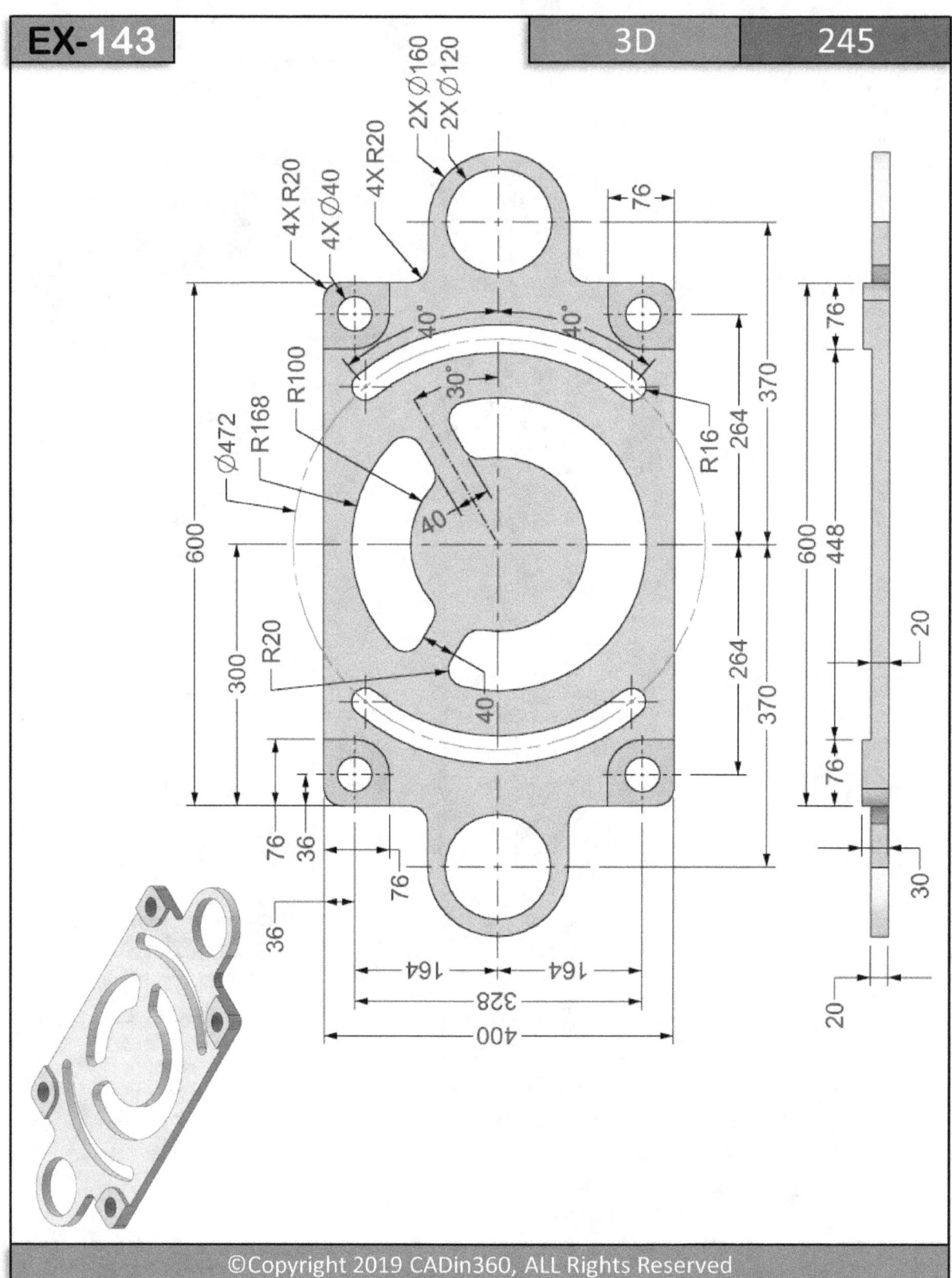

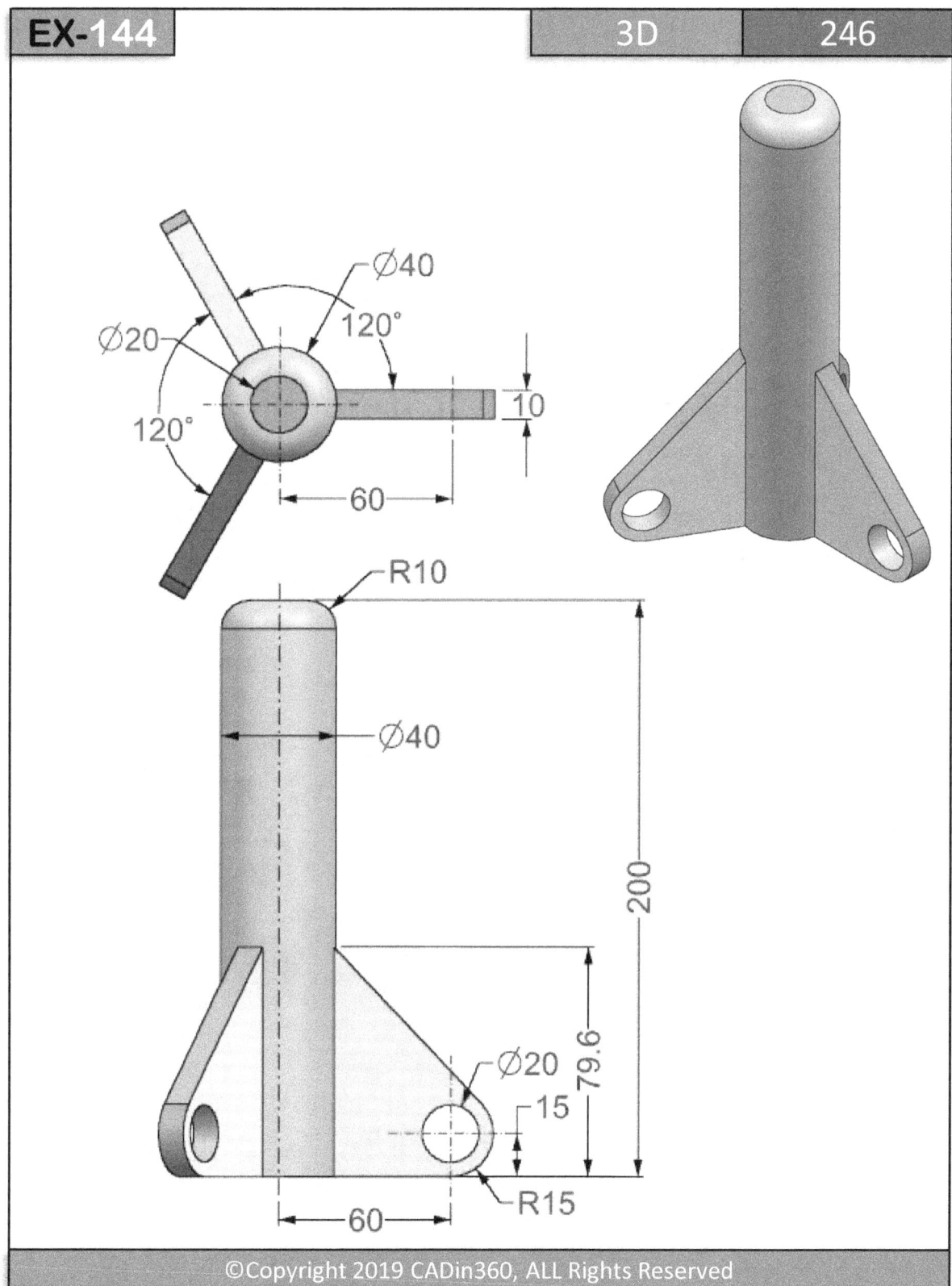

EX-144

3D

246

Ø40

120°

Ø20

120°

10

60

R10

Ø40

200

79.6

Ø20

15

R15

60

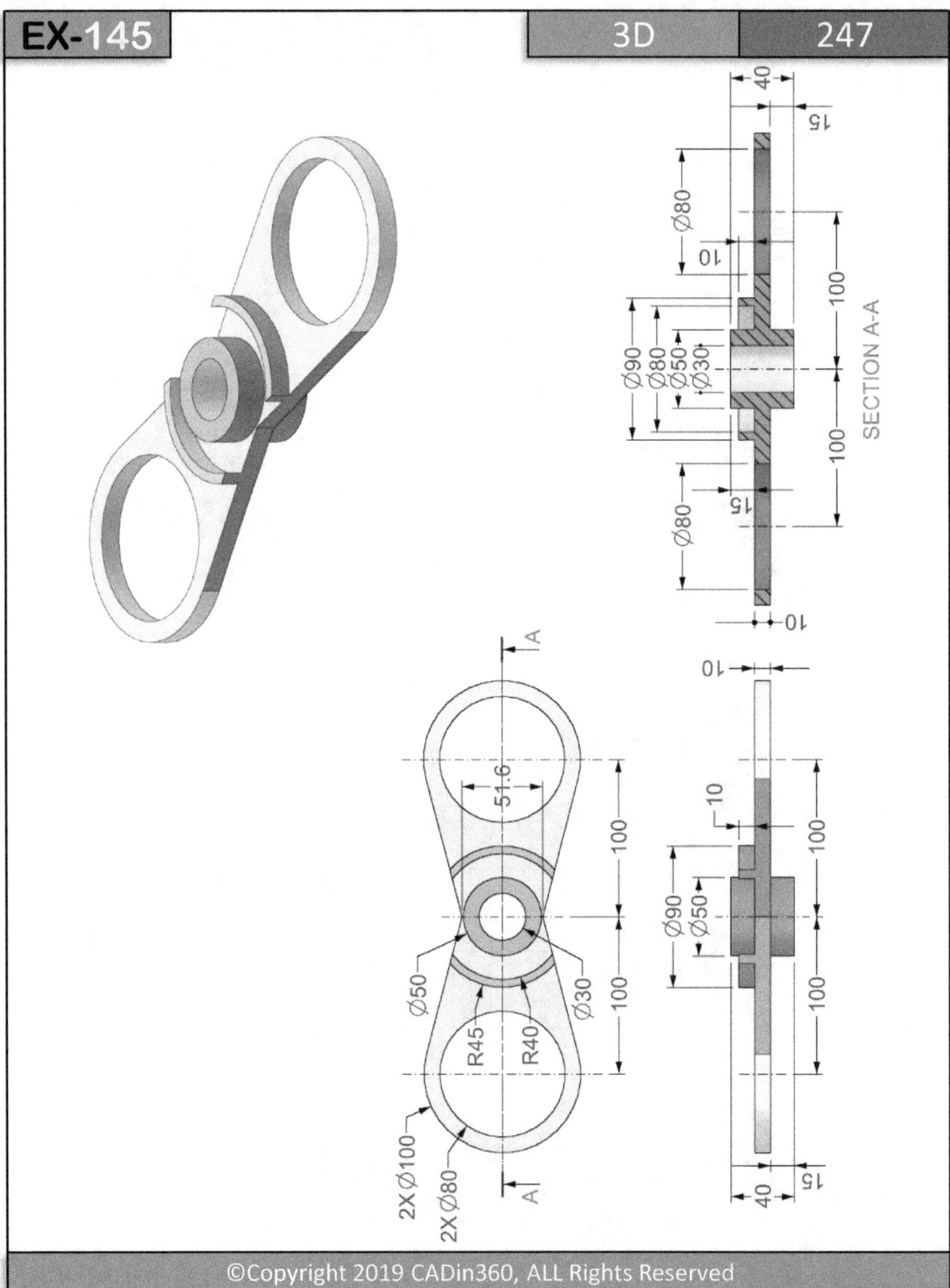

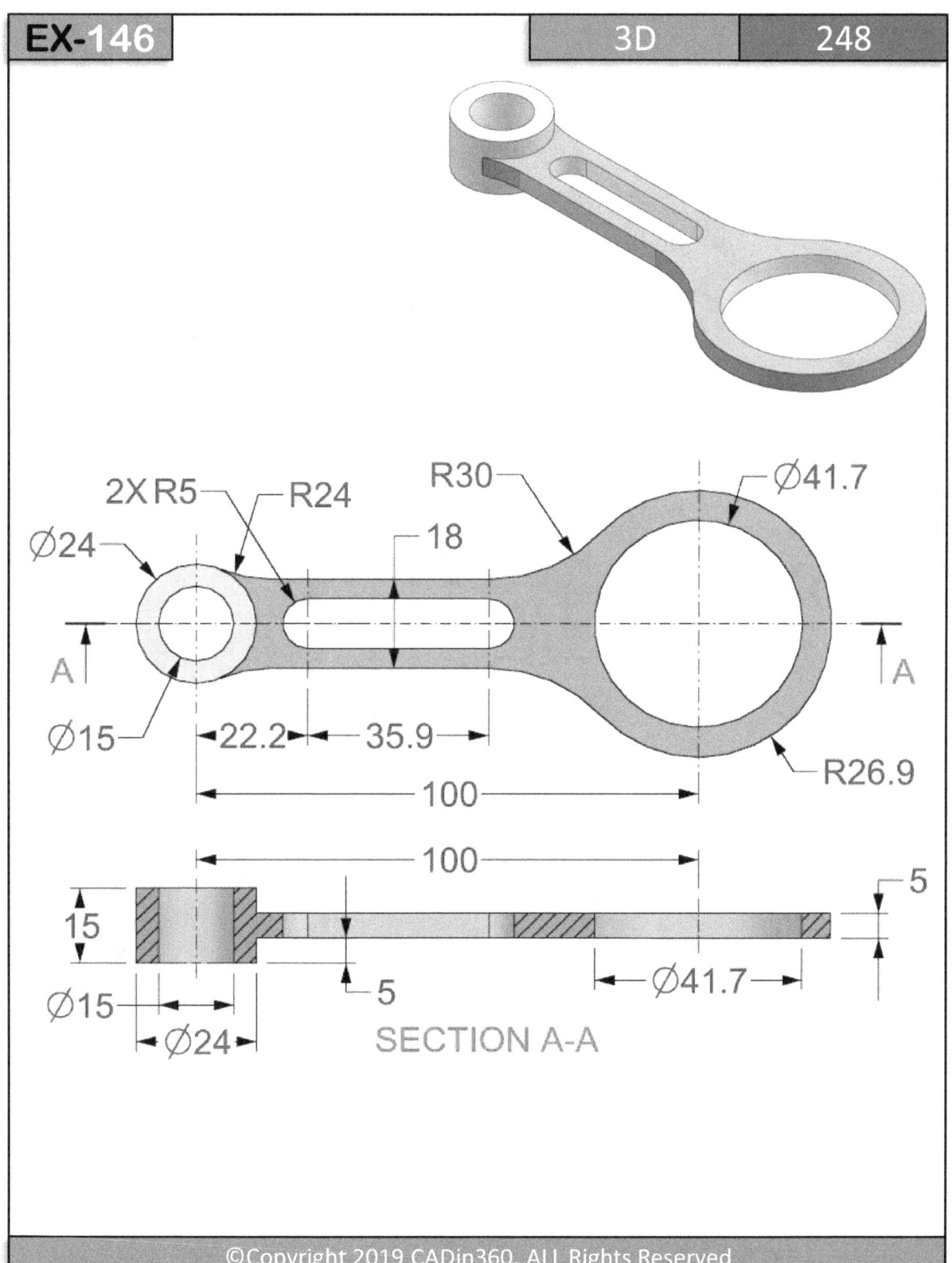

EX-146   3D   248

∅24
2X R5 — R24
18
R30 — ∅41.7

A

∅15
22.2   35.9
100

R26.9

100

15   5
∅41.7
∅15   5
∅24

SECTION A-A

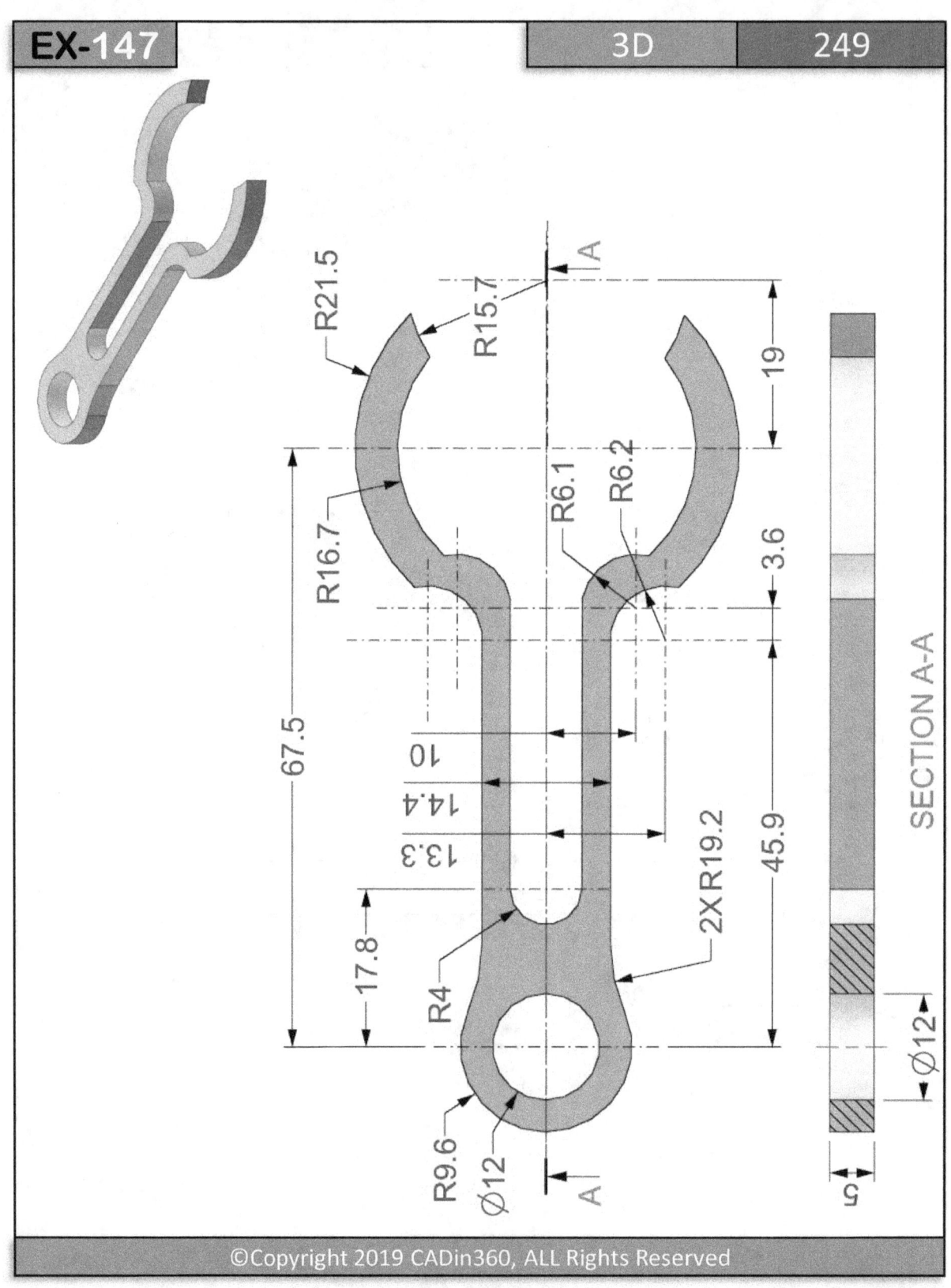

R21.5

R15.7

A

19

R6.1

R6.2

3.6

R16.7

67.5

10

14.4

13.3

45.9

2XR19.2

17.8

R4

R9.6

Ø12

A

SECTION A-A

Ø12

5

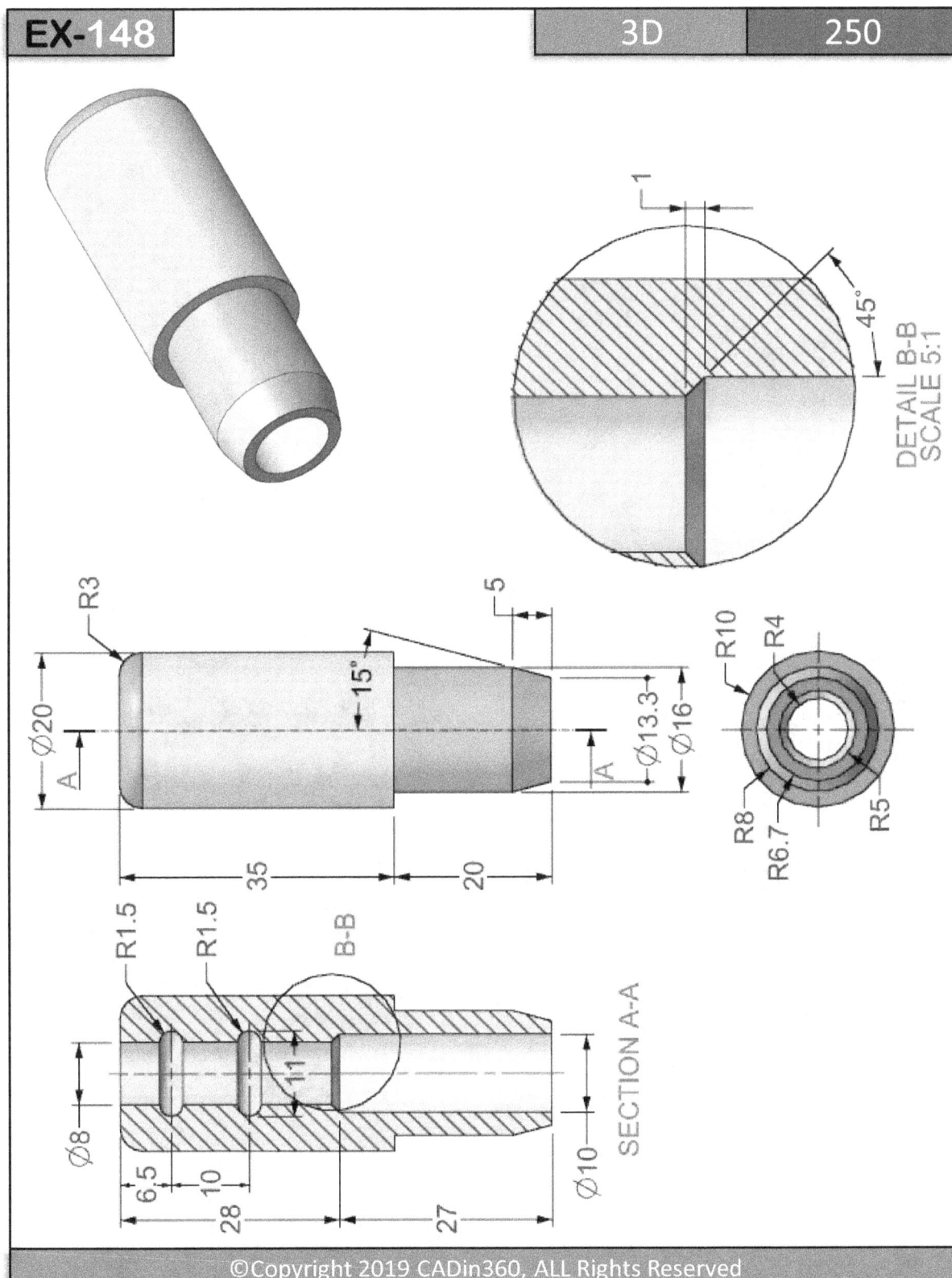

EX-148

3D

250

DETAIL B-B
SCALE 5:1

45°

1

R3

Ø20

A

15°

Ø13.3

Ø16

5

35

20

R10

R4

R8

R6.7

R5

R1.5

R1.5

B-B

Ø8

11

6.5

10

28

Ø10

27

SECTION A-A

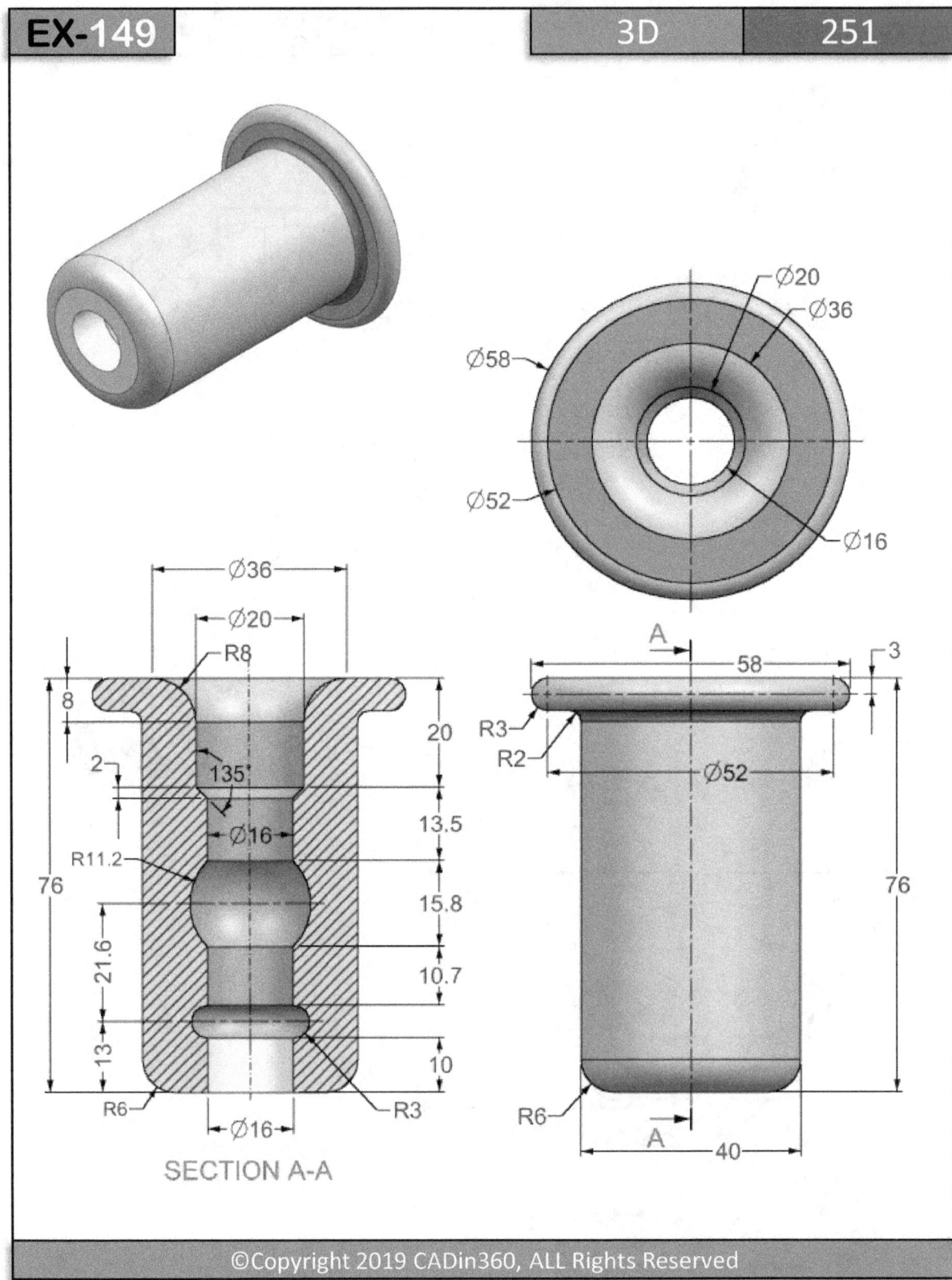

EX-149

Ø20
Ø36
Ø58
Ø52
Ø16

Ø36
Ø20
R8
8
20
2
135°
Ø16
13.5
R11.2
15.8
76
21.6
10.7
13
10
R6
R3
Ø16

SECTION A-A

A
58
3
R3
R2
Ø52
76
R6
A
40

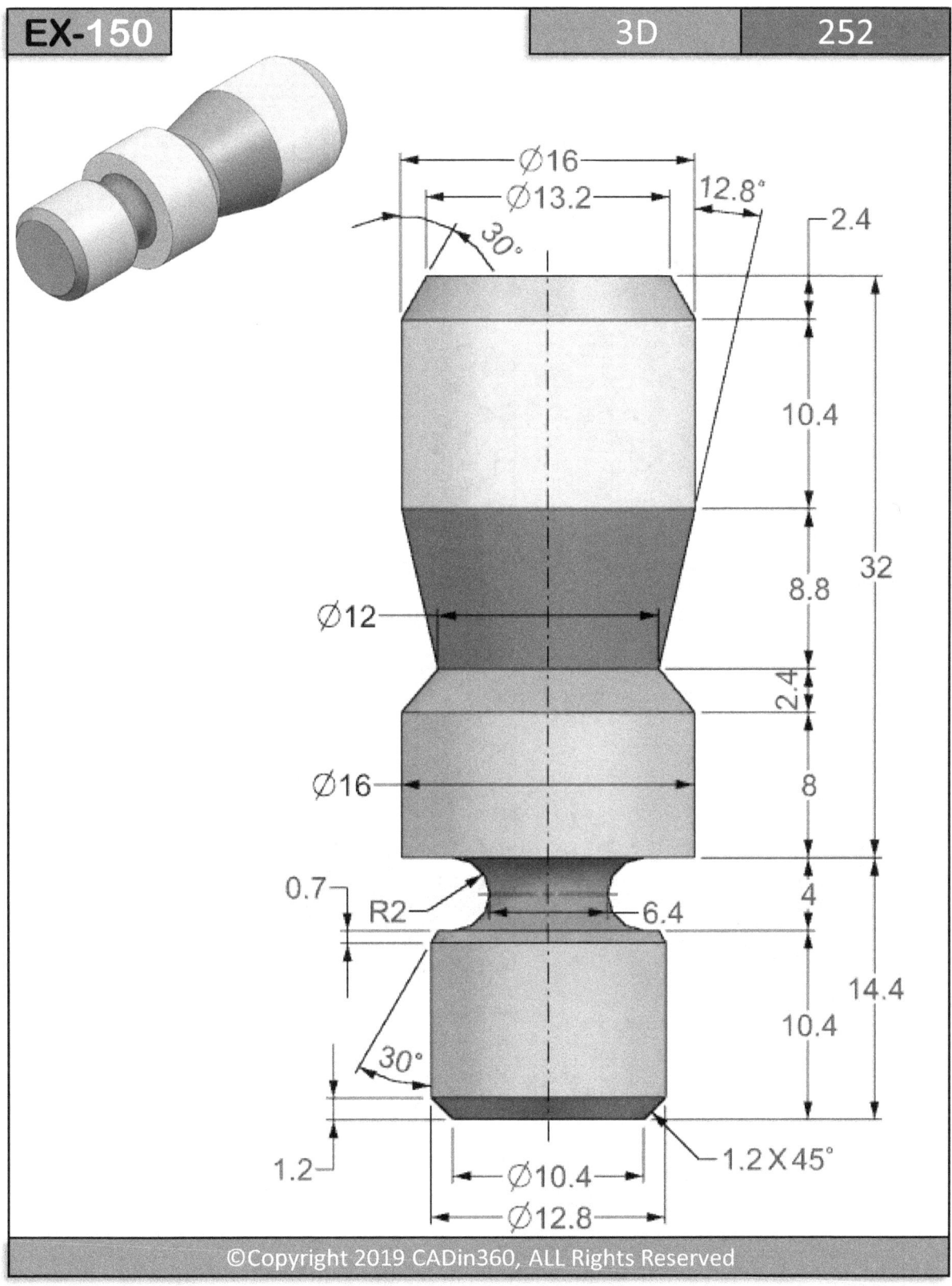

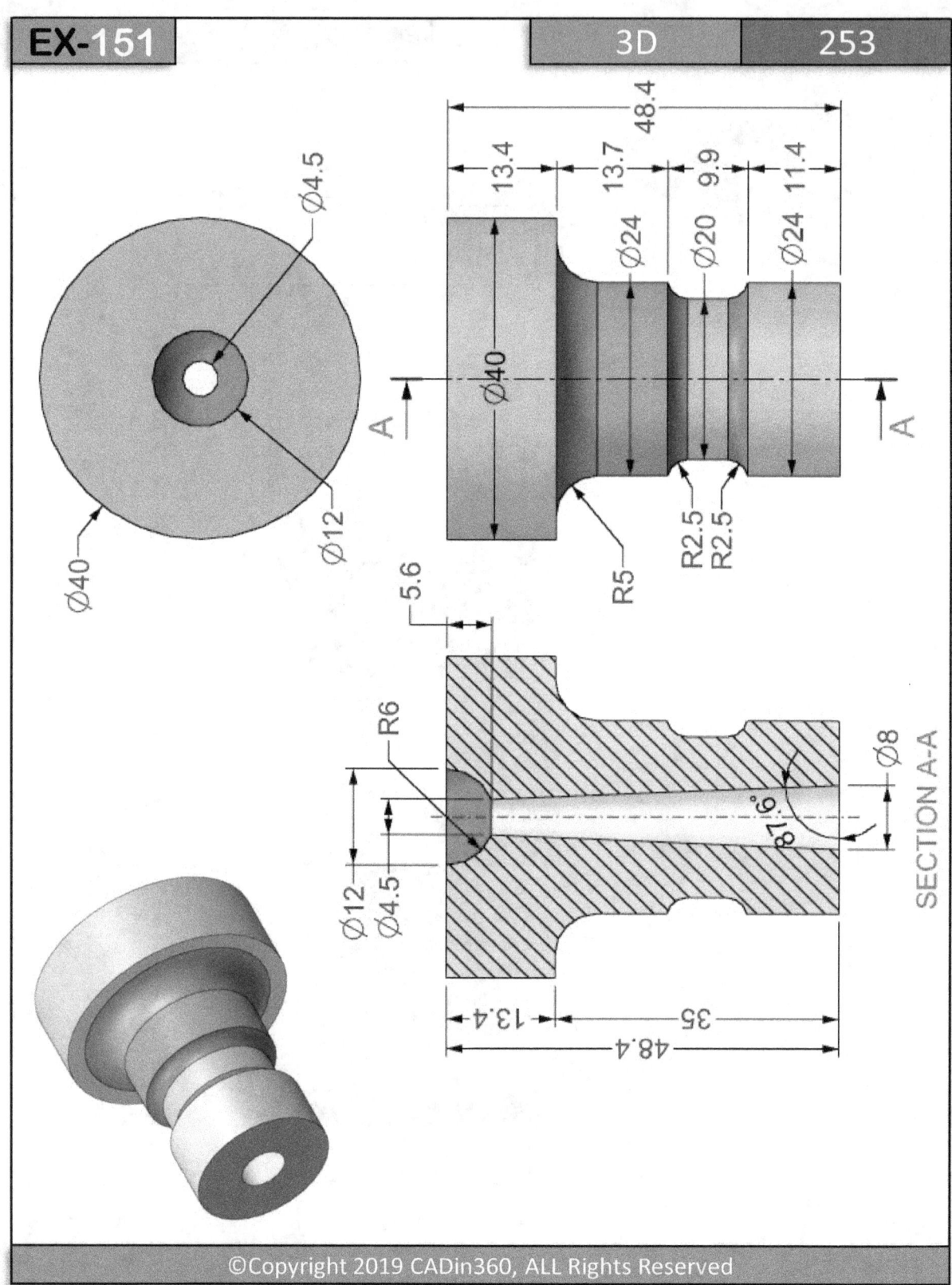

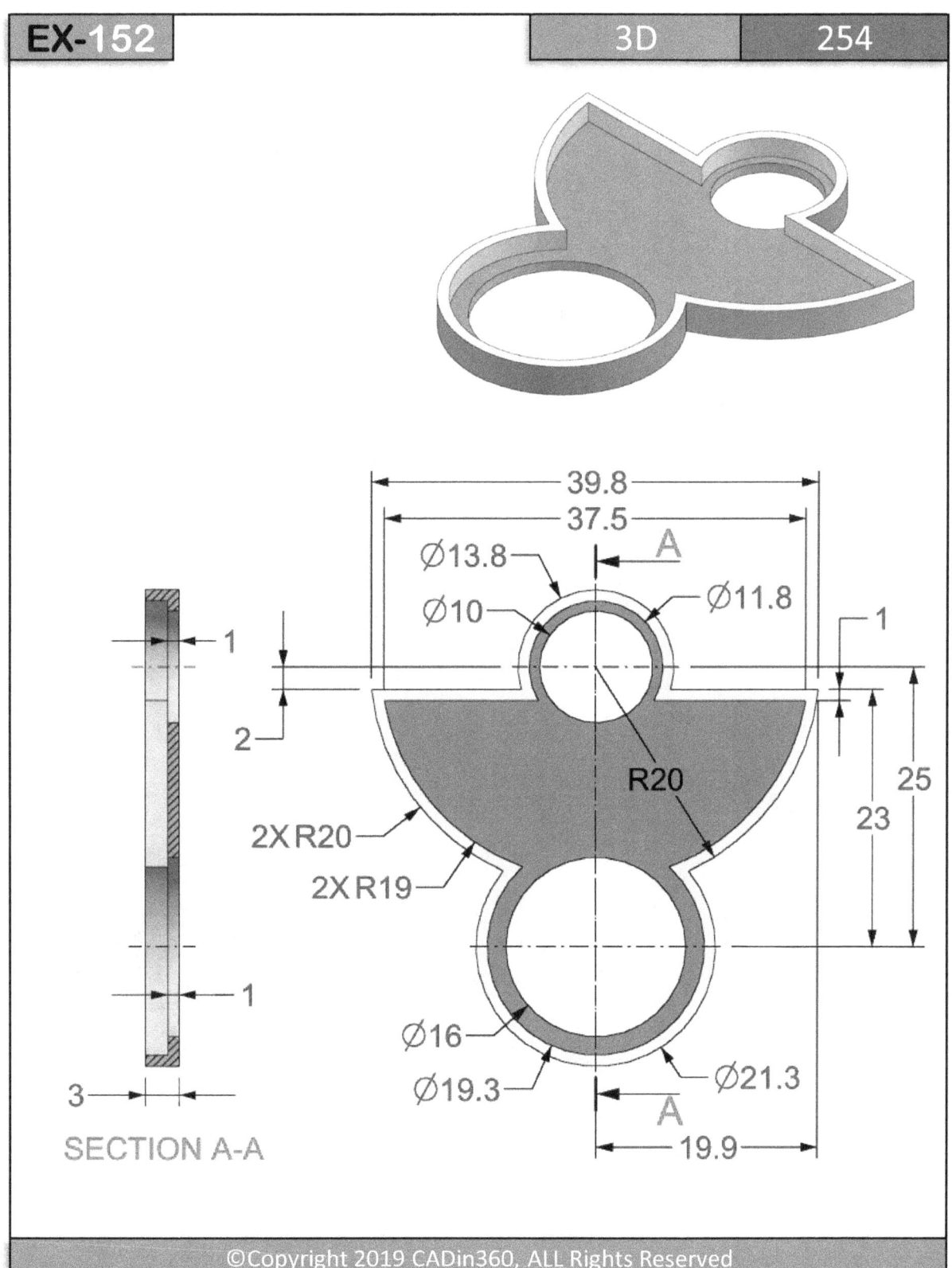

Ø13.8
Ø10
Ø11.8
39.8
37.5
A
1
2
R20
25
23
2X R20
2X R19
1
Ø16
Ø19.3
Ø21.3
A
19.9

1
3
SECTION A-A

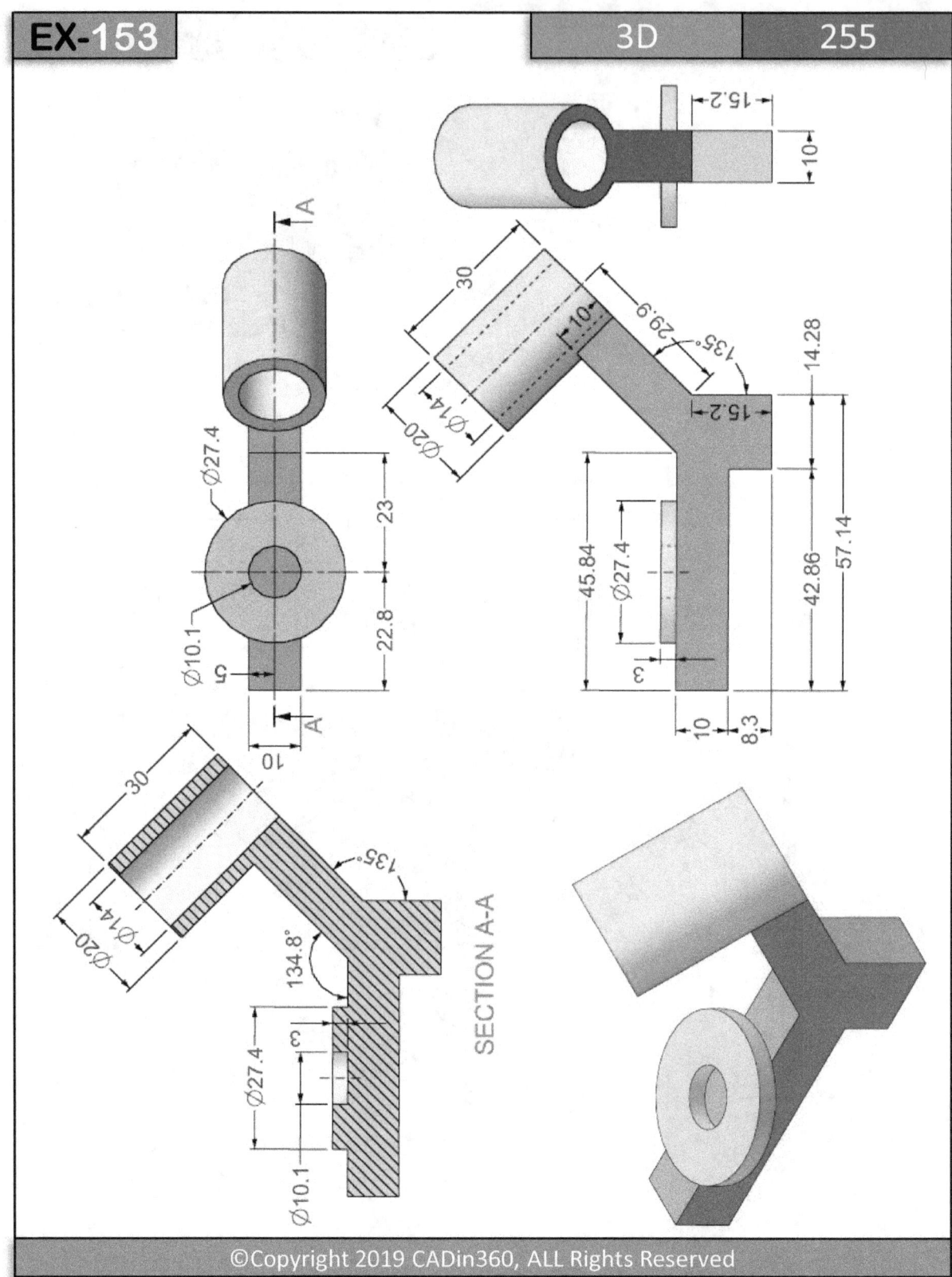

⌀27.4

A

A

23

22.8

⌀10.1

5

10

30

⌀20

⌀14

10

⌀20

30

⌀14

134.8°

⌀27.4

3

⌀10.1

135°

SECTION A-A

15.2

10

30

10

29.9

135°

15.2

14.28

45.84

⌀27.4

42.86

57.14

3

10

8.3

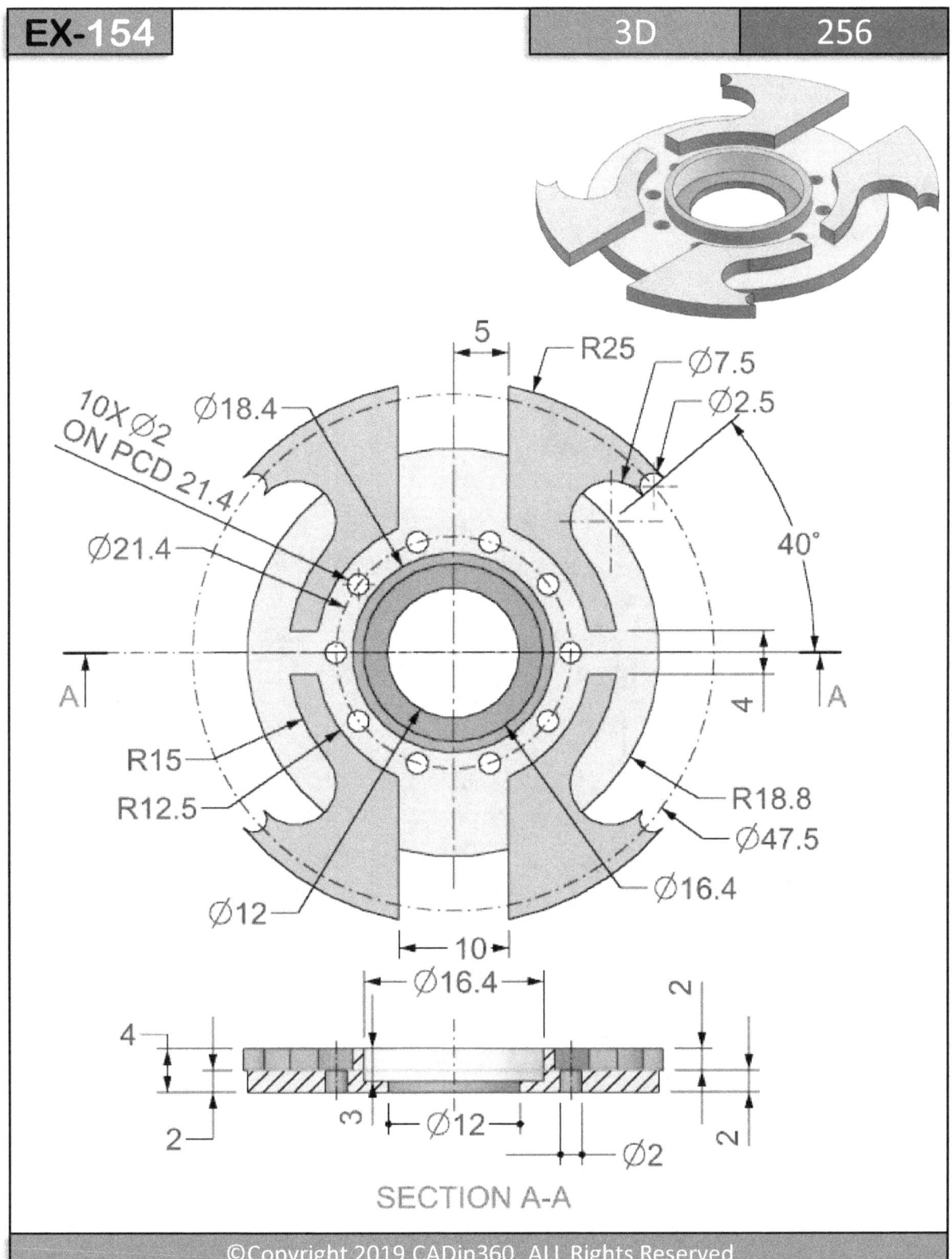

EX-154 3D 256

5
R25
Ø7.5
Ø2.5
10X Ø2
ON PCD 21.4
Ø18.4
40°
Ø21.4
4
A
A
R15
R12.5
R18.8
Ø47.5
Ø12
Ø16.4
10
Ø16.4
2
4
3
2
Ø12
Ø2
2

SECTION A-A

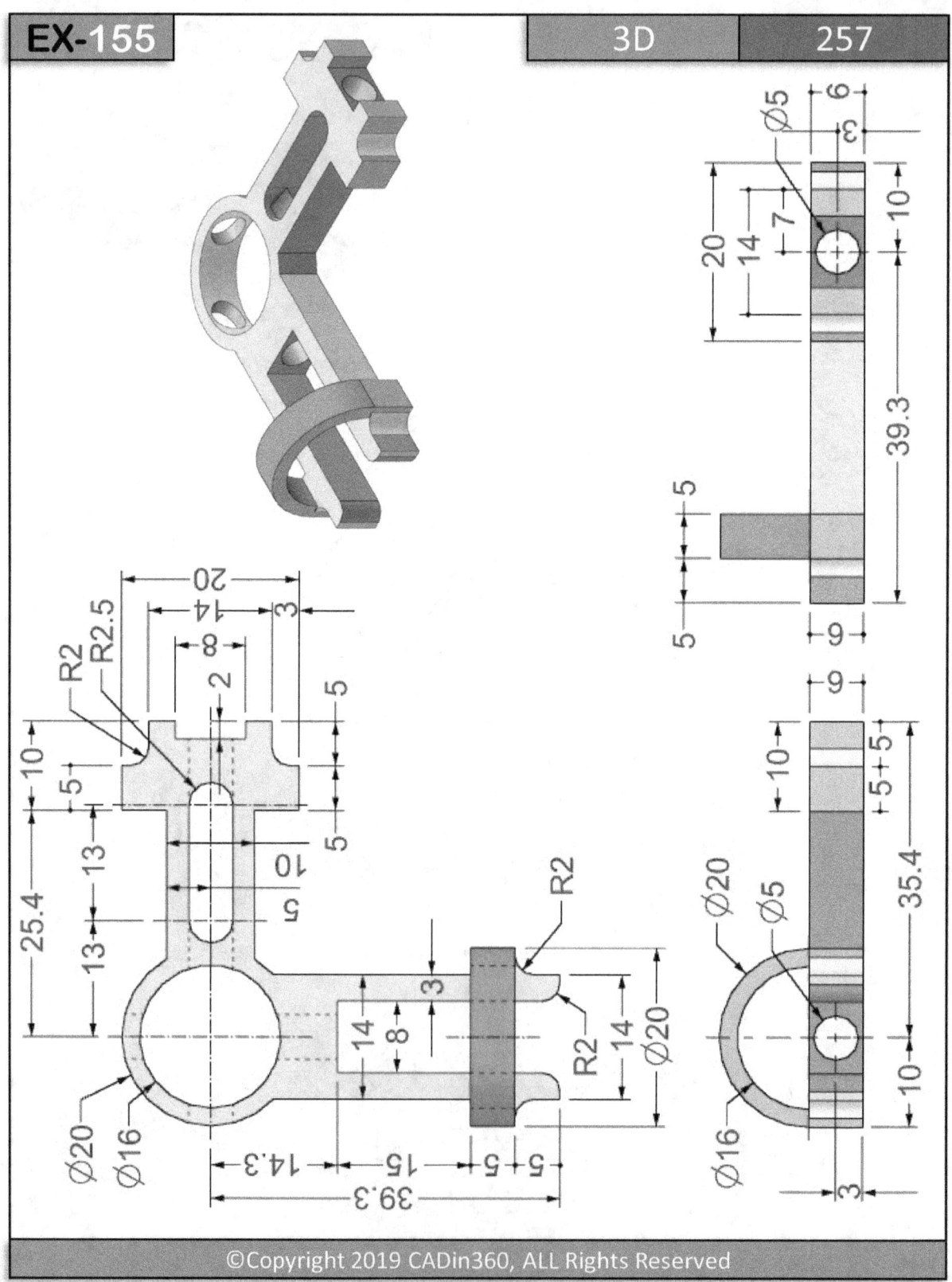

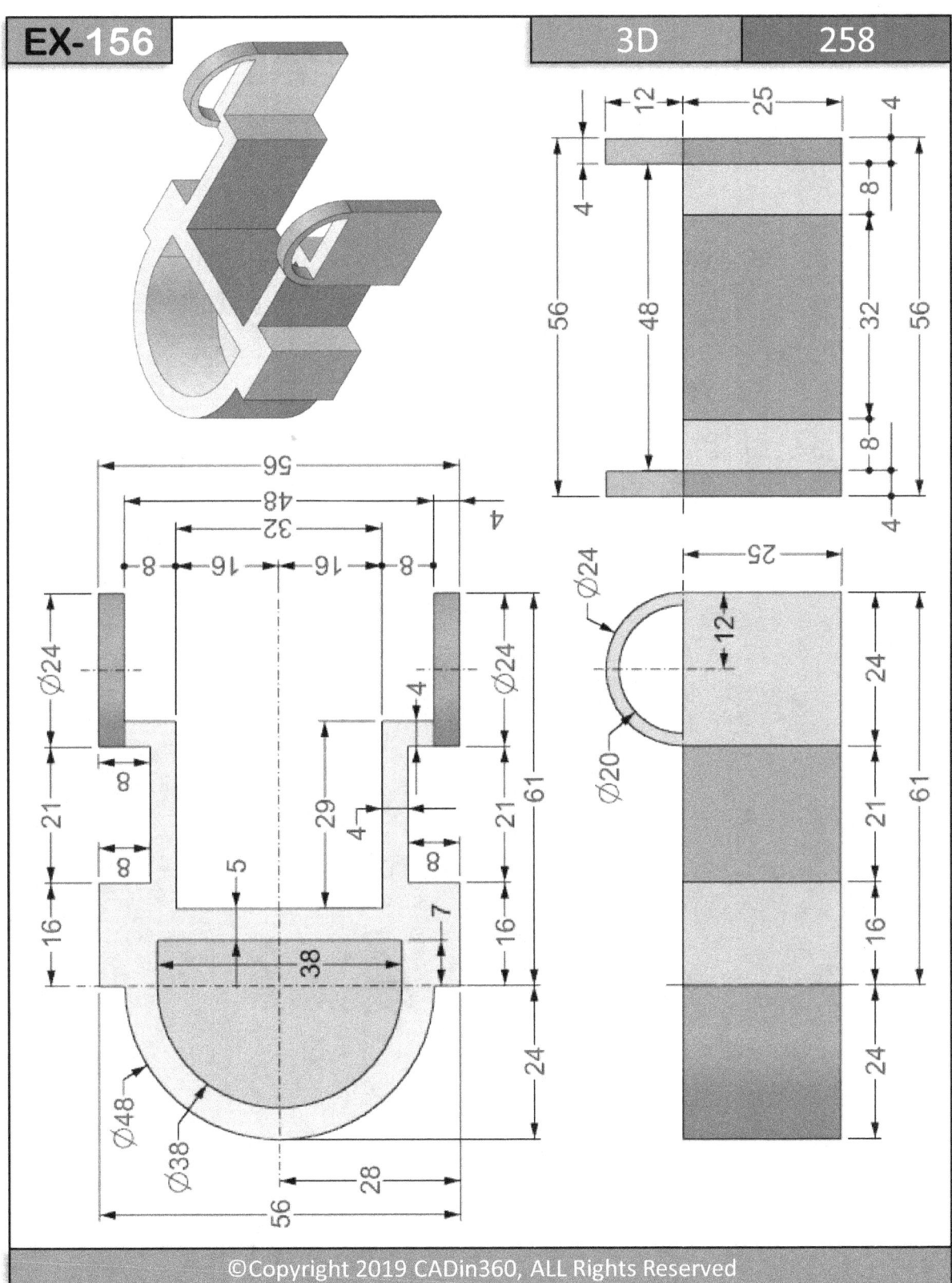

EX-156

3D

258

©Copyright 2019 CADin360, ALL Rights Reserved

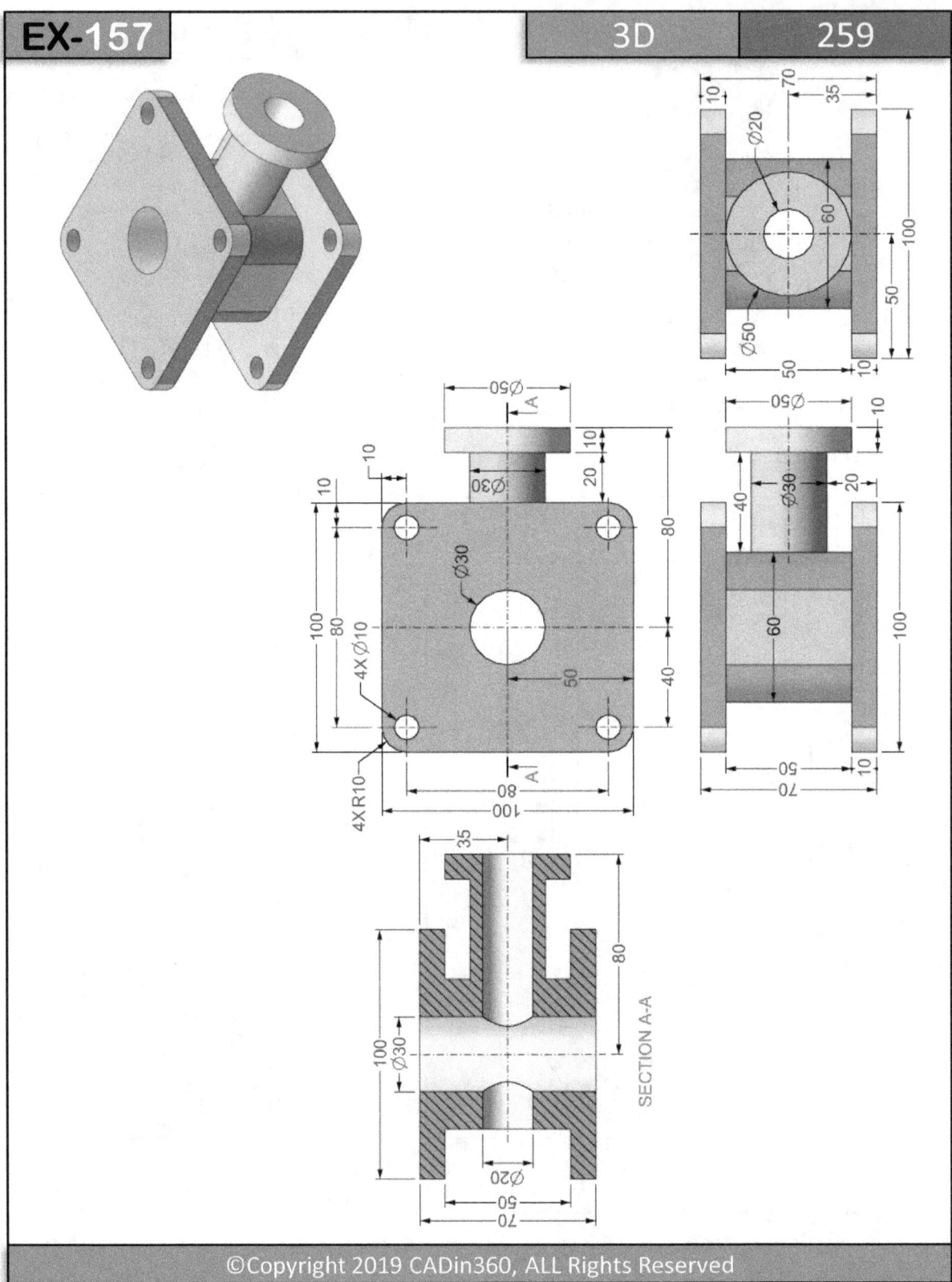

SECTION A-A

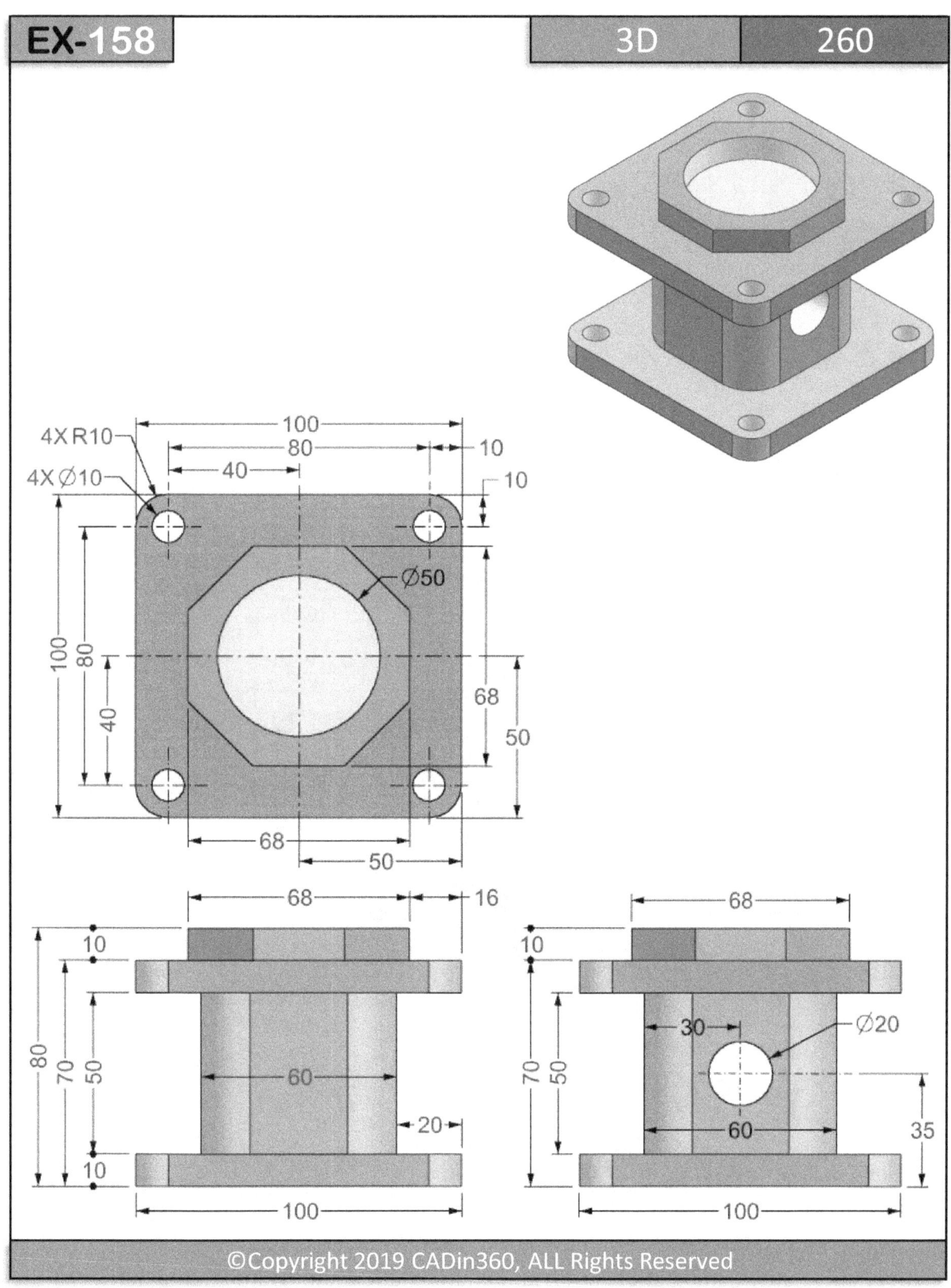

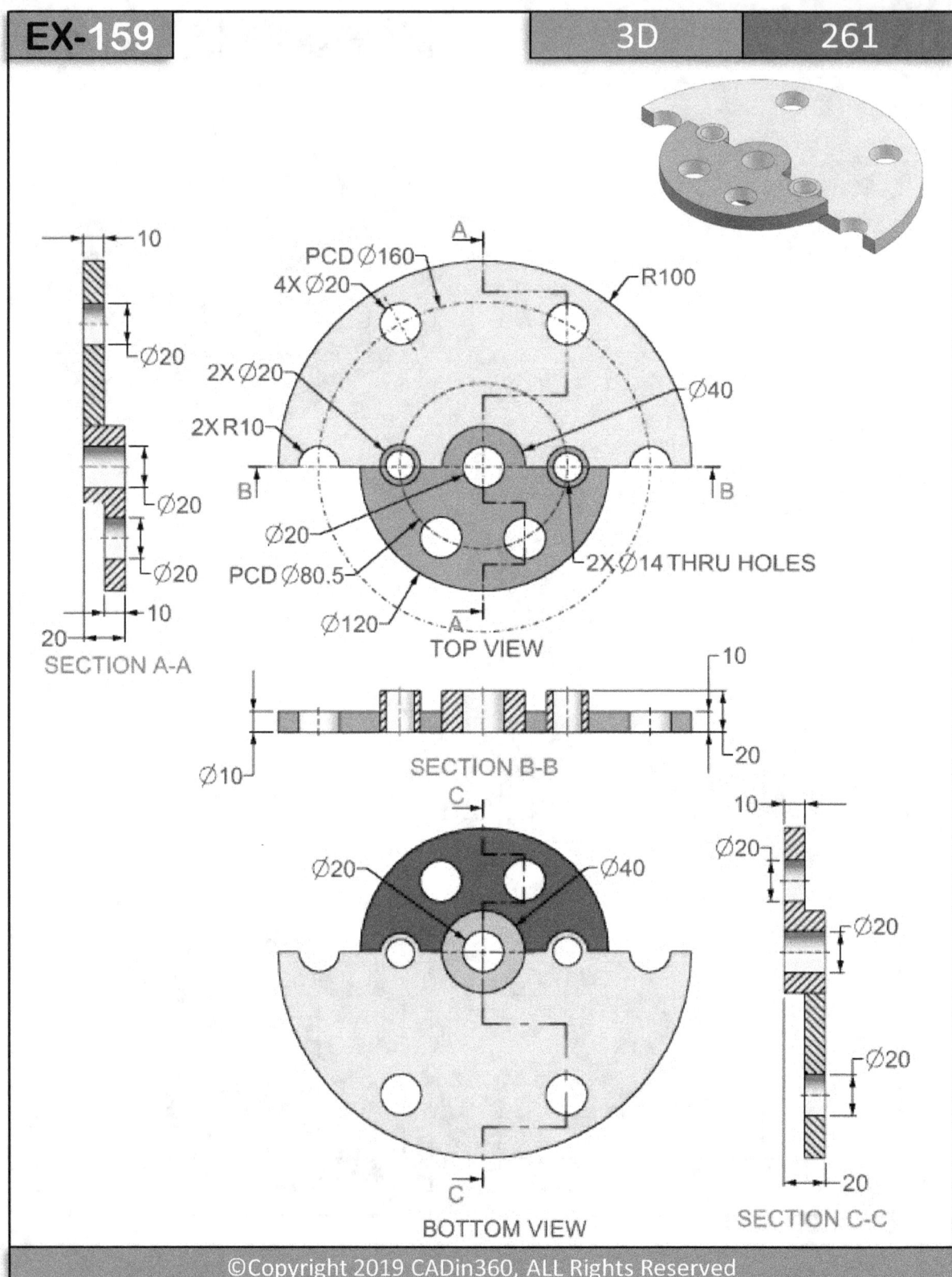

10

Ø20

Ø20

Ø20

10

20

SECTION A-A

PCD Ø160

4X Ø20

2X Ø20

2X R10

Ø20

PCD Ø80.5

Ø120

A

B

B

R100

Ø40

2X Ø14 THRU HOLES

A

TOP VIEW

10

20

Ø10

SECTION B-B

C

Ø20

Ø40

C

BOTTOM VIEW

10

Ø20

Ø20

Ø20

20

SECTION C-C

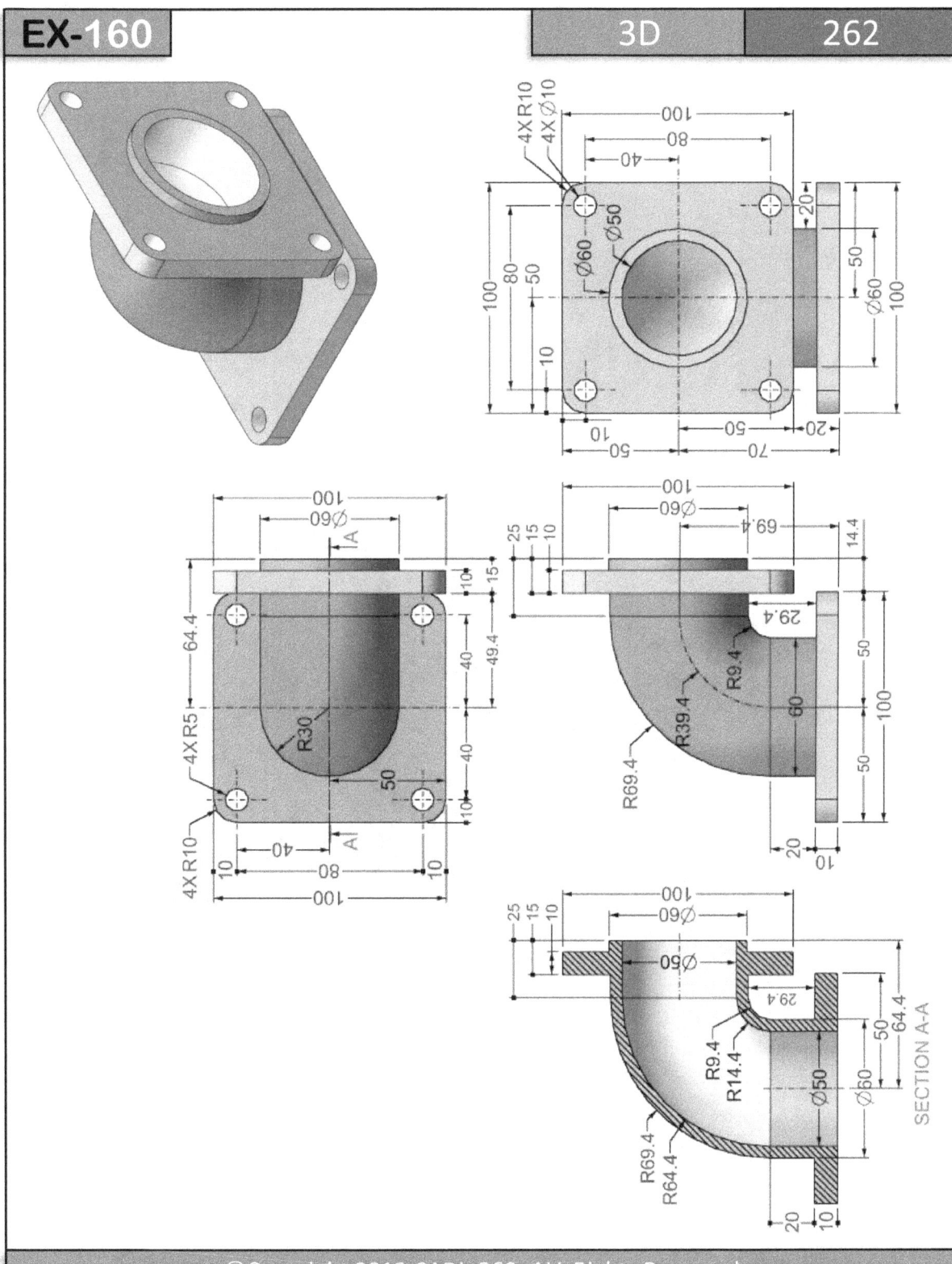

EX-160

3D

262

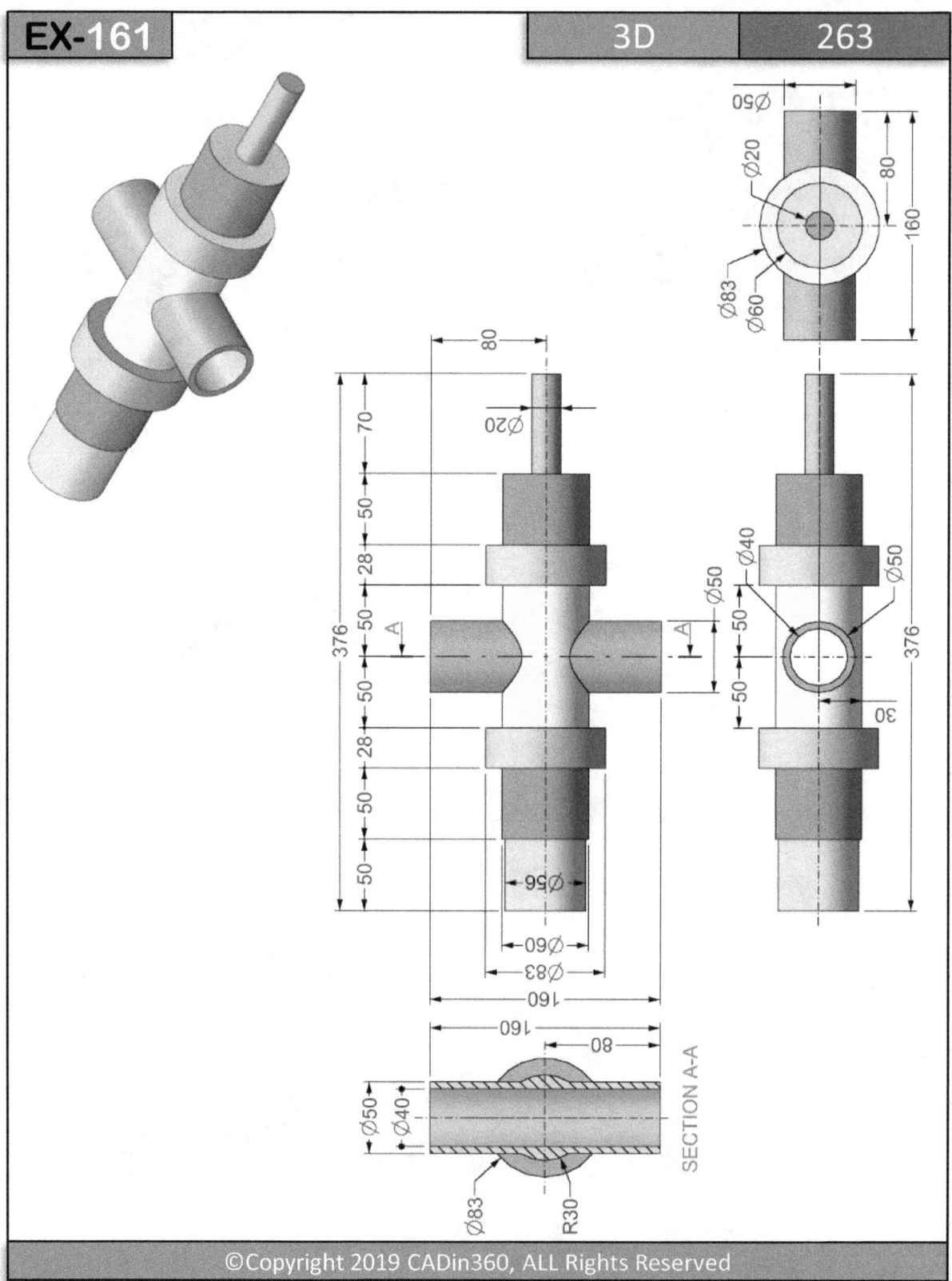

EX-161

3D   263

Ø50
Ø20
Ø83
Ø60
80
160

80
Ø20
70
50
28
376
50
A
50
28
50
50
Ø56
Ø60
Ø83
160

A
Ø50
Ø40
Ø50
50
50
376
30

160
80
Ø50
Ø40
Ø83
R30
SECTION A-A

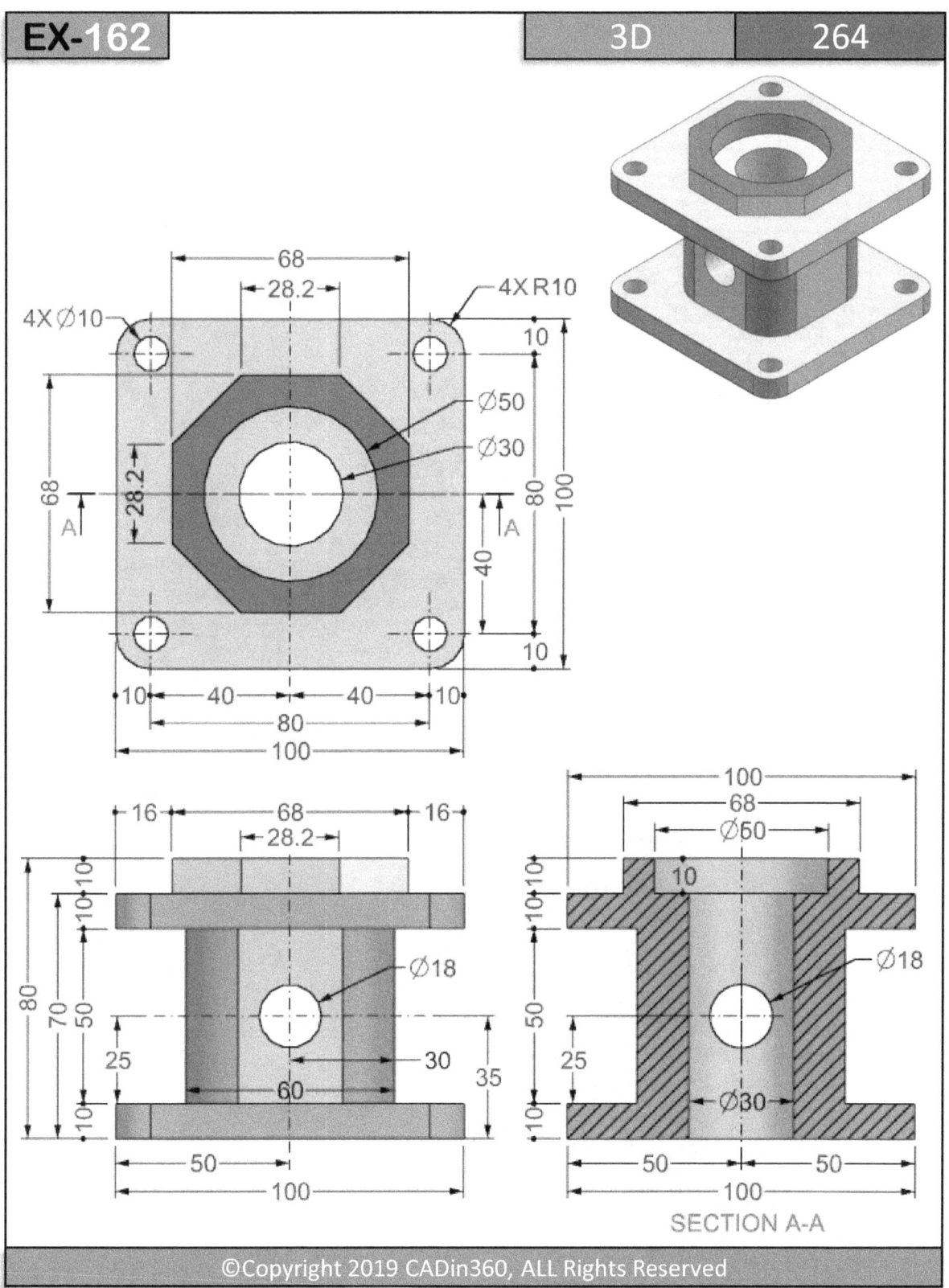

EX-162

3D 264

4X Ø10

68
28.2
4X R10
10
Ø50
Ø30
68
28.2
80
100
A
A
40
10
10 40 40 10
80
100

16 68 16
28.2
10 10
Ø18
80
70
50
25
30
35
10
60
50
100

100
68
Ø50
10 10
10
50
Ø18
50
25
Ø30
10
50 50
100
SECTION A-A

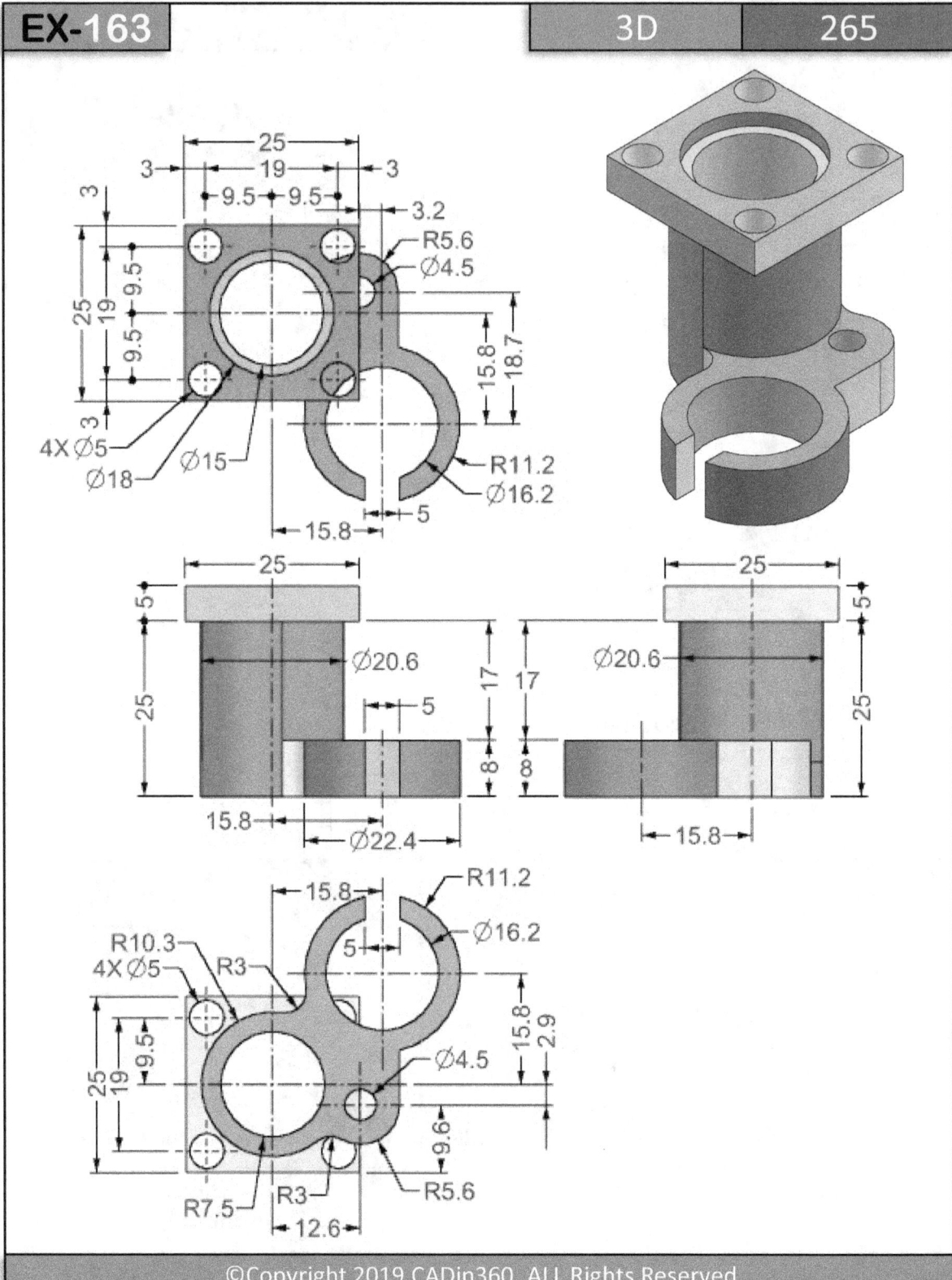

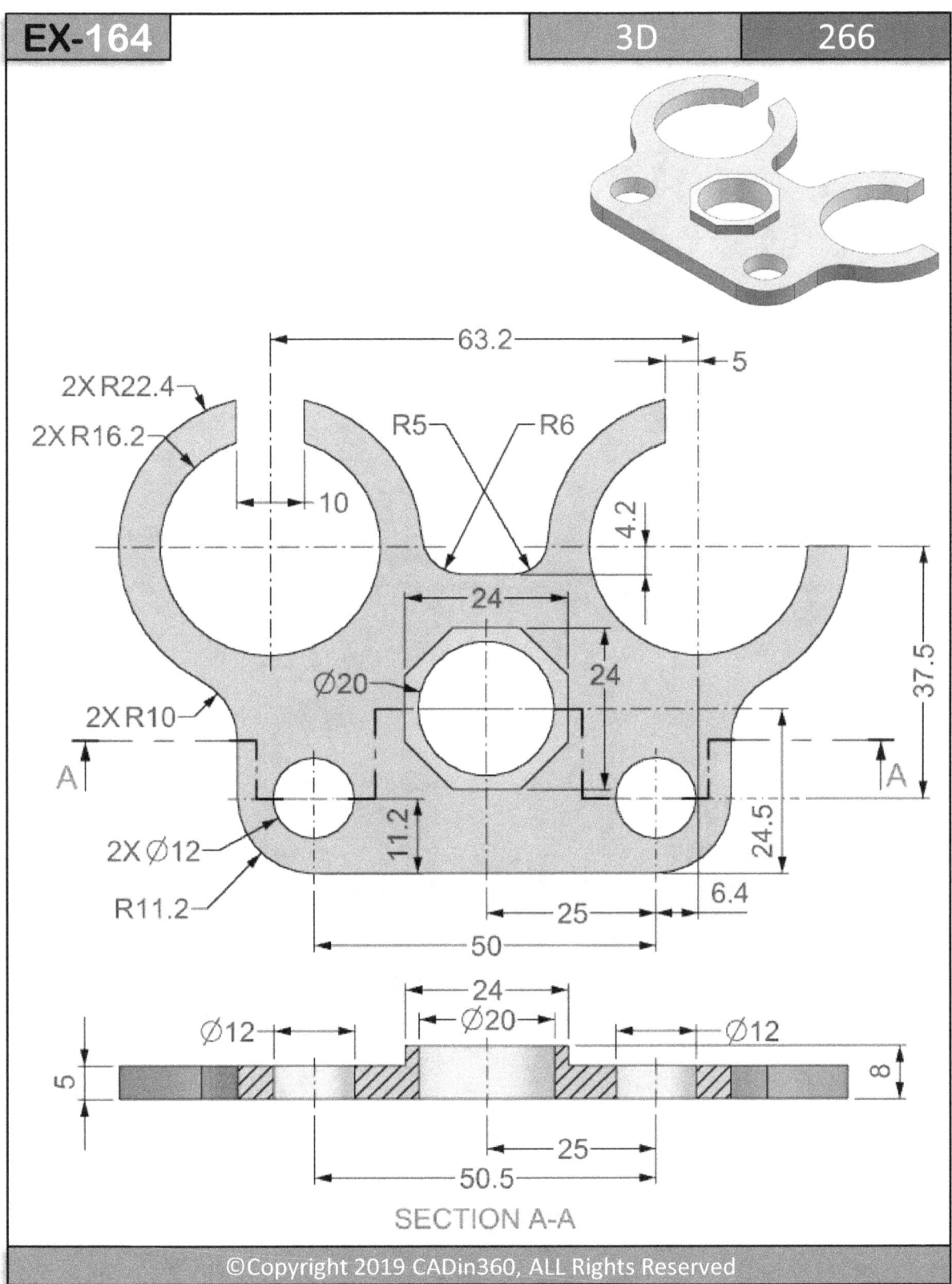

2X R22.4
2X R16.2
63.2
5
R5　R6
10
4.2
Ø20
24
24
2X R10
37.5
A
2X Ø12
R11.2
11.2
24.5
6.4
25
50

24
Ø12
Ø20
Ø12
5
8
25
50.5

SECTION A-A

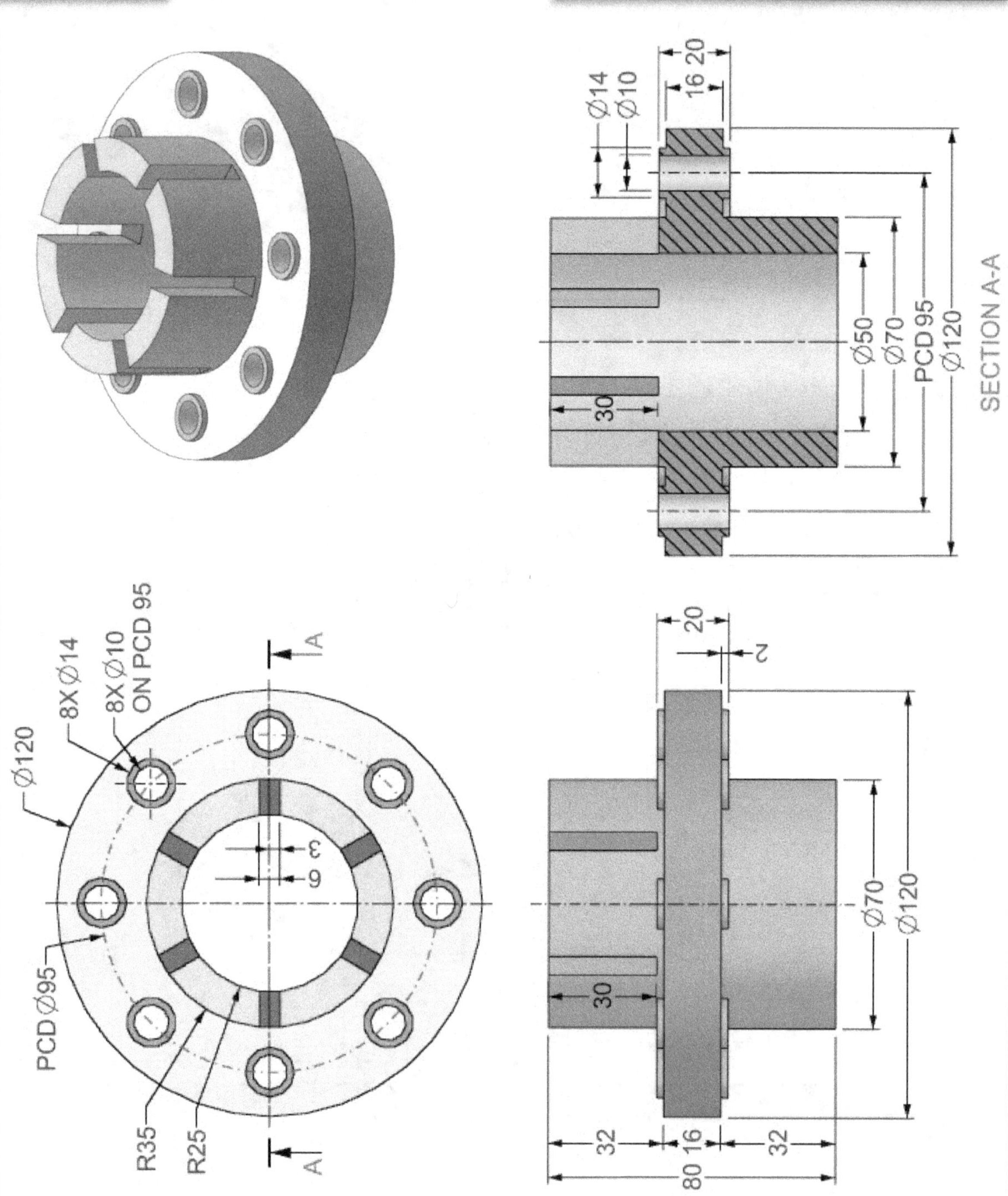

SECTION A-A

Ø14
Ø10
16 20
Ø50
Ø70
PCD 95
Ø120
30

8X Ø14
8X Ø10
ON PCD 95
Ø120
PCD Ø95
R35
R25
3
6
A
A

20
2
Ø70
Ø120
30
32
16
80
32

EX-166

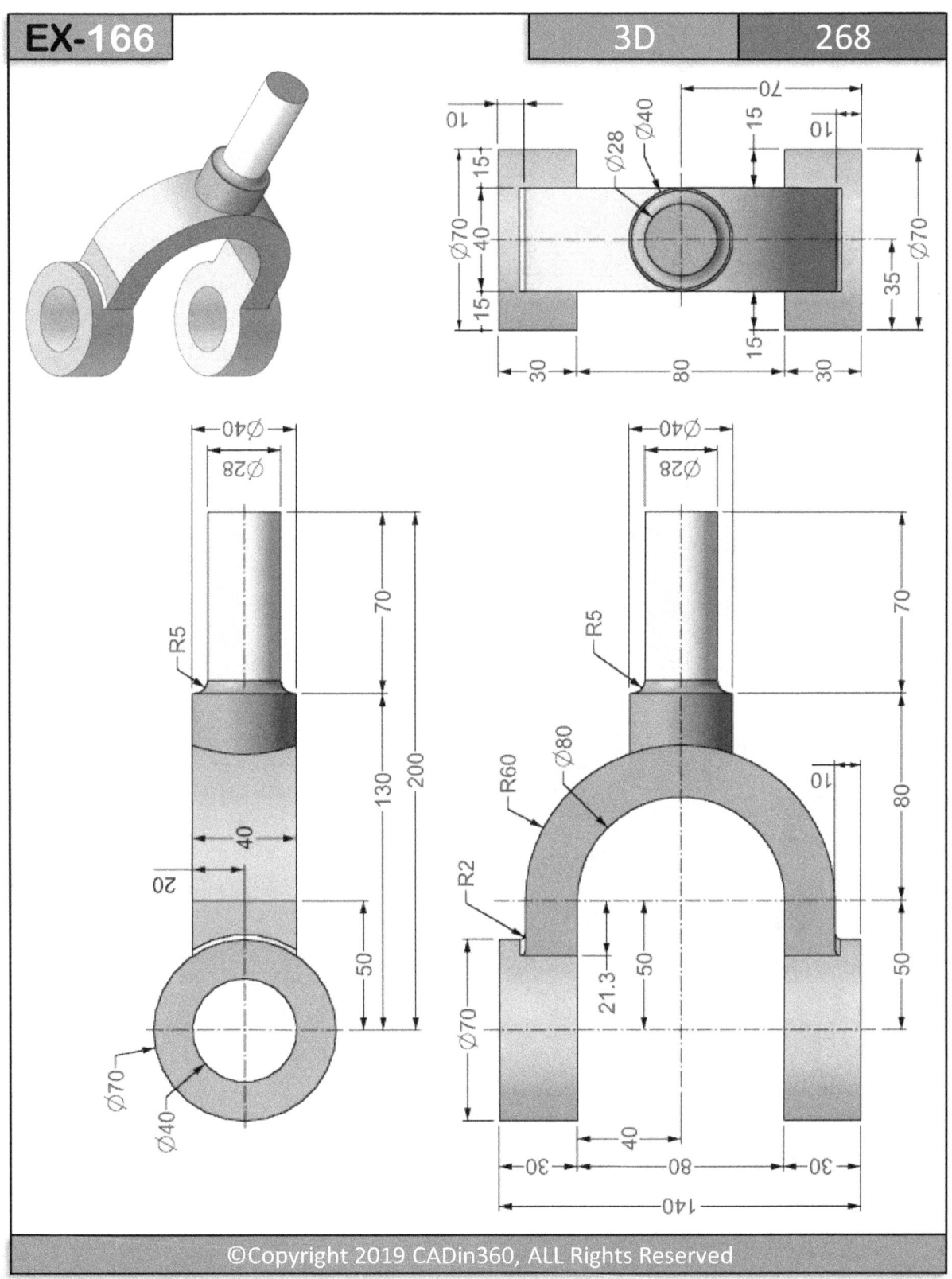

Ø40
Ø28
R5
70
130
200
40
20
50
Ø70
Ø40

10
15
Ø70
40
15
15
Ø28
Ø40
70
15
10
Ø70
35
15
30
80
30

Ø40
Ø28
R5
R60
Ø80
R2
70
80
50
10
21.3
50
Ø70
30
40
80
30
140

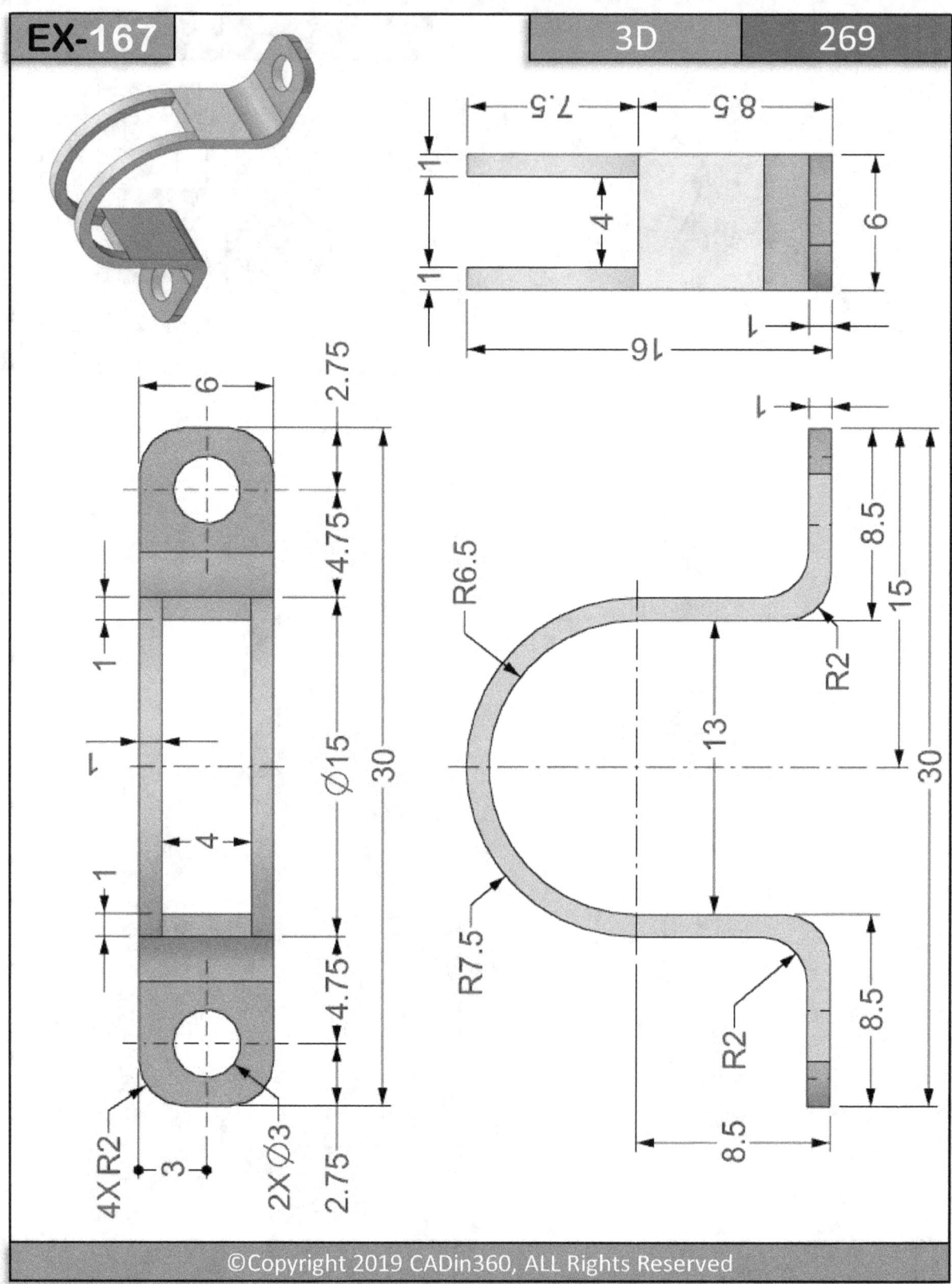

EX-167

3D

269

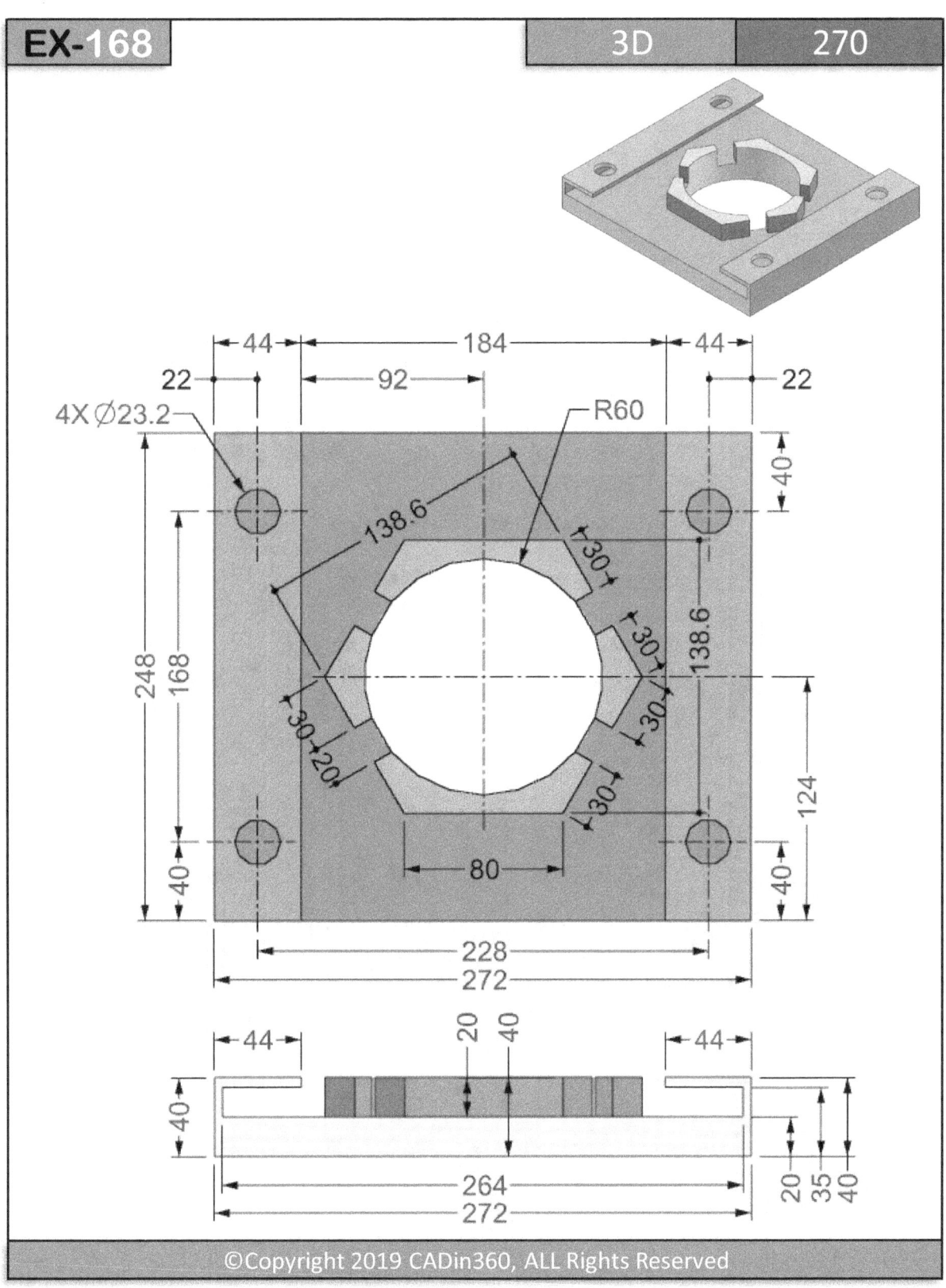

70
29
19
20.5
25
20.5
15
40
30
20
10
80
5
40
25
25
5
5
5
5
20
2X R15
20.5
15
20.5
35
2X Ø15
30
70
35
20
19
29
70

4X R15
40
4X Ø15
15
20
30
15
20
25
30
25
40
40
80
70
35

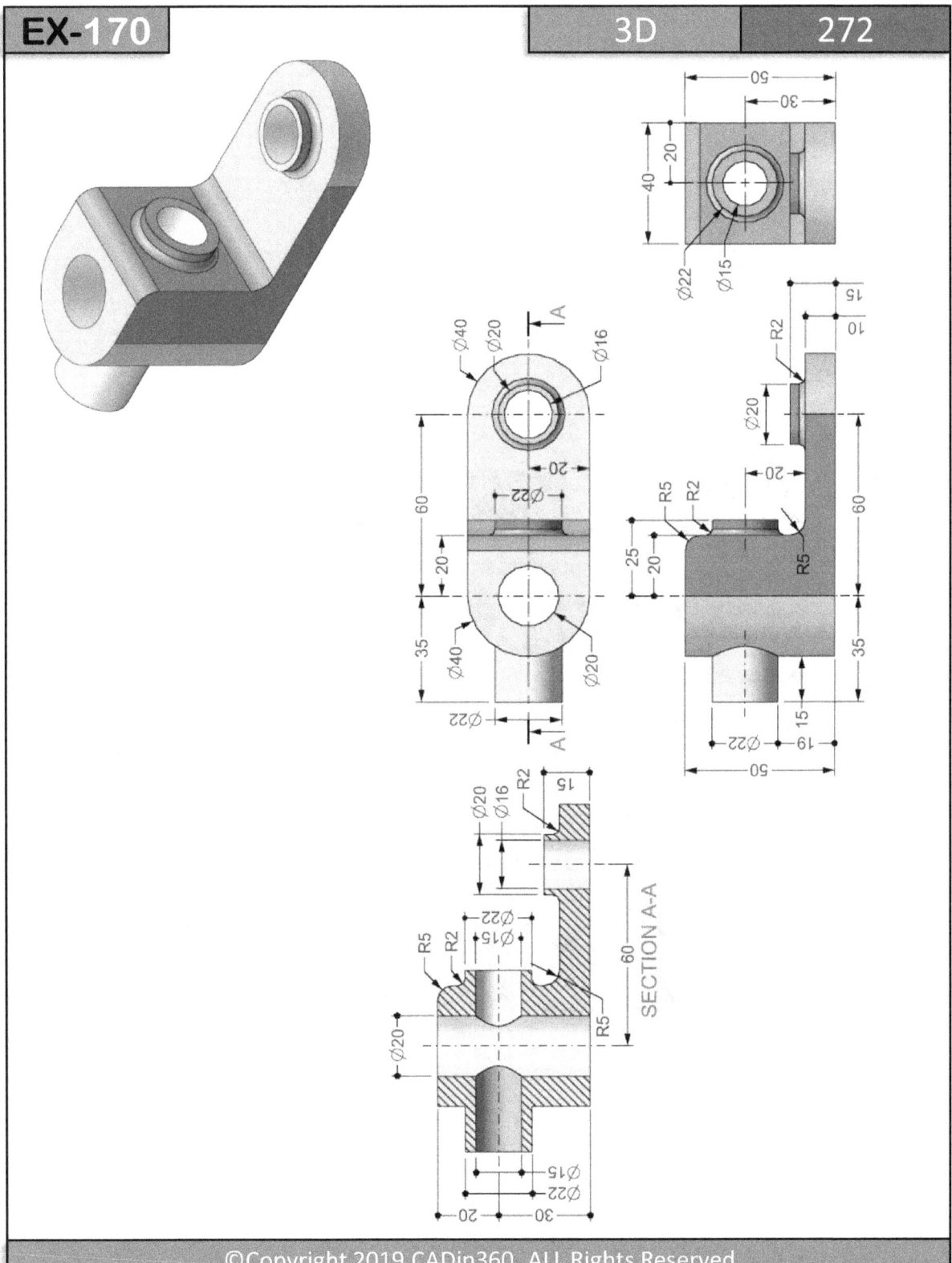

SECTION A-A

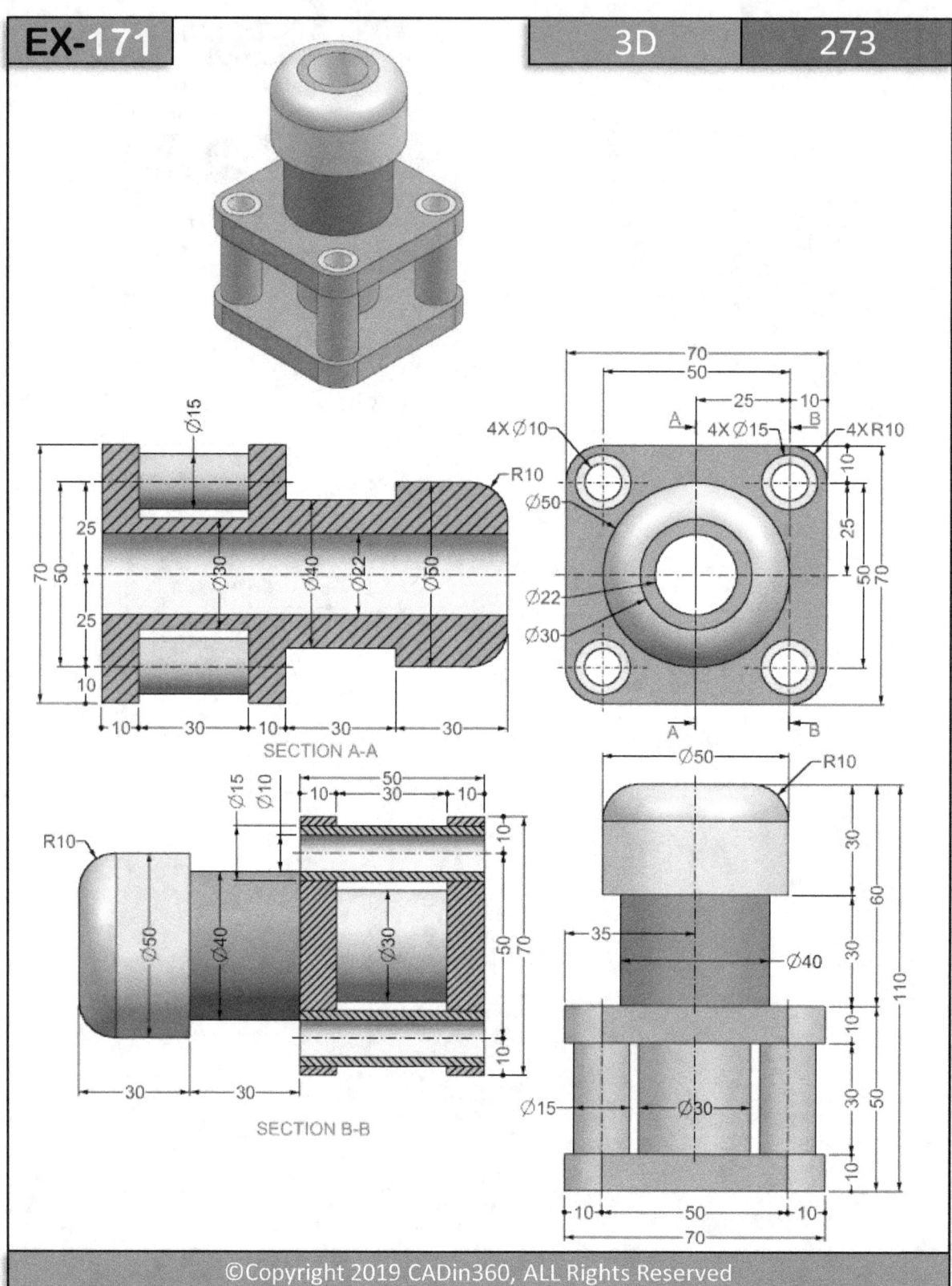

SECTION A-A

SECTION B-B

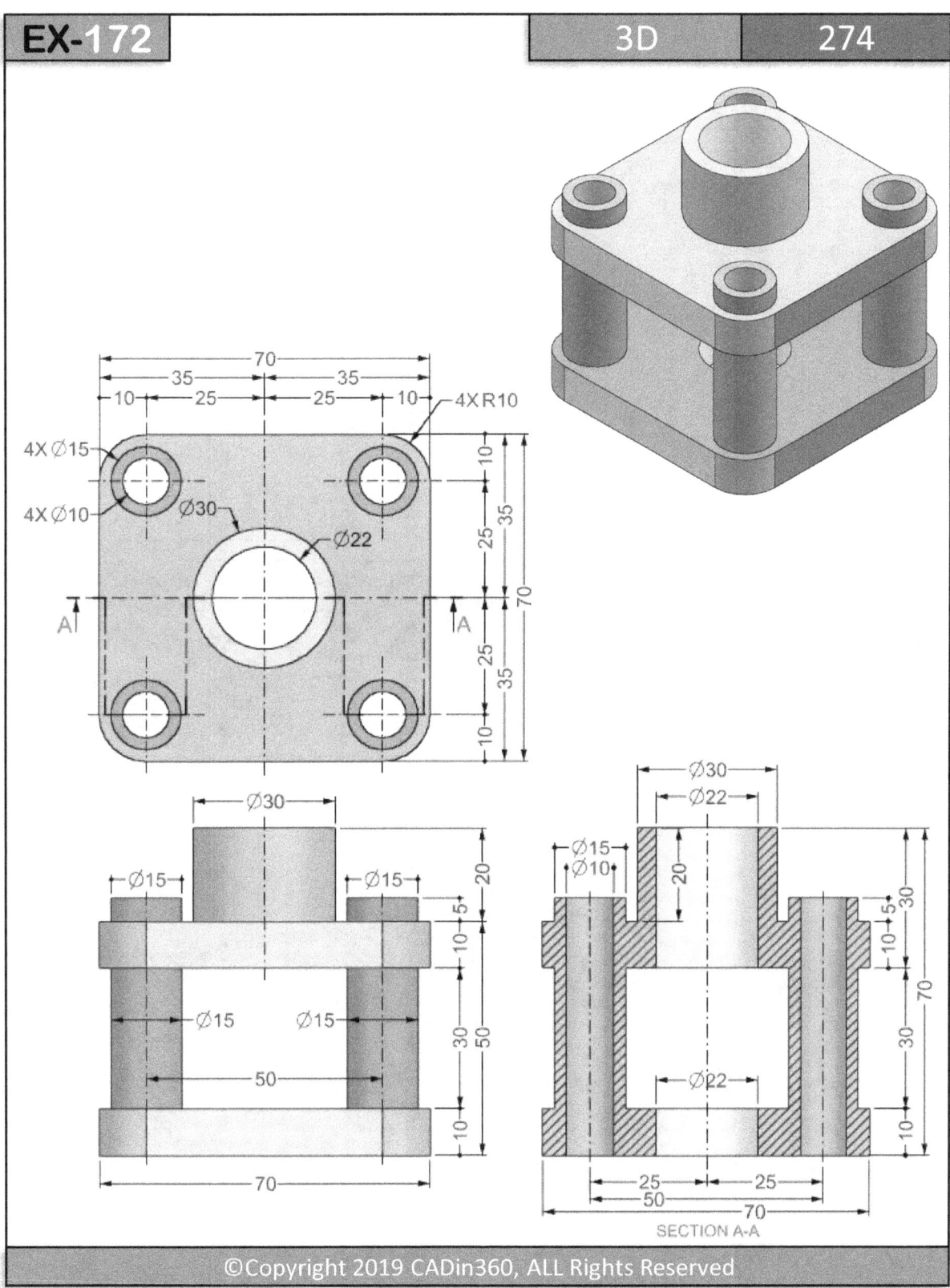

SECTION A-A

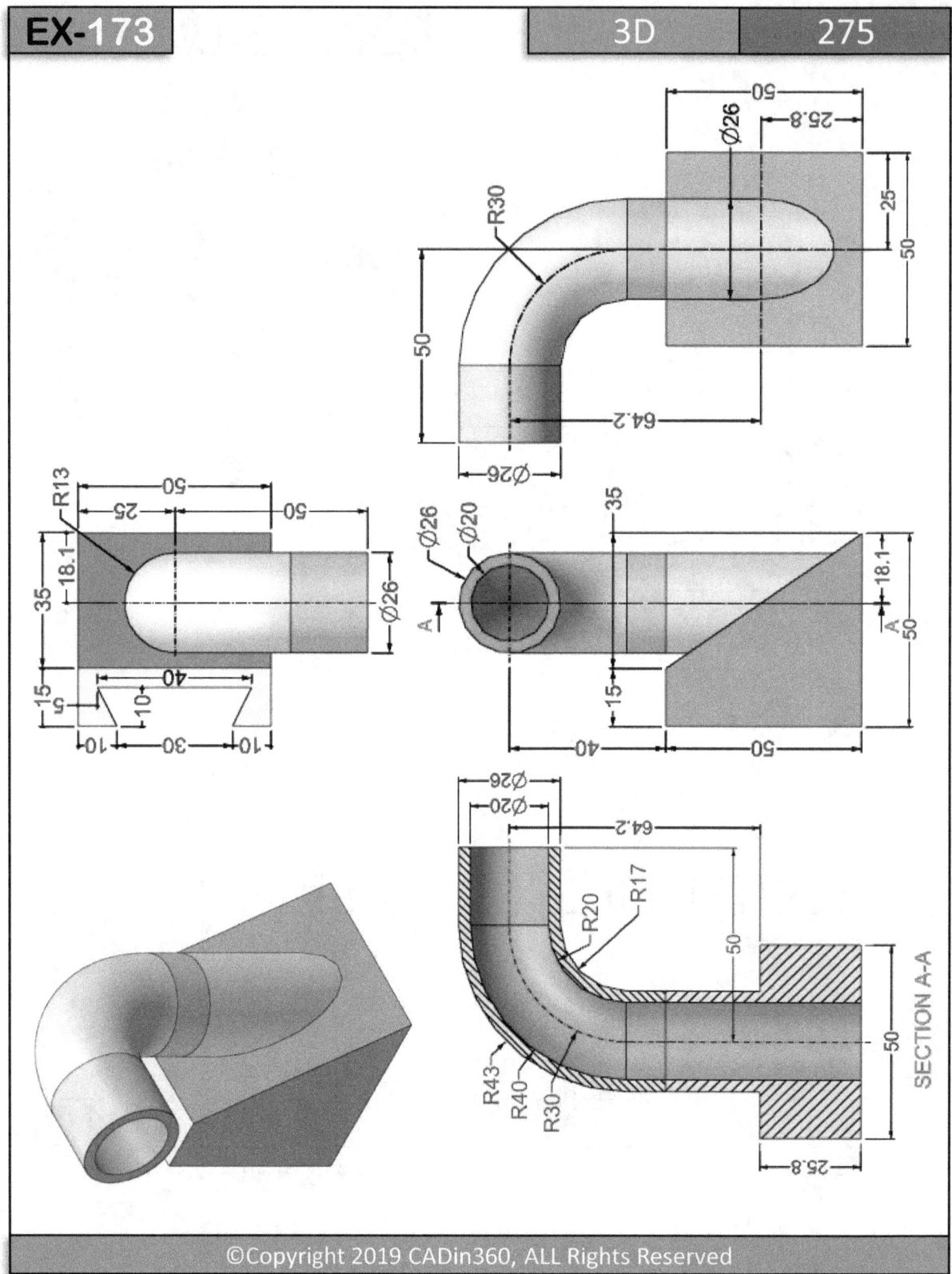

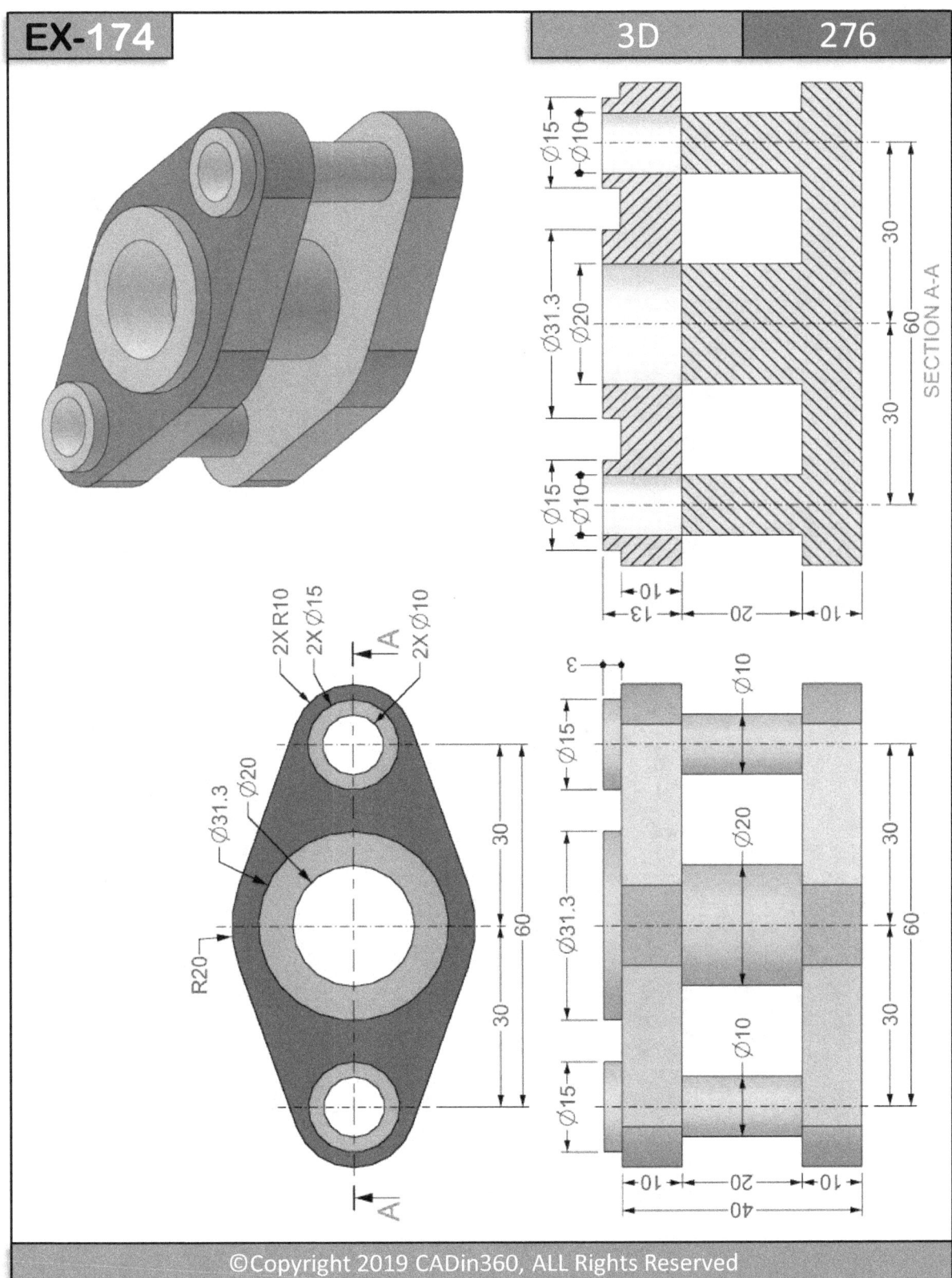

EX-174

3D   276

2X R10
2X Ø15
A
2X Ø10
Ø31.3
Ø20
R20
30
60
30
A

SECTION A-A

Ø15
Ø10
Ø31.3
Ø20
Ø15
Ø10
30
60
30
10
13   20   10

3
Ø15
Ø10
Ø31.3
Ø20
Ø15
Ø10
30
60
30
10   20   10
40

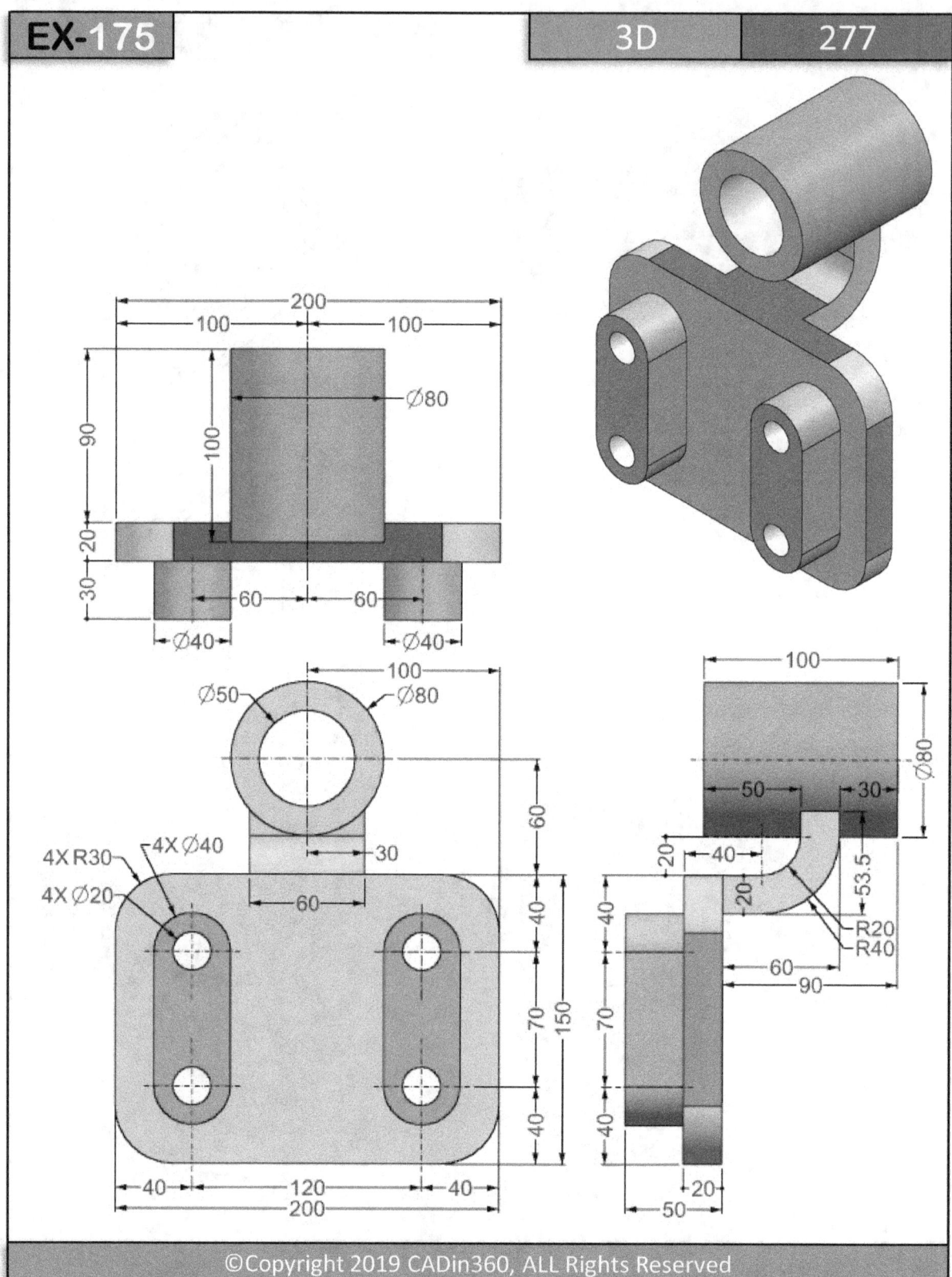

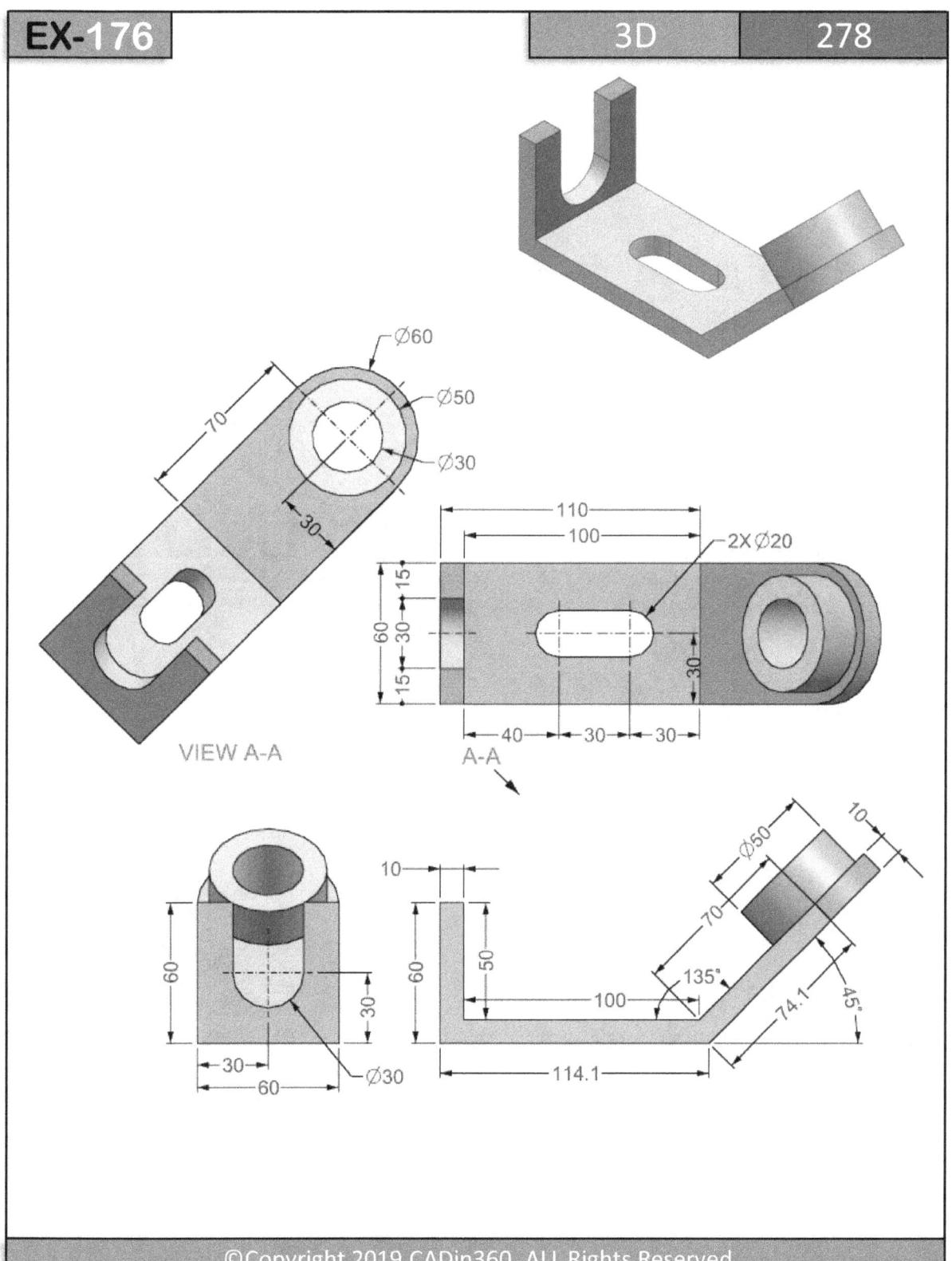

Ø60
Ø50
Ø30
70
30

VIEW A-A

110
100
2X Ø20
60
15
30
15
30
40
30
30
A-A

60
30
60
Ø30
30

10
60
50
10
Ø50
70
135°
100
45°
74.1
114.1

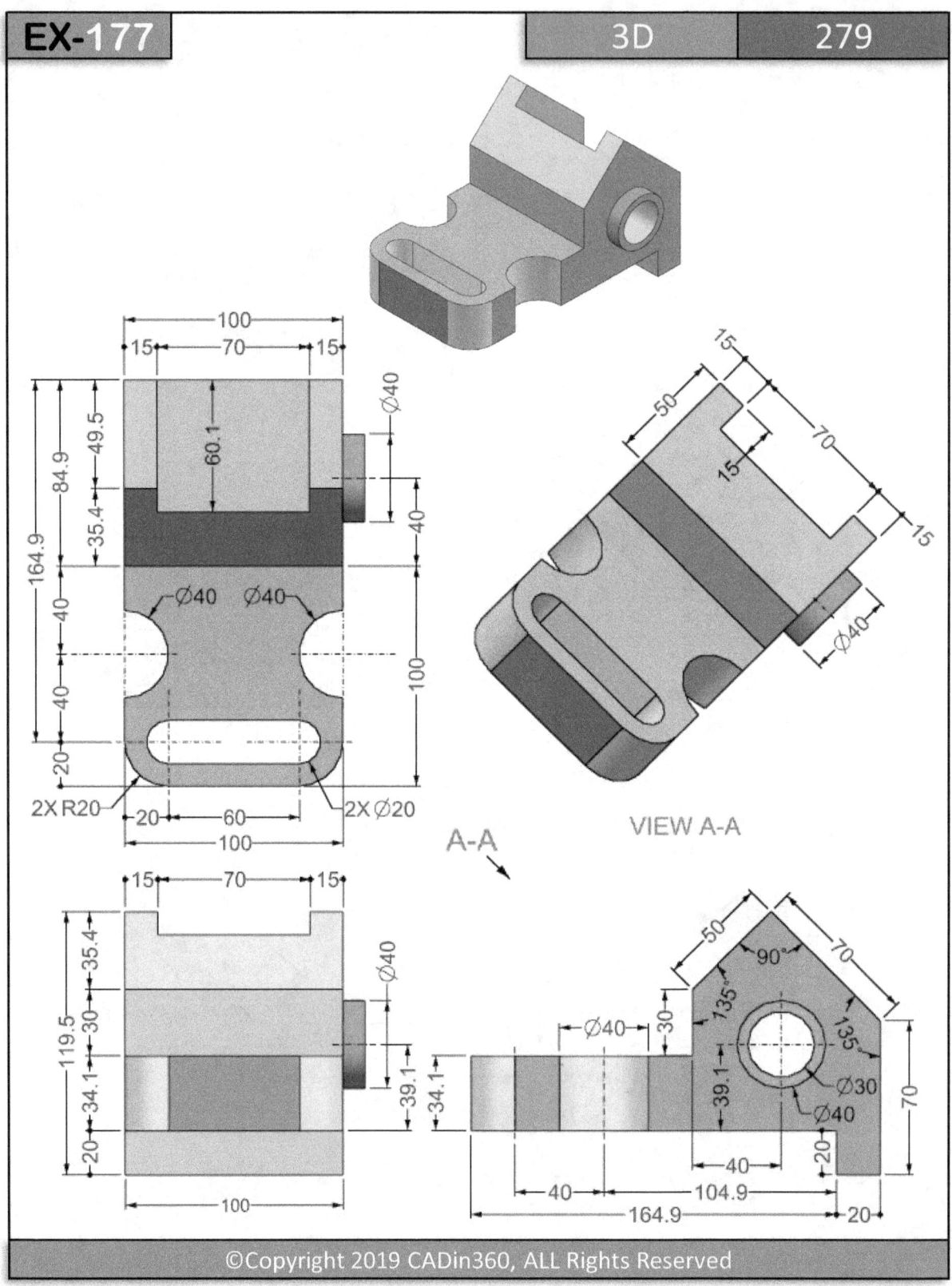

100
15 | 70 | 15
Ø40
60.1
49.5
84.9
35.4
164.9
40
Ø40   Ø40
40
100
40
20
2X R20
20 | 60
2X Ø20
100

15 | 70 | 15
35.4
Ø40
30
119.5
39.1
34.1
20
100

A-A

VIEW A-A

15
50
15
70
15
Ø40

Ø40
30
135°
90°
50
70
39.1
135°
Ø30
Ø40
40
104.9
20
70
40
164.9
20
34.1

8X R10

8X Ø10

30

20

20

20

60

30

20

20

20

100

60

40

40

40

20

10

20

40

40

100

80

40

20

10

SECTION B-B

30

60

2X R10

20 10 20

SECTION A-A

2X R10

20

40

20

40

30

100

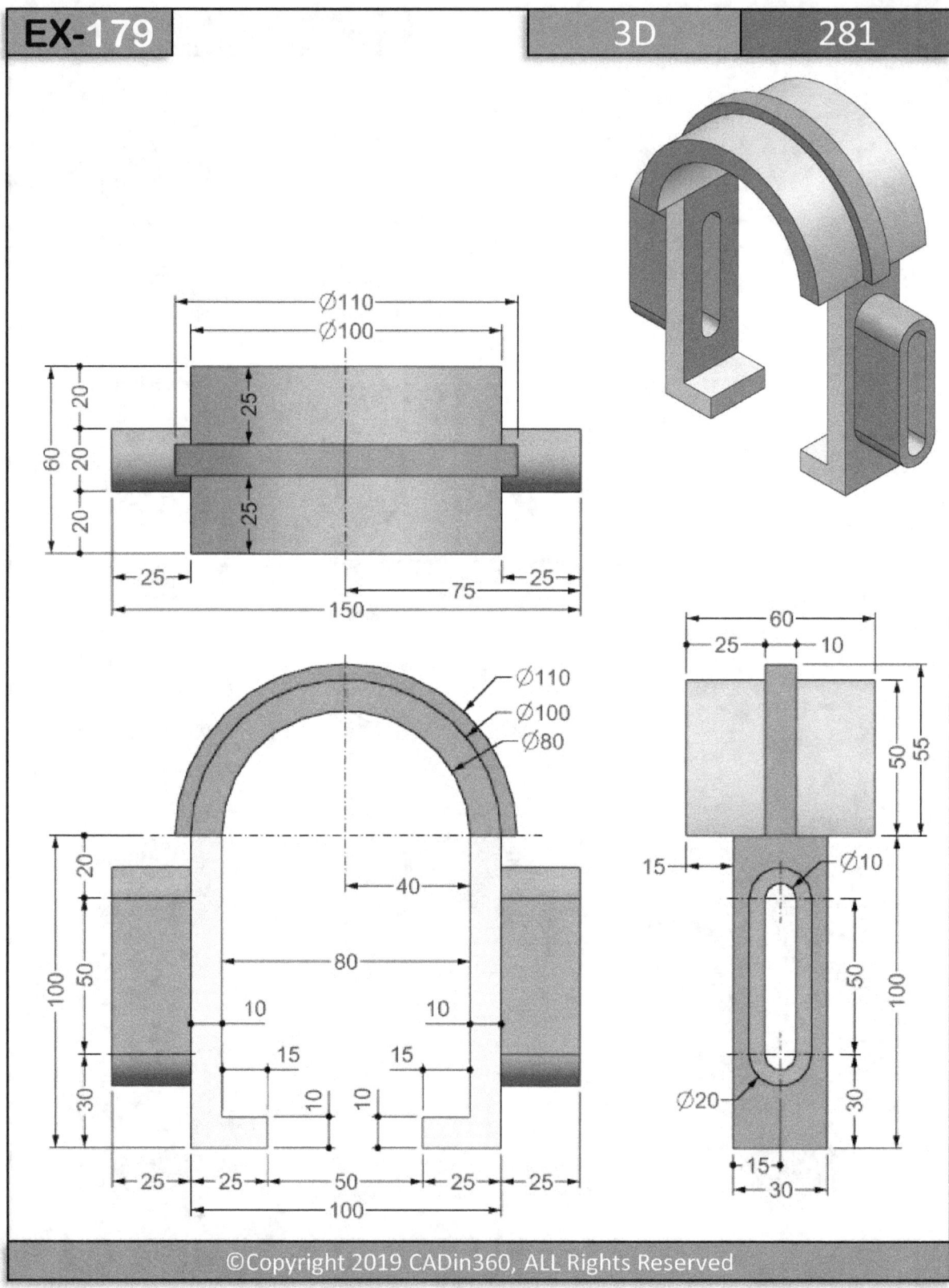

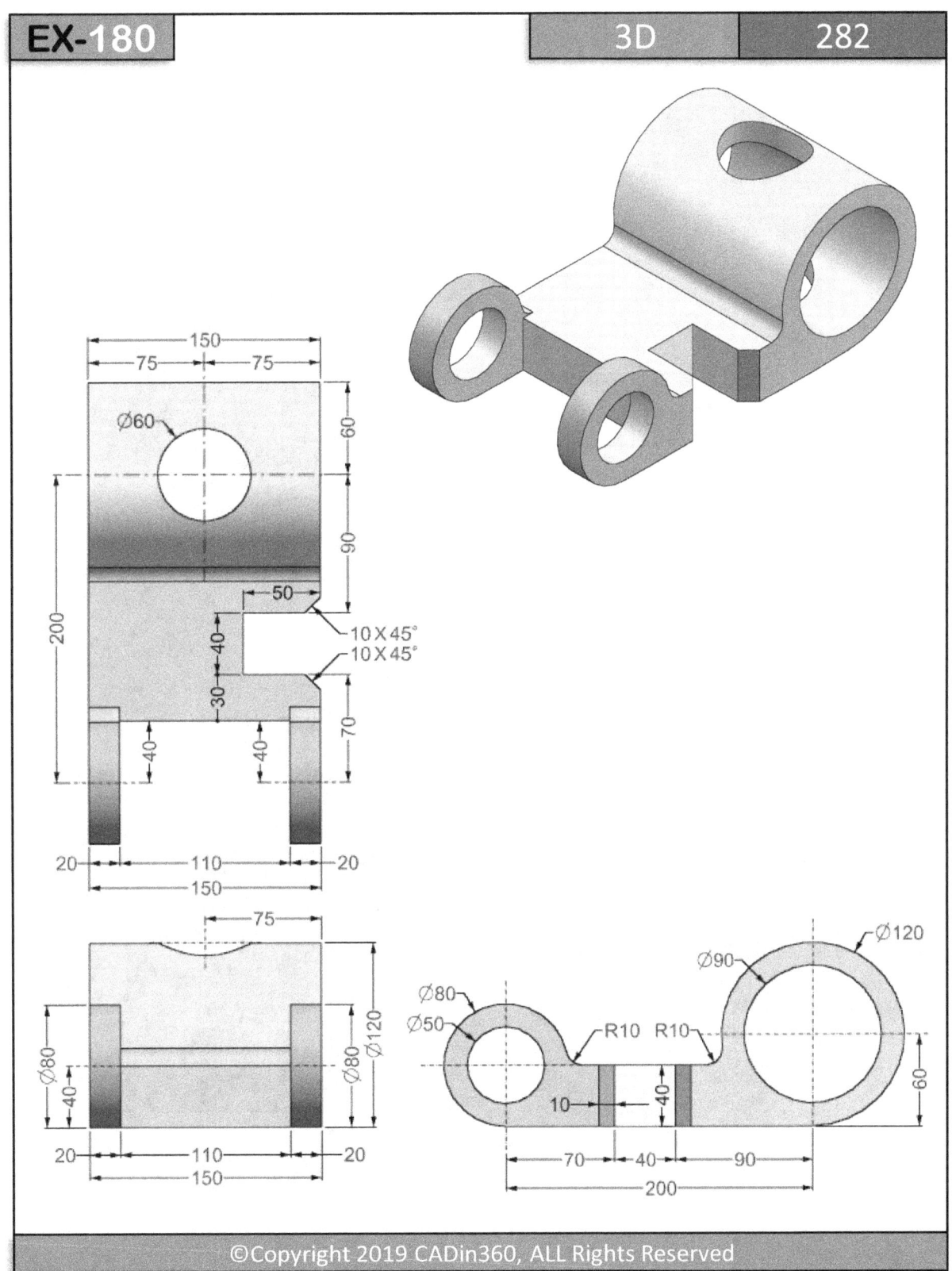

150
75   75
Ø60
200
90
60
50
10 X 45°
40
10 X 45°
30
70
40   40
20   110   20
150

75
Ø80
Ø120
Ø80
40
20   110   20
150

Ø120
Ø90
Ø80
Ø50
R10   R10
10
40
60
70   40   90
200

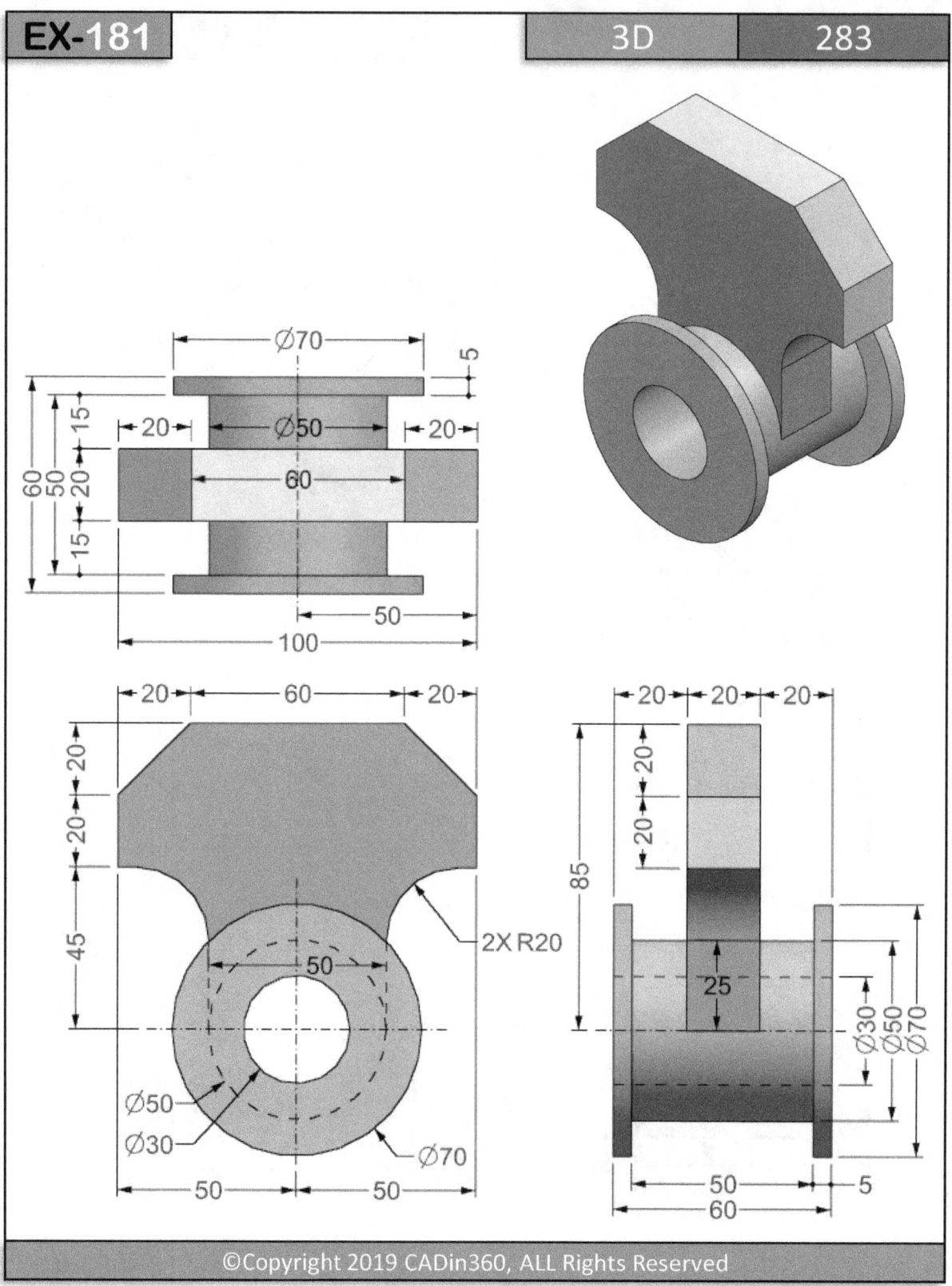

65 | 60

30

160

65

R95

R87.5

2X R20

2X R12.5

R62.5

R55

37.5

7.5

12.5 — 40

75

90

52.5

R20

5

90

4X Ø15

2X R15

Ø30

40 | 40 | 40

30

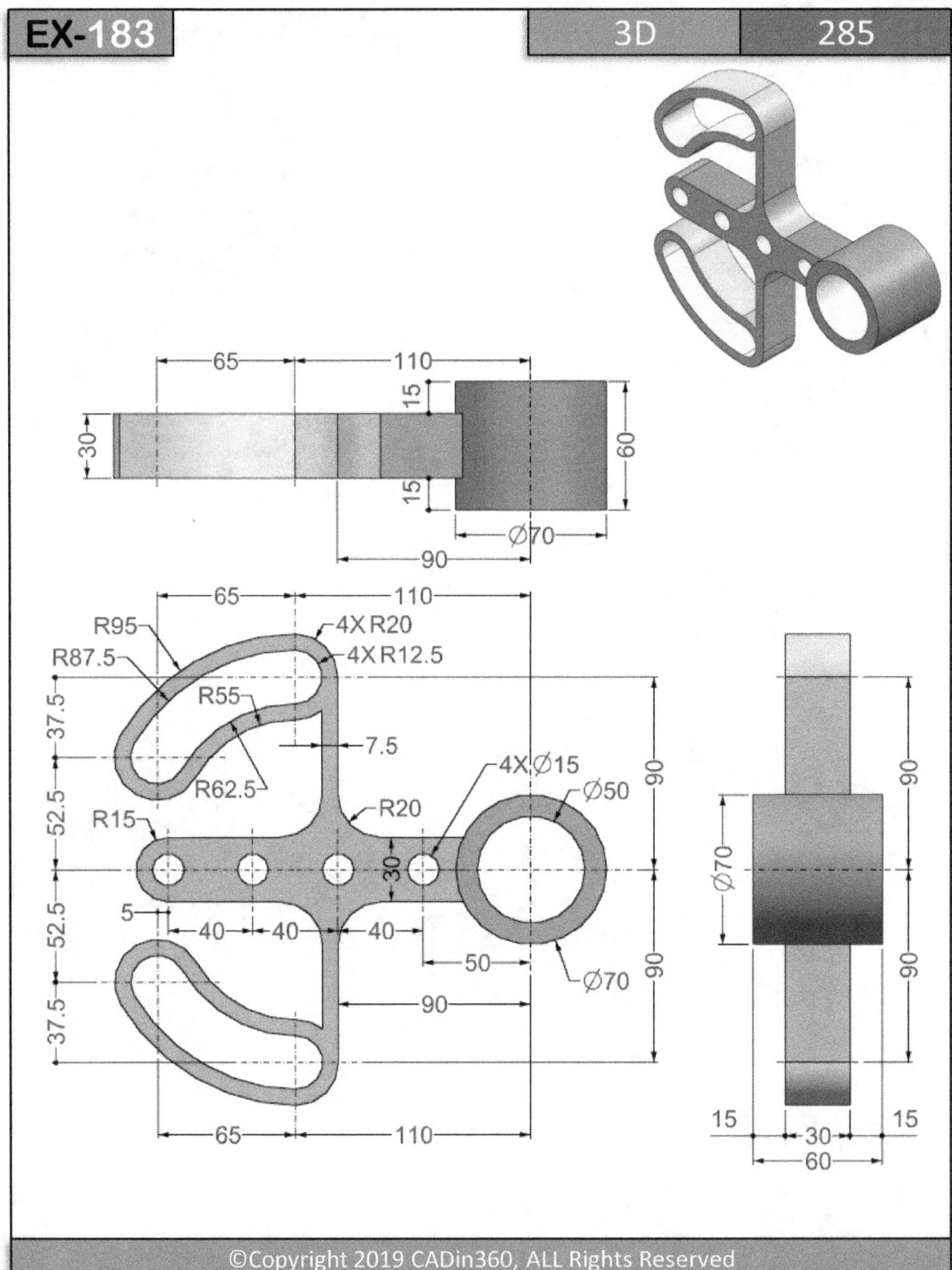

65
110
15
30
15
60
Ø70
90

65
110
R95
4X R20
4X R12.5
R87.5
37.5
R55
7.5
52.5
R62.5
R20
4X Ø15
Ø50
R15
90
30
5
40 40 40
52.5
50
90
37.5
Ø70
65
110

90
Ø70
90
15
30
15
60

100
15
70
15
24
22
24
5
15
35
30
2X R15
2X Ø15

22
2X R7.5
2X R5
6
39
2X R50
58.9
51.4
68.9
10
30
30
10
15
70
15
100

35
Ø15
5
15
61.4
51.4
5
41.4
68.9
5
25
10
5
15
30

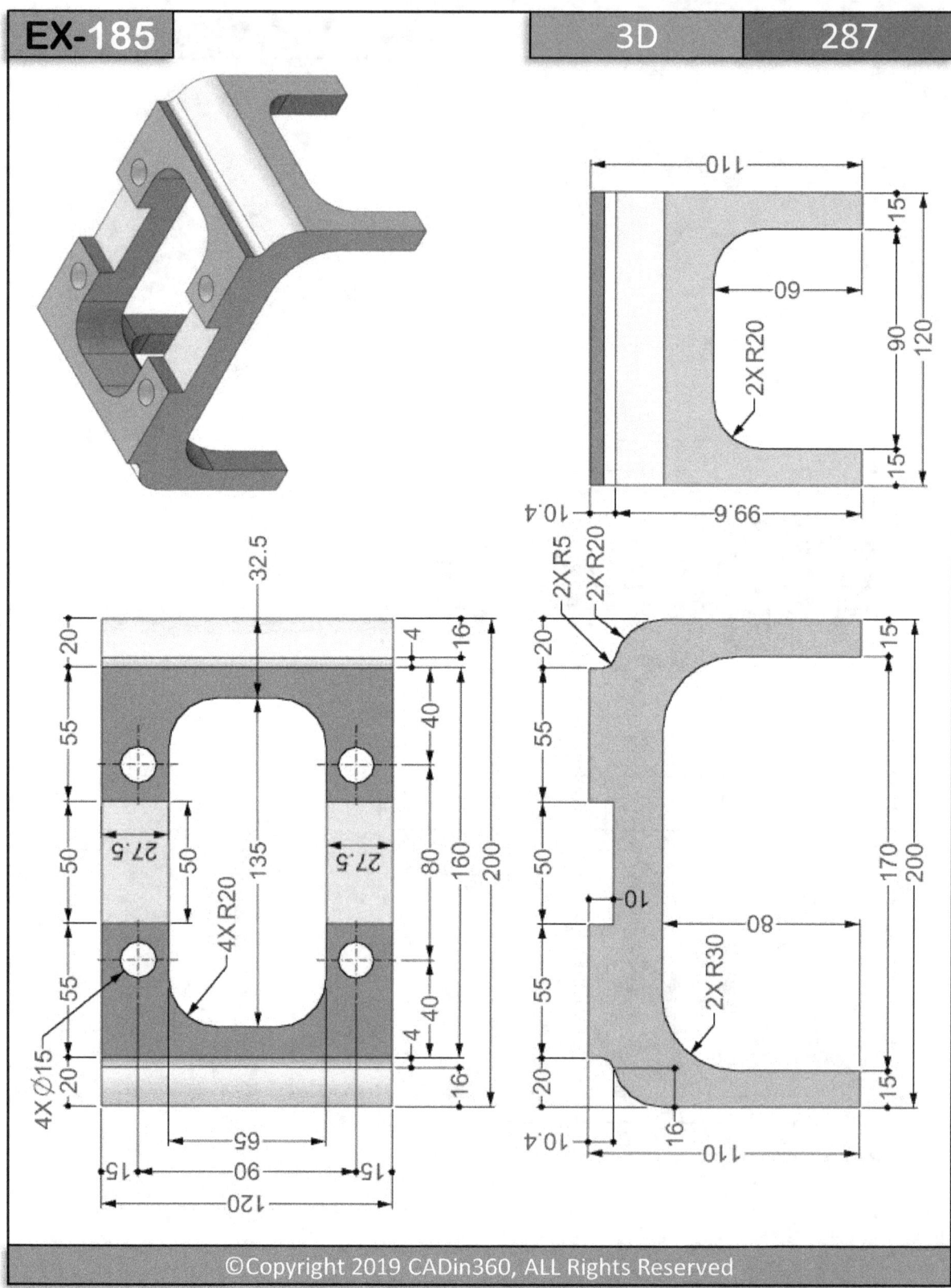

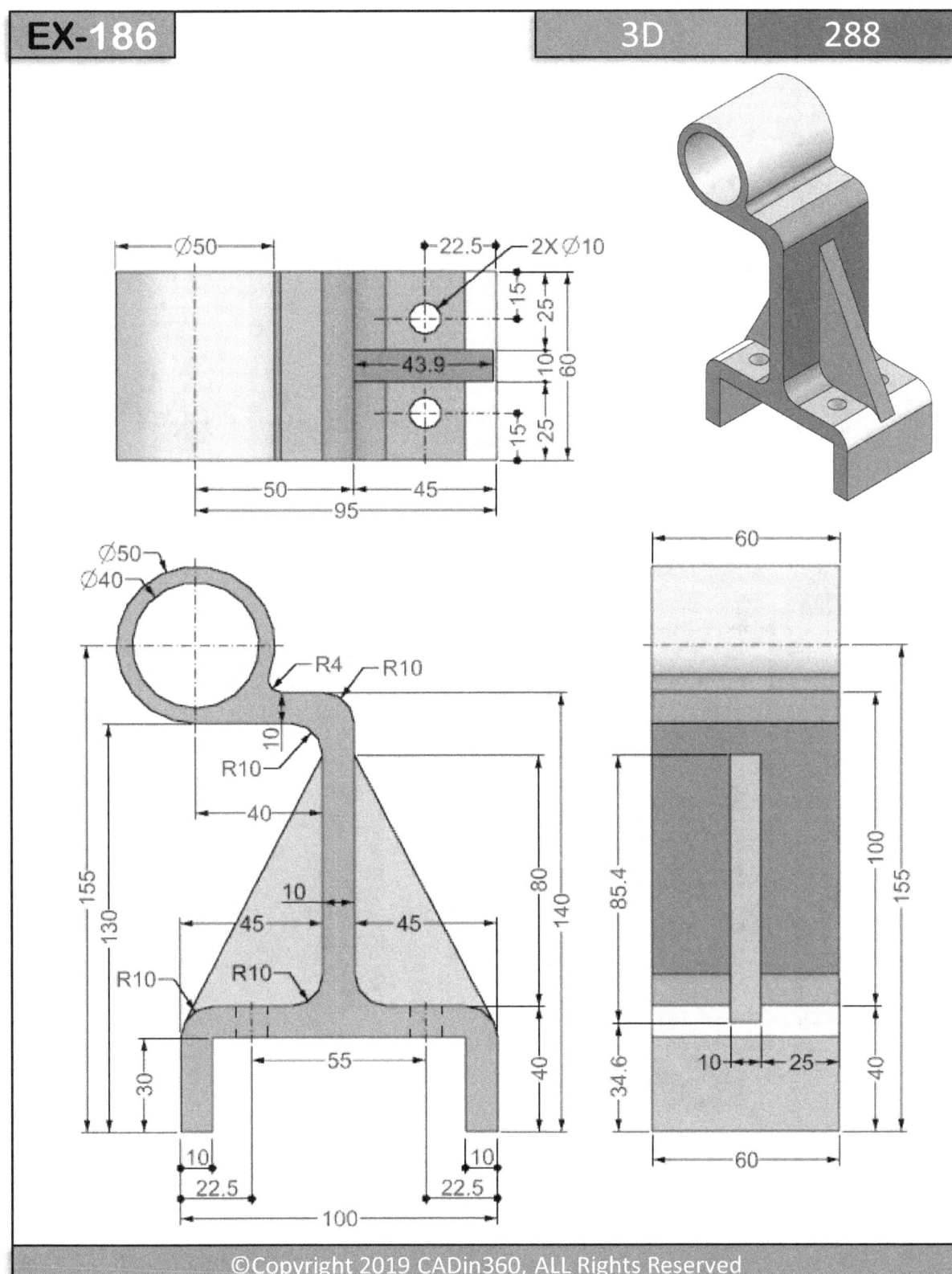

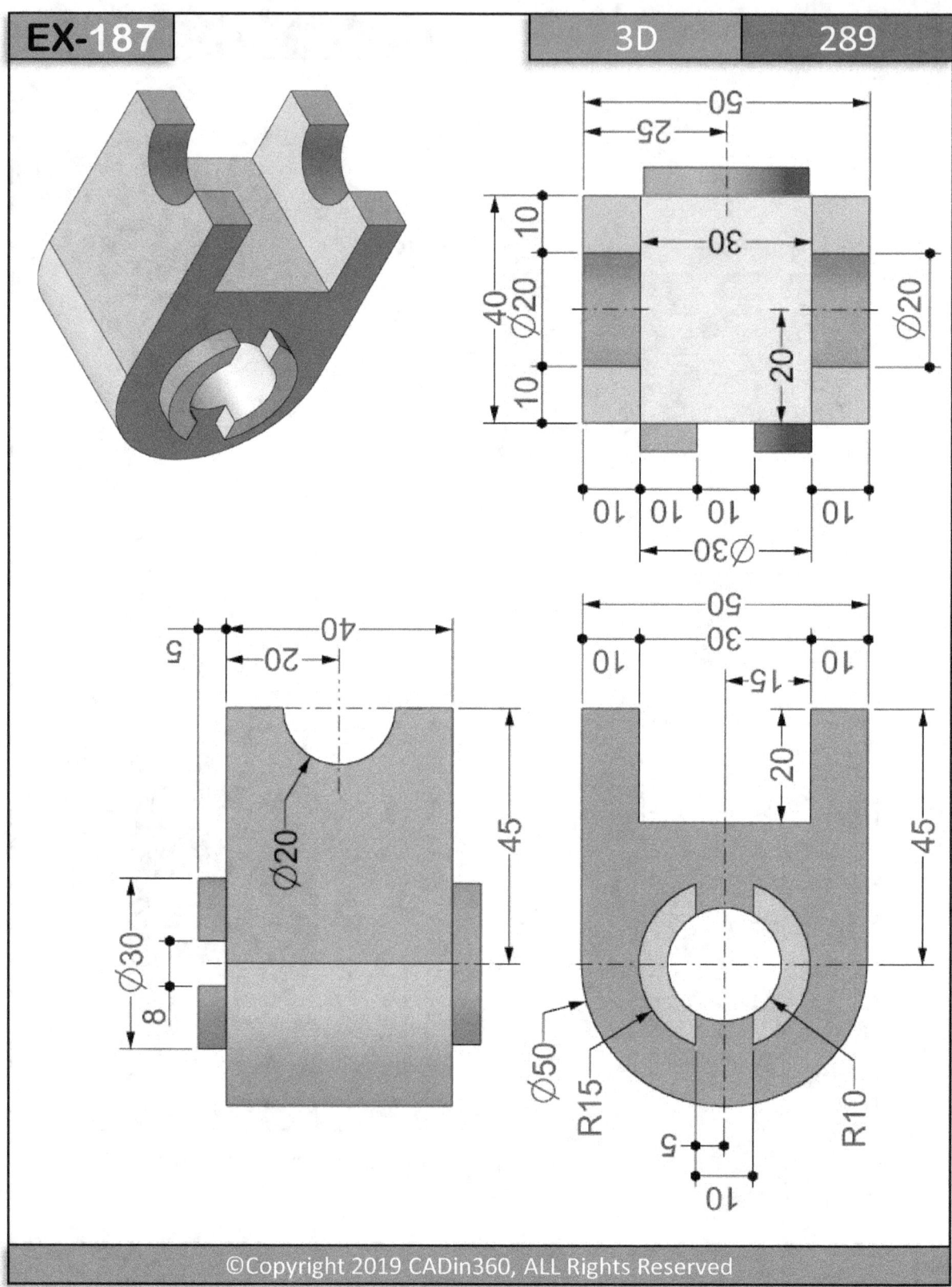

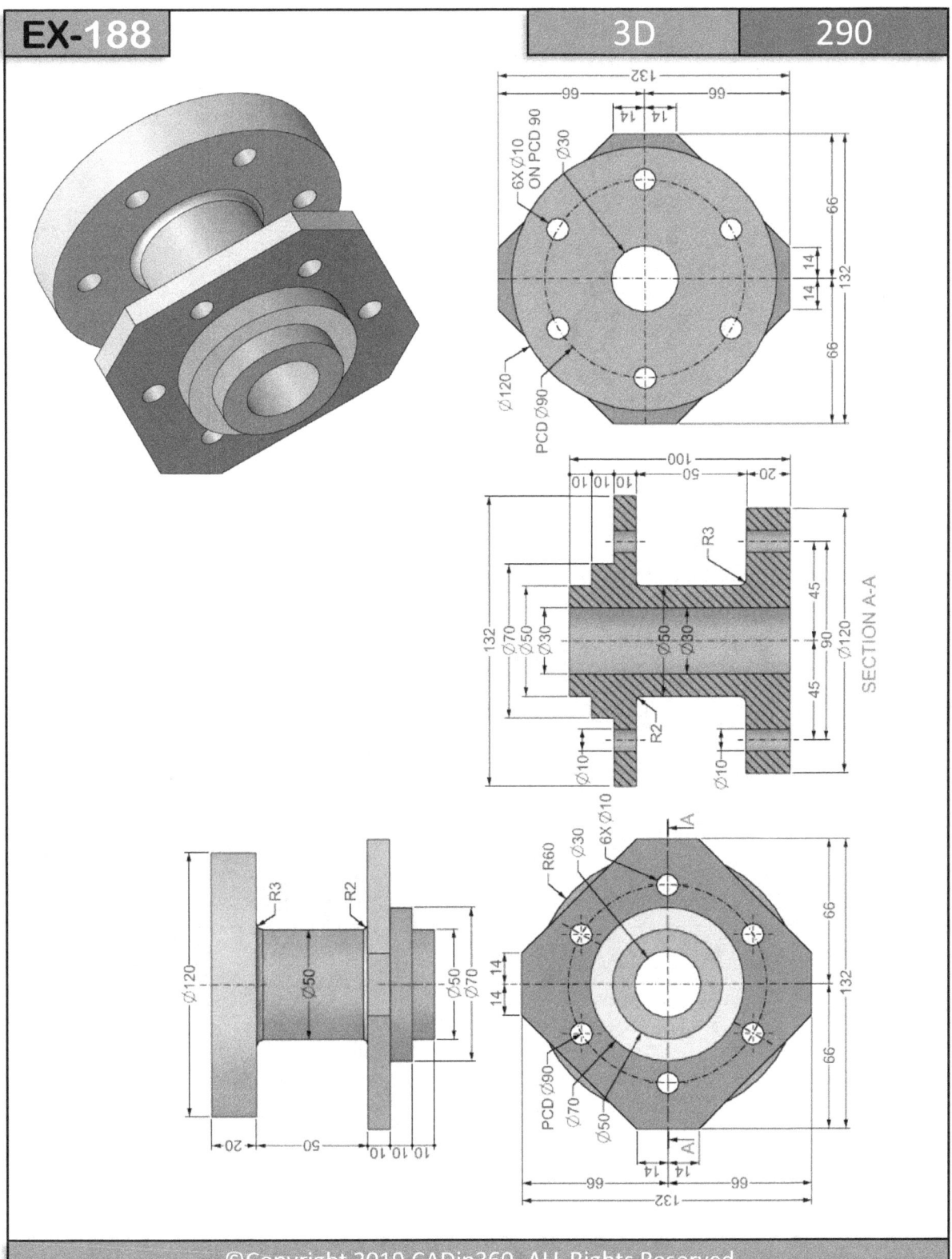

EX-188

3D 290

SECTION A-A

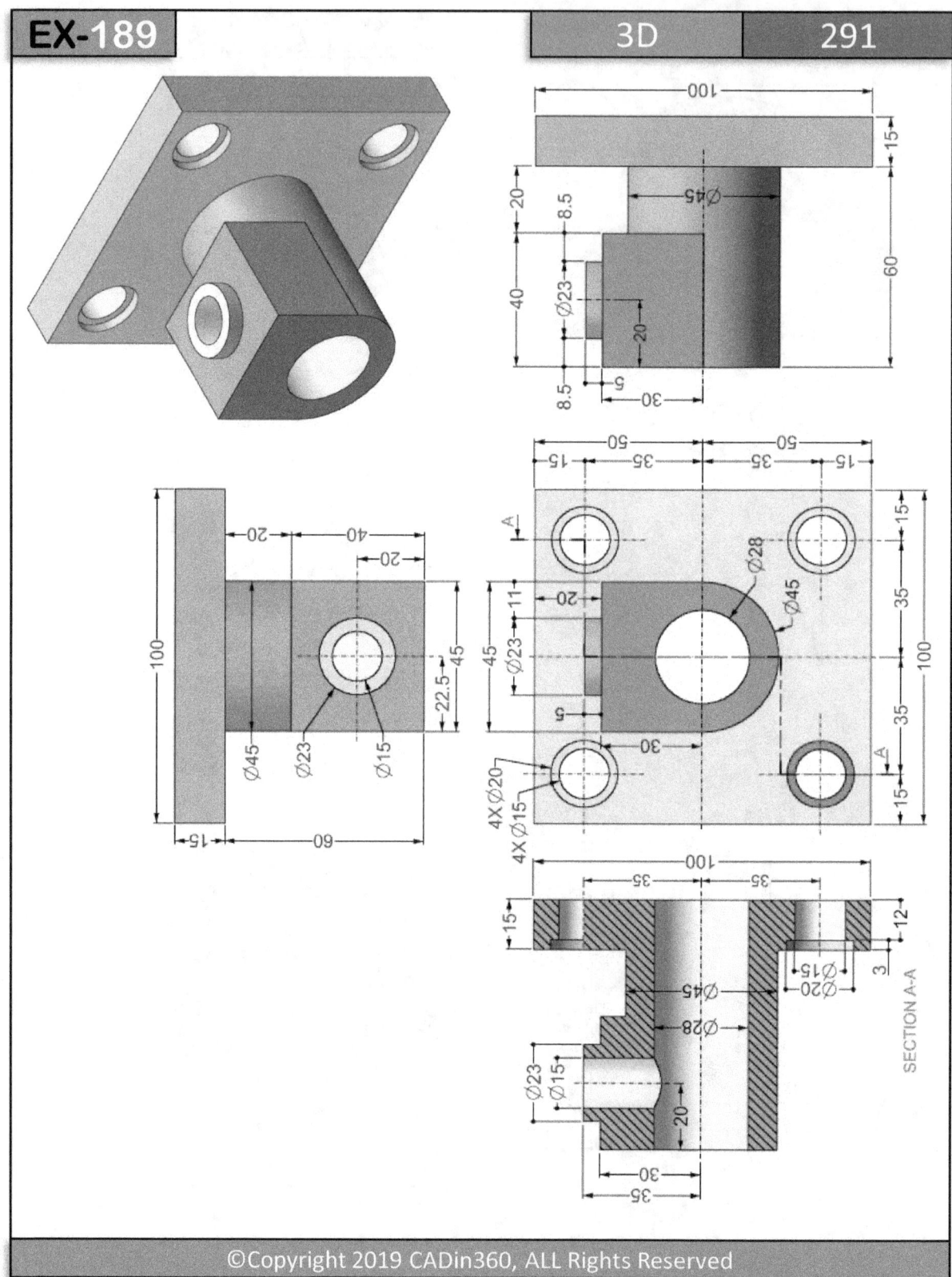

SECTION A-A

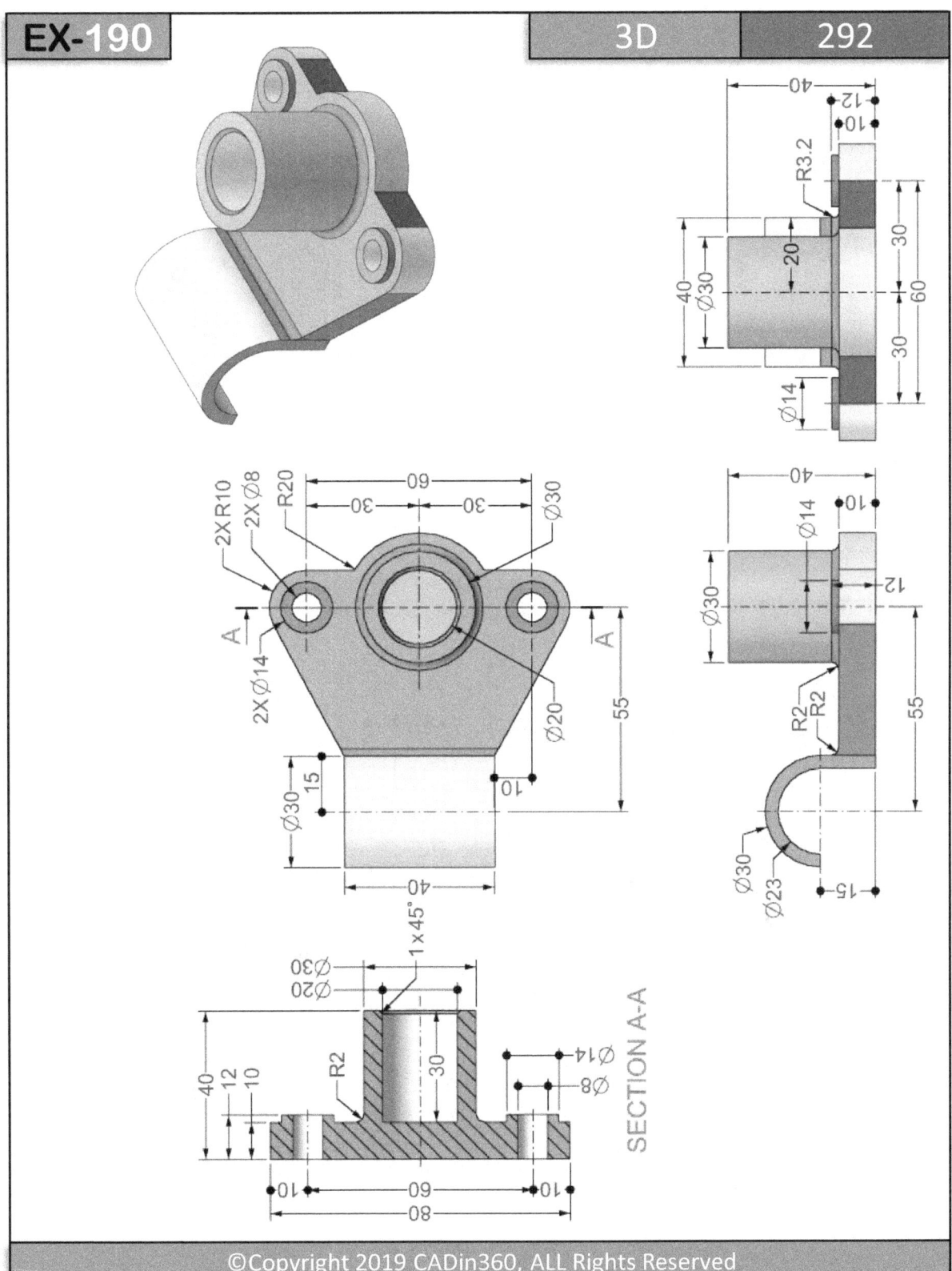

SECTION A-A

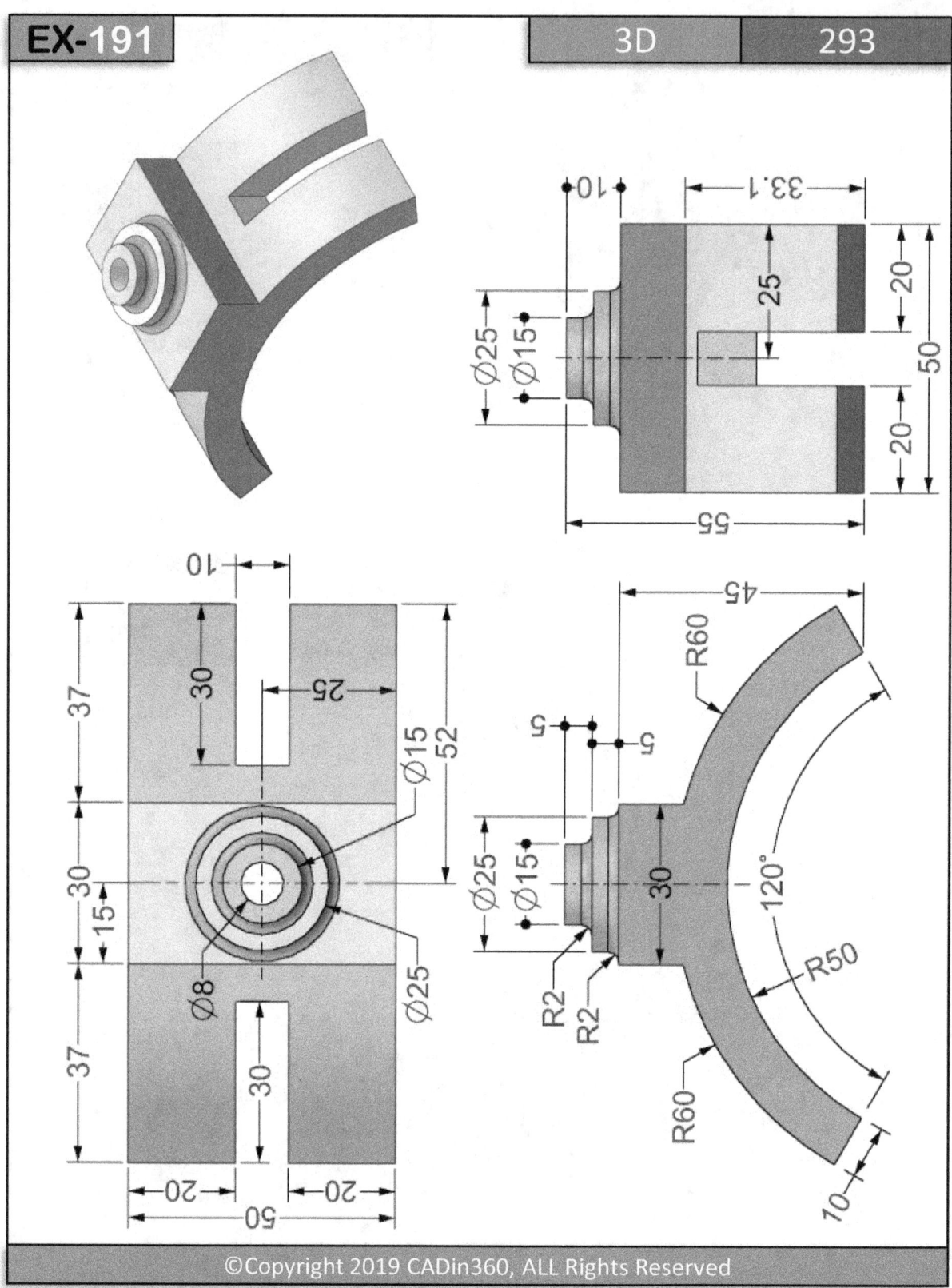

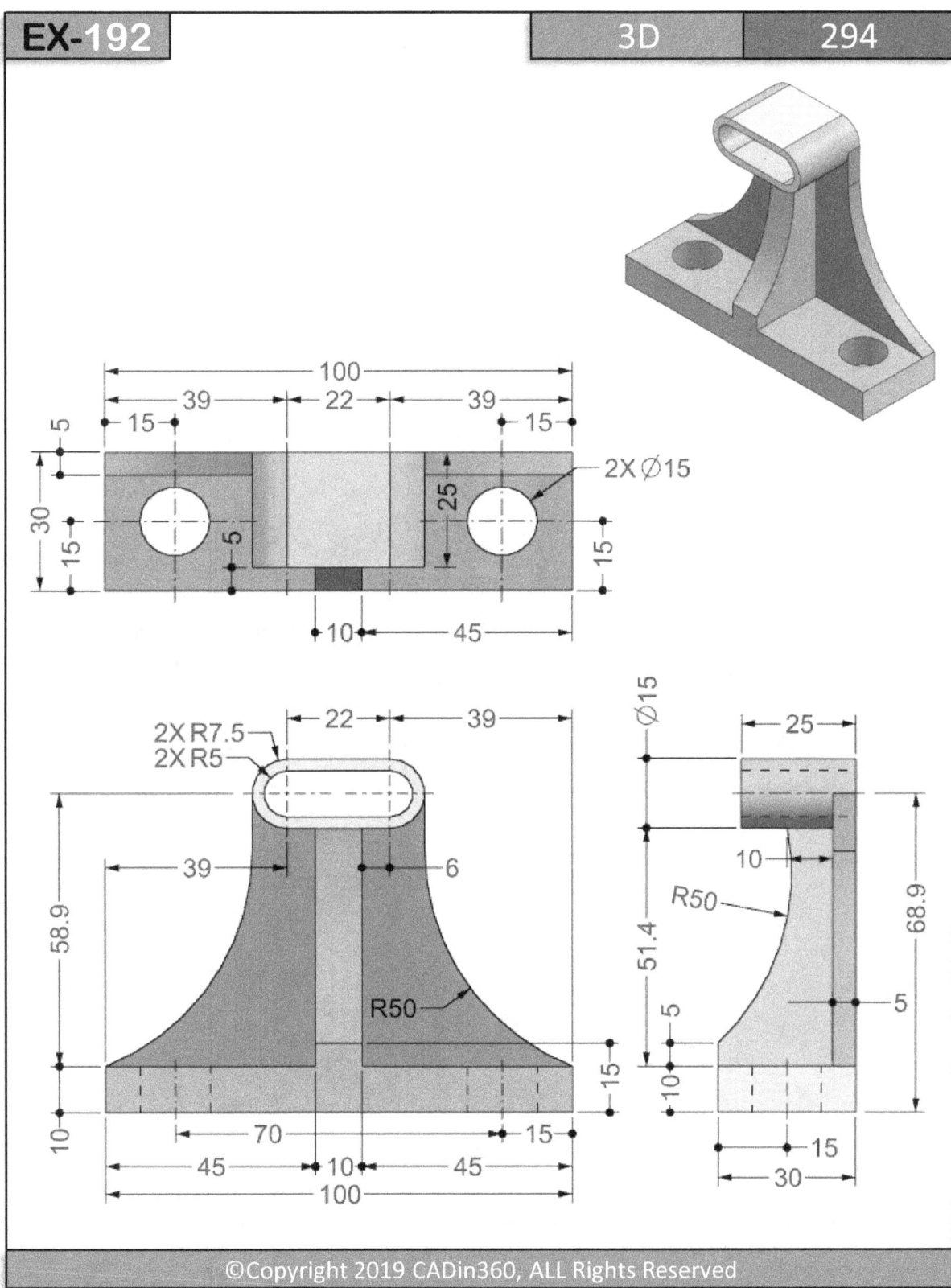

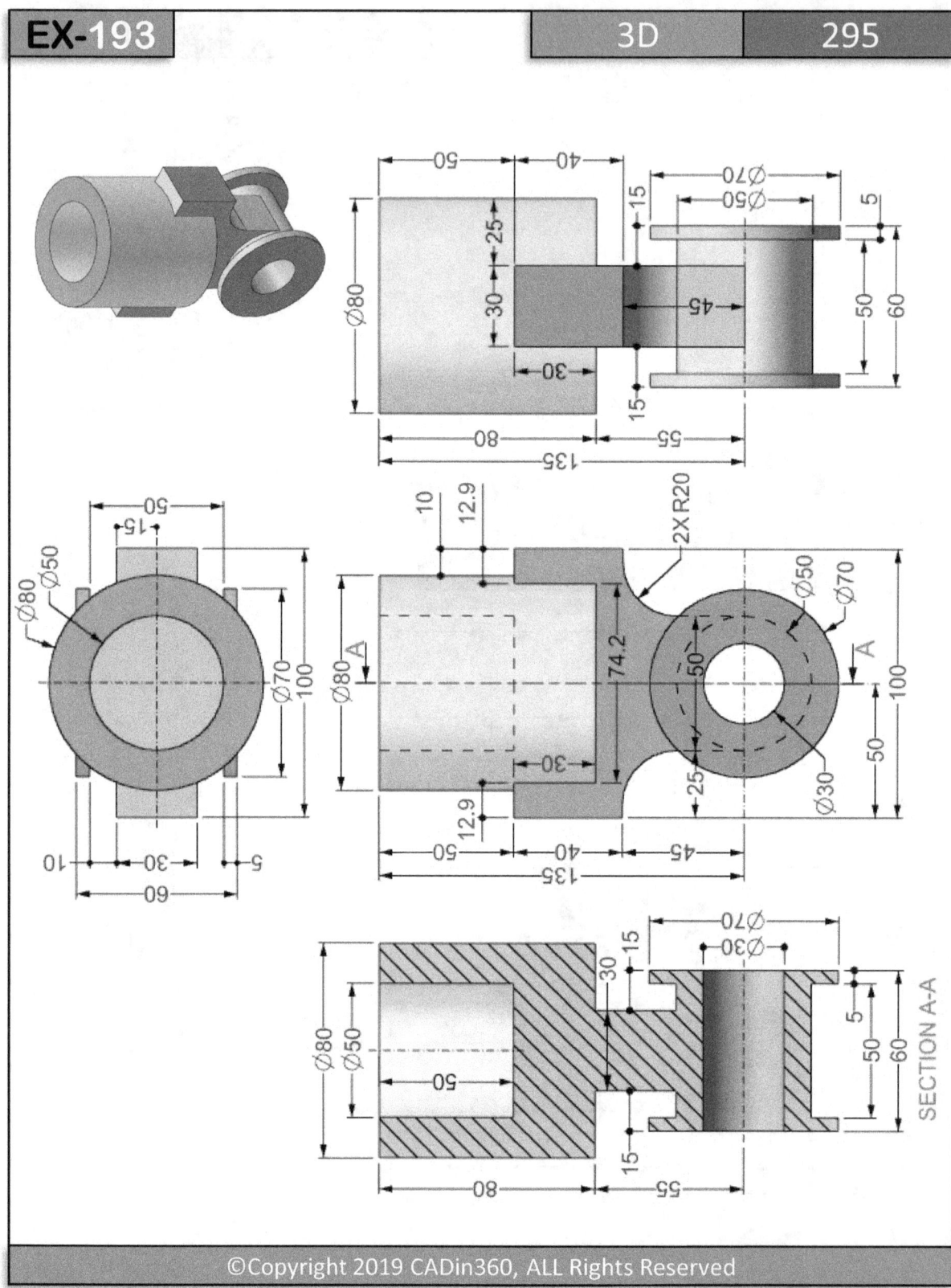

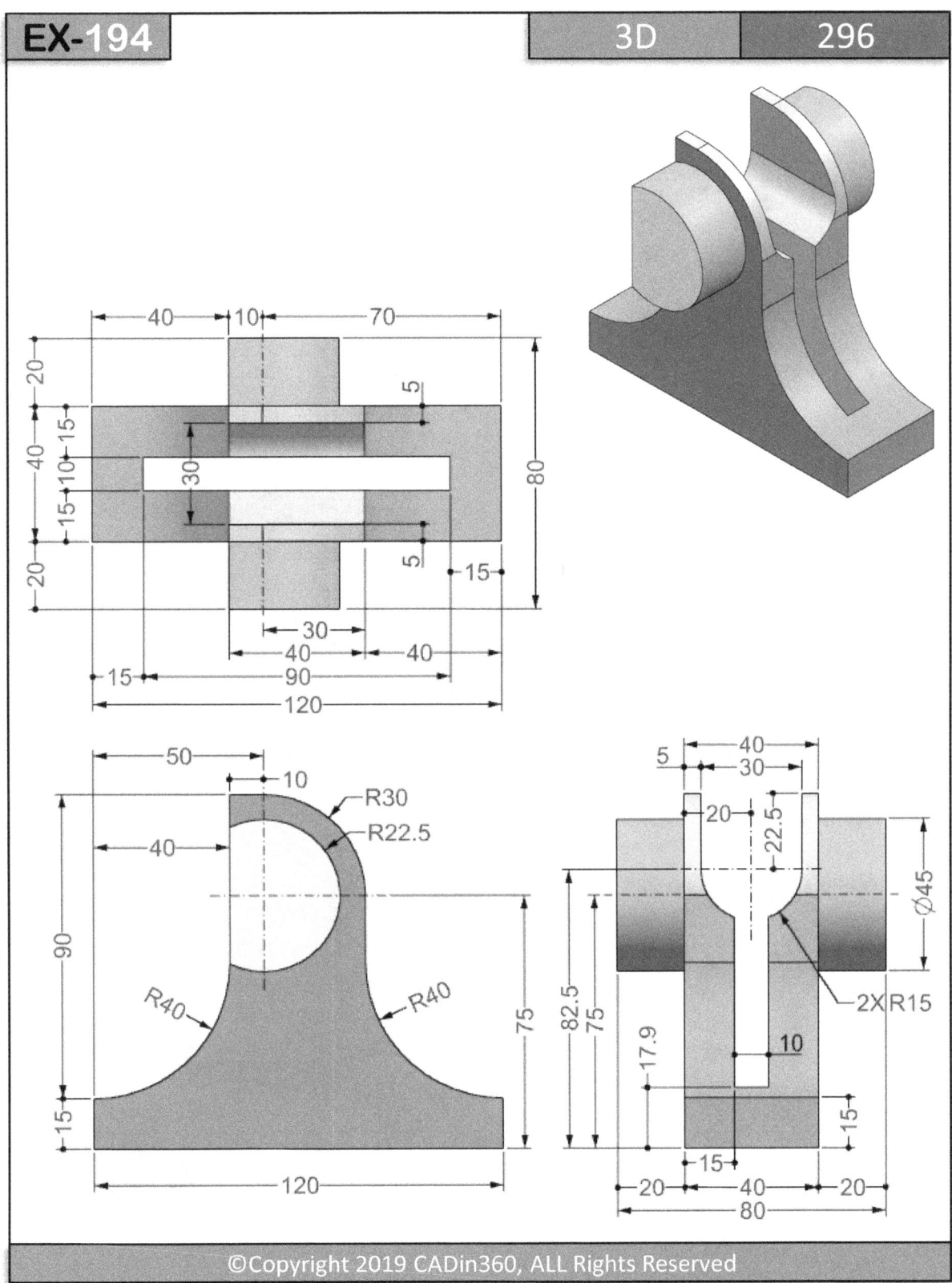

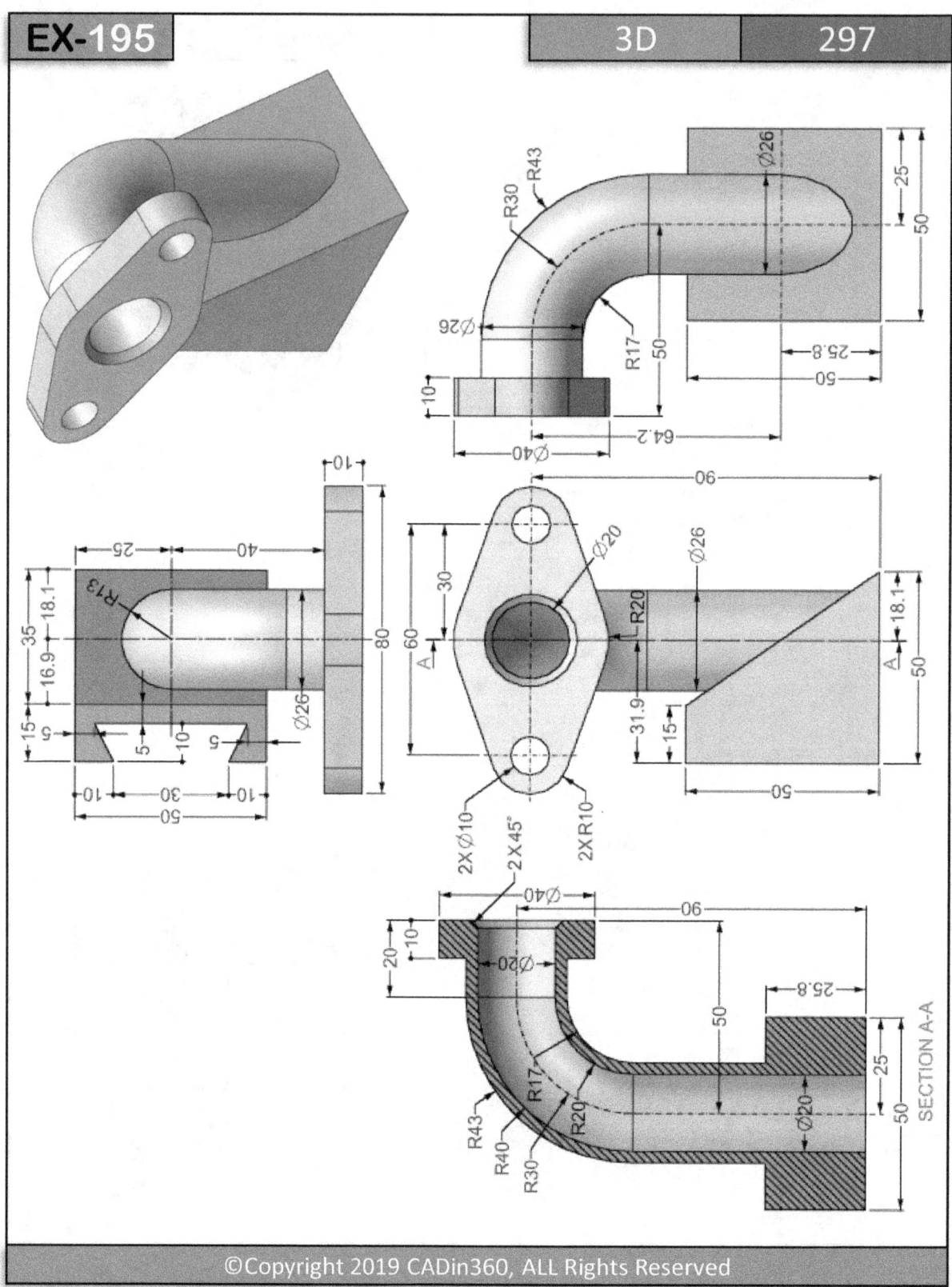

R30
R43
Ø26
25
50
Ø26
R17
50
25.8
50
10
64.2
Ø40

10
25
40
R13
35
18.1
16.9
80
Ø26
15
5
5
10
5
10
30
10
50

30
60
A
Ø20
Ø20
90
R20
Ø26
A
18.1
50
31.9
15
50

2X Ø10
2 X 45°
2XR10
Ø40
20
10
Ø20
90
50
R17
R20
25.8
25
50
R43
R40
R30
Ø20
50
SECTION A-A

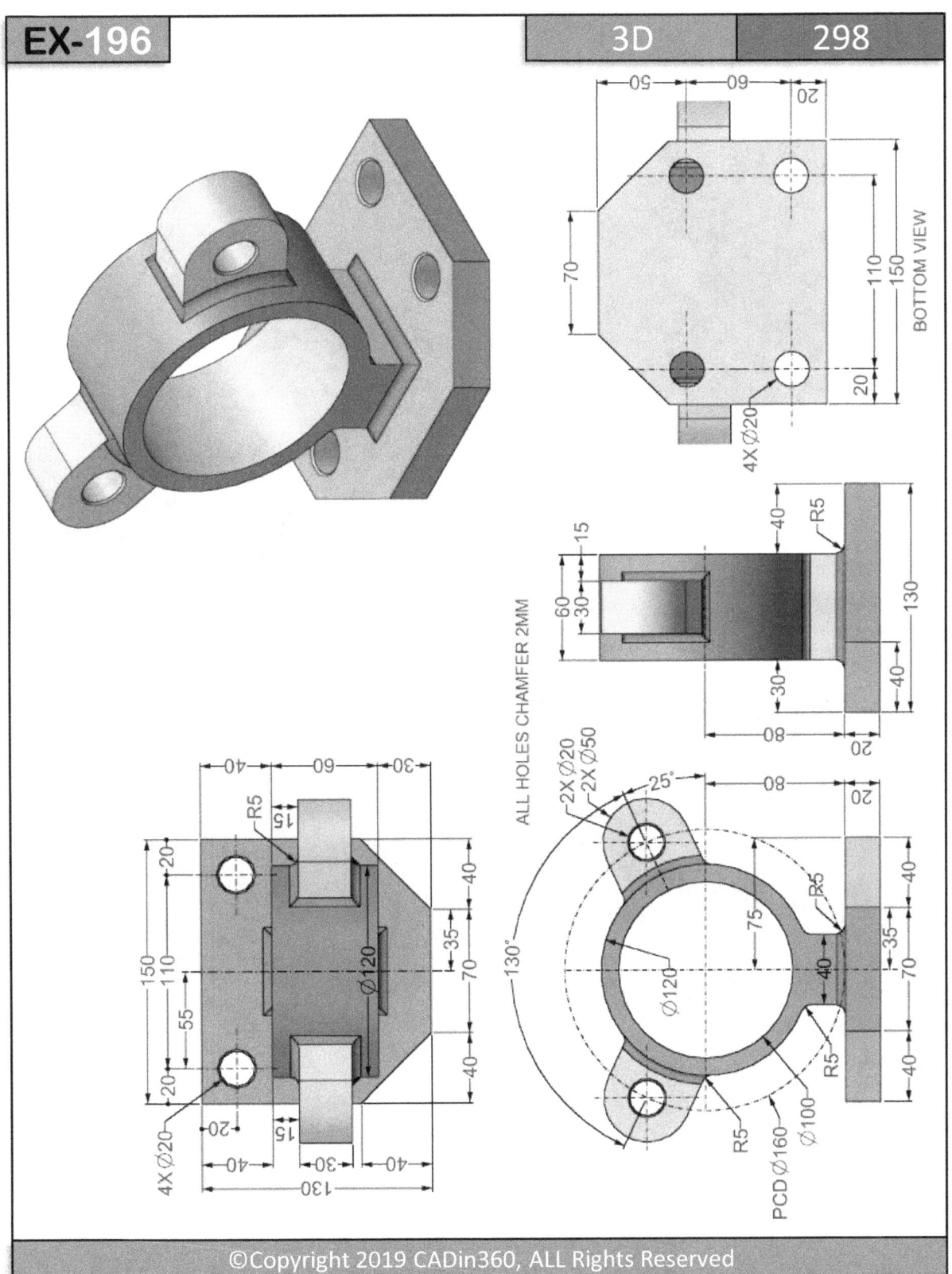

EX-196

BOTTOM VIEW

50
60
20
70
110
150
20
4X Ø20

15
60
30
40
R5
130
30
40
80
20

ALL HOLES CHAMFER 2MM

2X Ø20
2X Ø50
25°
80
130°
75
R5
Ø120
40
Ø100
R5
R5
PCD Ø160
20
40
35
70
40

40
60
30
R5
15
20
150
110
55
20
Ø120
40
35
70
40
4X Ø20
20
15
40
30
40
130

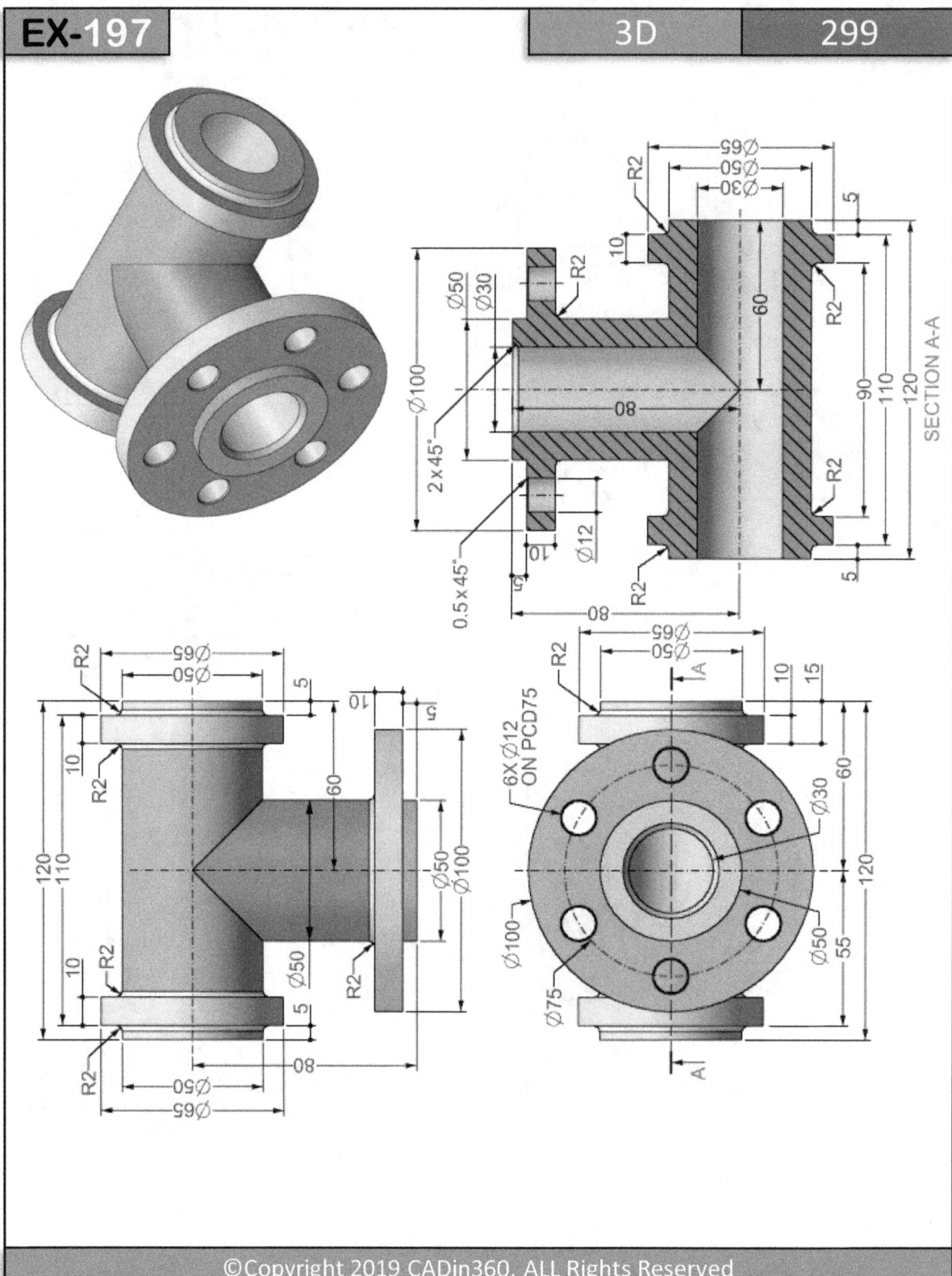

SECTION A-A

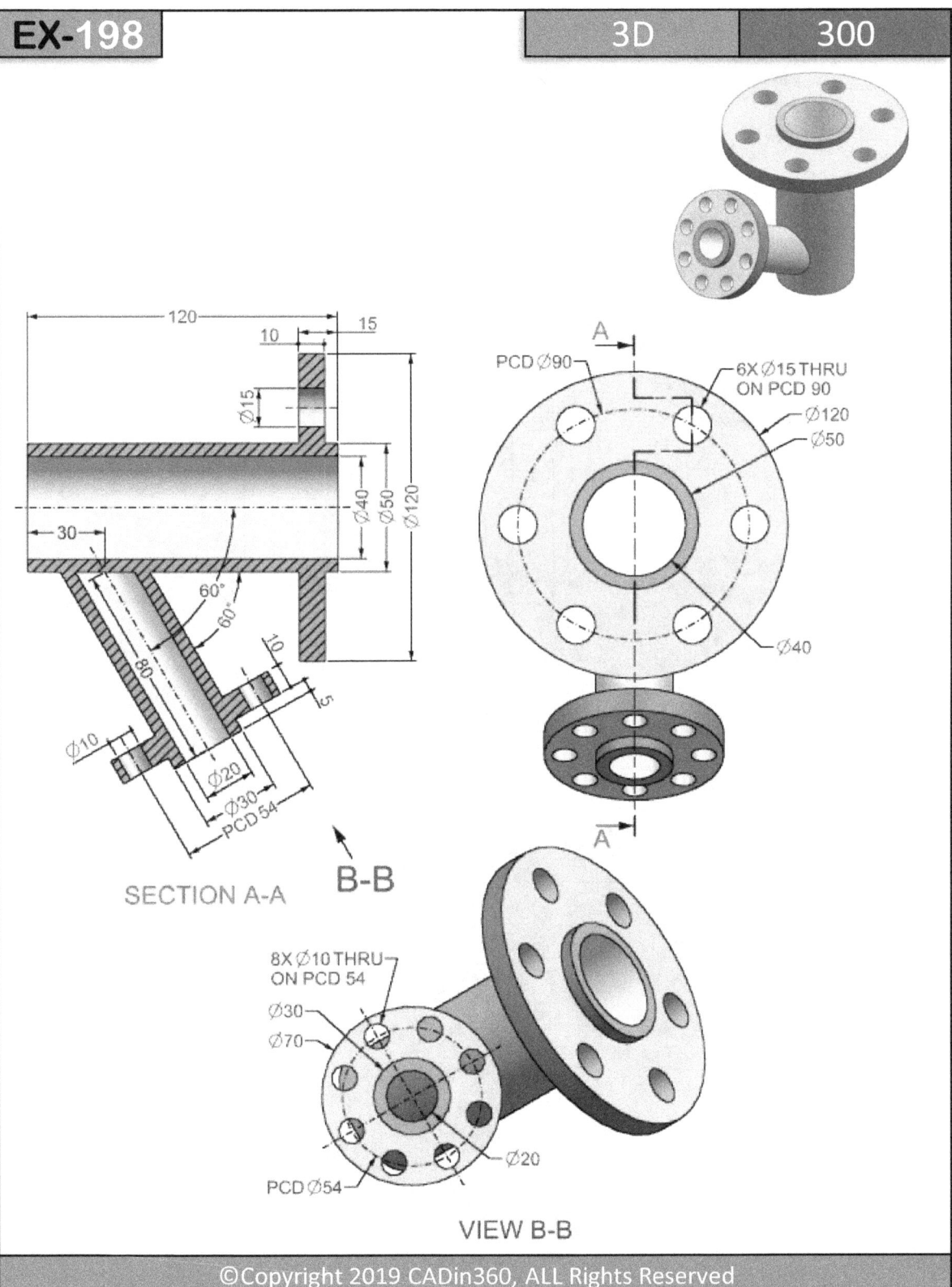

120
10 15
Ø15
Ø40 Ø50 Ø120
30
60°
60°
80
10
5
Ø10
Ø20
Ø30
PCD 54

SECTION A-A

B-B

A

PCD Ø90
6X Ø15 THRU ON PCD 90
Ø120
Ø50
Ø40

A

8X Ø10 THRU ON PCD 54
Ø30
Ø70
Ø20
PCD Ø54

VIEW B-B

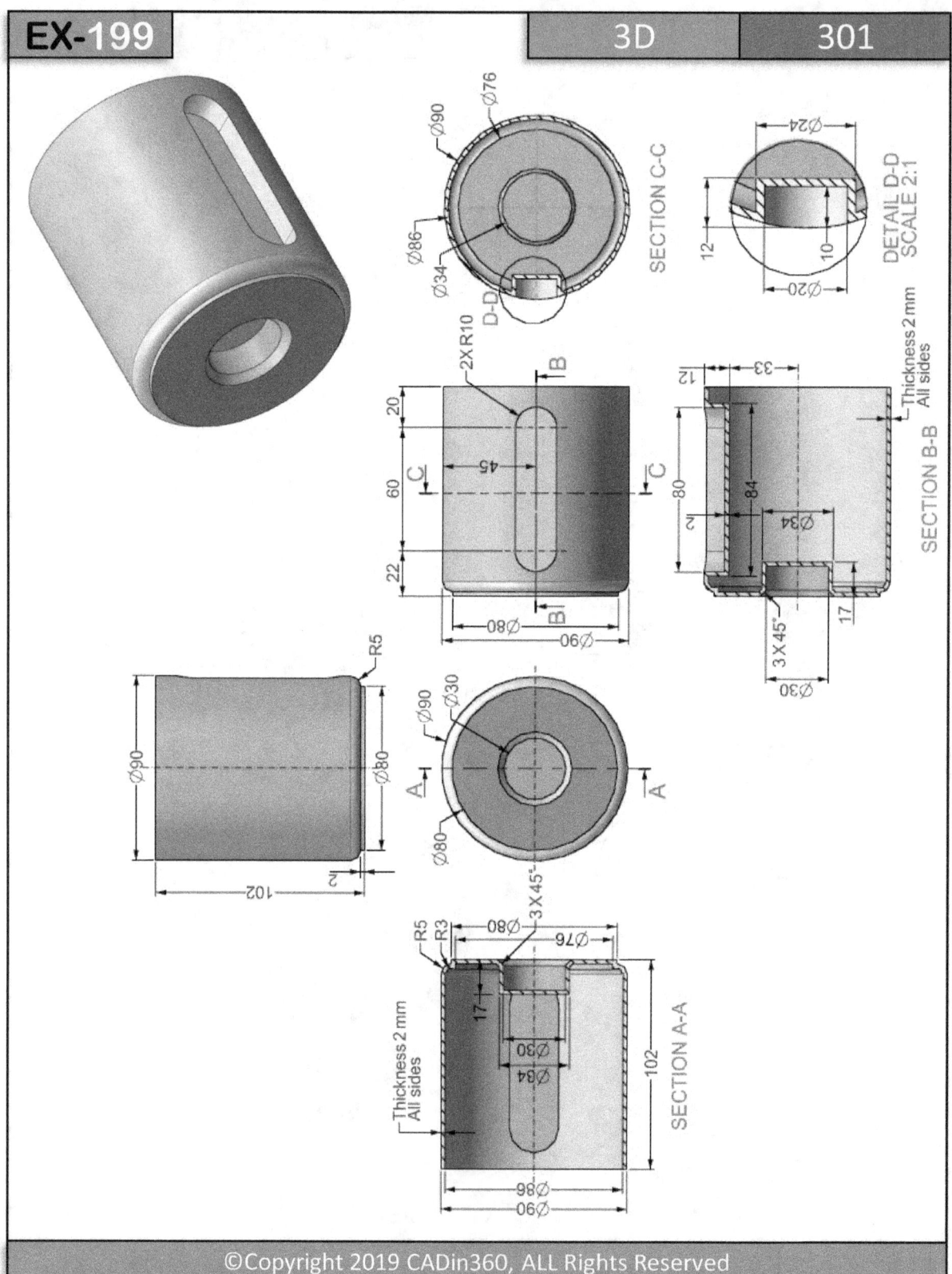

Ø76
Ø90
Ø86
Ø34
SECTION C-C
D-D

Ø24
12
10
Ø20
DETAIL D-D
SCALE 2:1

2X R10
20
45
60
22
C
C
B
B
Ø80
Ø90

12
33
Thickness 2 mm
All sides
80
84
2
Ø34
3 X 45°
17
Ø30
SECTION B-B

R5
Ø90
Ø80
2
102

Ø90
Ø30
A
A
Ø80

R5
R3
3 X 45°
Ø80
Ø76
17
Ø30
Ø84
102
Thickness 2 mm
All sides
Ø86
Ø90
SECTION A-A

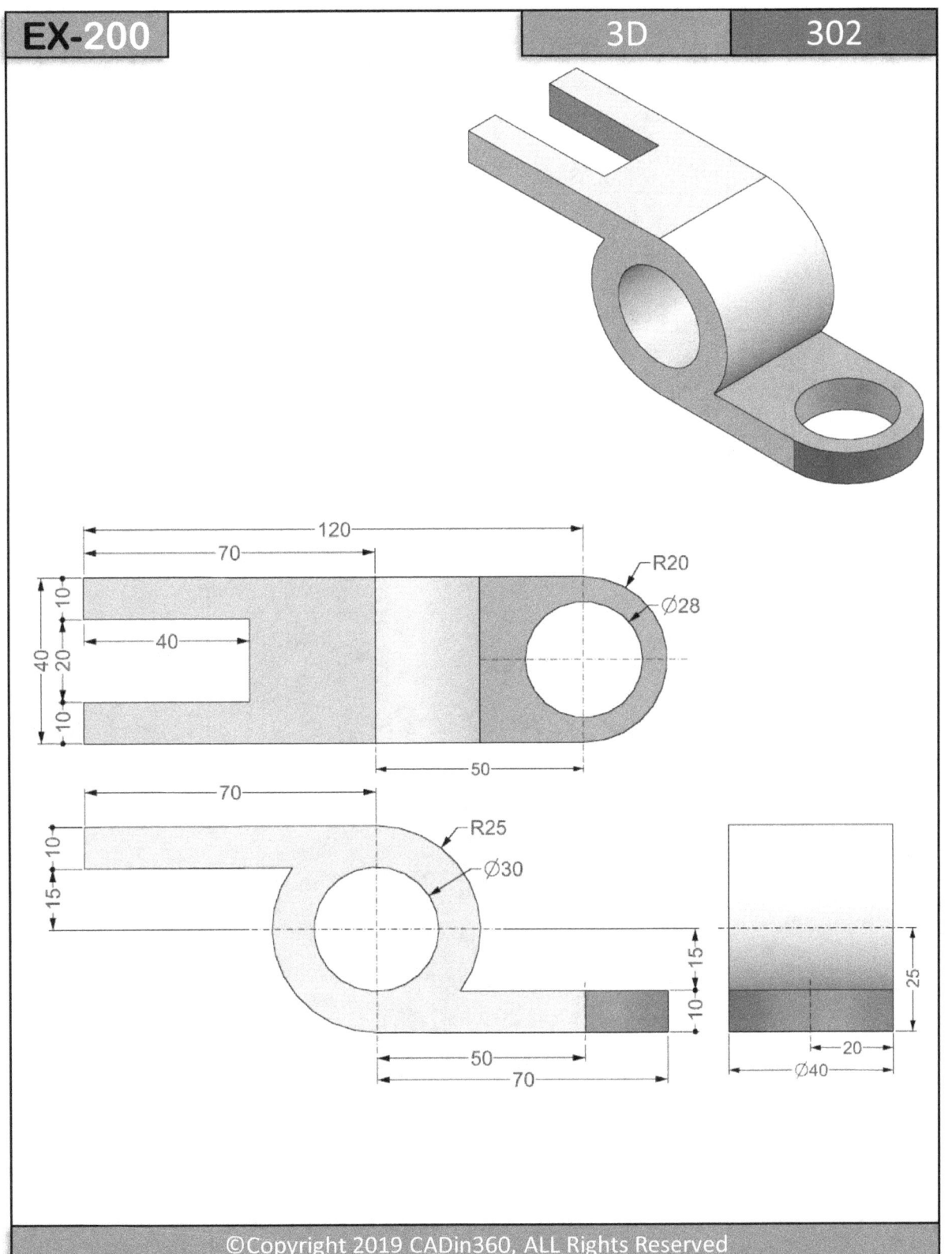

Other Books Authored by Sachidanand Jha

- ❖ 150 CAD Exercises
- ❖ AutoCAD Exercises
- ❖ CAD Exercises
- ❖ 50+ SolidWorks Exercises
- ❖ SolidWorks 200 Exercises
- ❖ Autodesk Inventor Exercises
- ❖ Catia Exercises
- ❖ Siemens NX Exercises
- ❖ Autodesk Fusion 360 Exercises
- ❖ Siemens Solid Edge Exercises
- ❖ PTC Creo Exercises